An Applied Guide to Water and Effluent Treatment Plant Design

Seán Moran

ELSEVIER

Elsevier
Radarweg 29, PO Box 211, 1000 AE Amsterdam, Netherlands
The Boulevard, Langford Lane, Kidlington, Oxford OX5 1GB, United Kingdom
50 Hampshire Street, 5th Floor, Cambridge, MA 02139, United States

Notices
Knowledge and best practice in this field are constantly changing. As new research and experience
broaden our understanding, changes in research methods, professional practices, or medical treatment
may become necessary.

Practitioners and researchers must always rely on their own experience and knowledge in evaluating and
using any information, methods, compounds, or experiments described herein. In using such information
or methods they should be mindful of their own safety and the safety of others, including parties for
whom they have a professional responsibility.

To the fullest extent of the law, neither the Publisher nor the authors, contributors, or editors, assume
any liability for any injury and/or damage to persons or property as a matter of products liability,
negligence or otherwise, or from any use or operation of any methods, products, instructions, or ideas
contained in the material herein.

British Library Cataloguing-in-Publication Data
A catalogue record for this book is available from the British Library

Library of Congress Cataloging-in-Publication Data
A catalog record for this book is available from the Library of Congress

ISBN: 978-0-12-811309-7

For Information on all Elsevier publications
visit our website at https://www.elsevier.com/books-and-journals

Working together
to grow libraries in
developing countries

www.elsevier.com • www.bookaid.org

Publisher: Joe Hayton
Acquisition Editor: Kostas Marinakis
Editorial Project Manager: Leticia Lima
Production Project Manager: Kamesh Ramajogi
Cover Designer: Matthew Limbert

Typeset by MPS Limited, Chennai, India

An Applied Guide to
Water and Effluent
Treatment
Plant Design

Contents

Preface xix
Acknowledgments xxiii

**CHAPTER 1 Introduction: The Nature of Water and Effluent
            Treatment Plant Design** 1
  What is this Book About? 1
  A Brief History of Water Treatment Plant Design...................... 2
  The Literature of Water Treatment Plant Design 3
  Interaction with Other Engineering Disciplines 3
  Hydraulic Calculations 4
  Water Chemistry.. 4
  Water Biology... 4
  Economics... 4
  Materials Selection 5
  The Importance of Statistics in Water and Effluent Treatment... 5
      Statistics in Sewage Treatment Plant Design and
      Performance 6
      Statistics and Discharge Consents.......................................7
  Contractual Arrangements in Water and Effluent Treatment 8
  Misconceptions in Water and Effluent Treatment Plant Design 10
      Chemical/Process Engineering Misconceptions 10
      Academic Misconceptions...10
      Sales Misconceptions 11
  Further Reading 12

SECTION 1 WATER ENGINEERING SCIENCE 13
CHAPTER 2 Water Chemistry... 15
  Hardness and Alkalinity 15
  Acids and Bases.. 15
      Buffering...17
      Scaling and Aggressiveness 19
      Oxidation and Reduction..21
      Dechlorination 23

Catalysis 23

Osmosis 23

Further Reading 24

CHAPTER 3   Biology **25**

Mind the (Biology/Biochemistry) Gap 25

Important Microorganisms 27

Bacteria 27

Protozoa and Rotifers 28

Algae 28

Fungi 30

Bacterial Growth 30

Biochemistry 31

Proteins 32

Fats 32

Carbohydrates 33

Nucleic Acids 33

Metabolism 34

The Biology of Aerobic Water Treatment 35

The Biology of Anaerobic Water Treatment 36

Stage 1: Hydrolysis 36

Stage 2: Volatile Fatty Acid Production 36

Stage 3: Conversion of Higher Fatty Acids to
Acetic Acid by Acetogenic Microorganisms 36

Stage 4: Conversion of Acetic Acid to Methane
and Carbon Dioxide by Methanogenic Bacteria 36

**CHAPTER 4   Engineering Science of Water Treatment Unit
Operations** **39**

Introduction 39

Coagulation/Flocculation 40

Coagulants 40

Flocculants 40

Sedimentation/Flotation 40

Sedimentation 40

Flotation 42

Filtration 43

Membrane Filtration/Screening 45

Depth Filtration 45

Mixing 46

Combining Substances 46

Promotion of Flocculation 46
Heat and Mass Transfer 47
Suspension of Solids 47
Gas Transfer 47
Adsorption 48
Disinfection and Sterilization 49
Dechlorination 51
Further Reading 51

CHAPTER 5   Fluid Mechanics **53**
Introduction 53
The Key Concepts 53
Pressure and Pressure Drop, Head and Headloss 54
Rheology 54
Bernoulli's Equations 56
Darcy-Wiesbach Equation 56
Less Accurate Explicit Equation 57
Stokes Law 57
Further Reading 58

SECTION 2 CLEAN WATER TREATMENT
ENGINEERING **59**

**CHAPTER 6   Clean Water Characterization and Treatment
Objectives** **61**
Impurities Commonly Found in Water 61
Characterization 62
Microbiological Quality 62
Color 62
Suspended Solids 63
Alkalinity 63
Iron 63
Manganese 64
Total Dissolved Solids 64
Treatment Objectives 64
Potable Water 65
Cooling Water 66
Ultrapure Water 66
Further Reading 67

CHAPTER 7   Clean Water Unit Operation Design: Physical Processes **69**

Introduction 69

Mixing 70

    High Shear 70

    Low Shear 70

Flocculation 70

Settlement 71

    Clarifier Design 71

Flotation 73

Filtration 74

    Coarse Surface Filtration 75

    Fine Depth Filtration 76

    Fine Surface Filtration: Membranes 87

Distillation 96

    Pretreatment 96

    Types of Distillation Process 96

Gas Transfer 97

Stripping 98

Physical Disinfection 98

    UV Irradiation 98

Novel Processes 99

    Electrodialysis 99

    Membrane Distillation 99

    Forward Osmosis 100

Further Reading 100

CHAPTER 8   Clean Water Unit Operation Design: Chemical Processes **101**

Introduction 101

Drinking Water Treatment 102

    pH and Aggressiveness Correction 102

    Coagulation 104

    Precipitation 105

    Softening 105

    Iron and Manganese Removal by Oxidation 107

Boiler and Cooling Water Treatment 108

    Cooling Water Treatment 108

    Boiler Water Treatment 109

CHAPTER 9 **Clean Water Unit Operation Design: Biological Processes** **111**
    Introduction 111
    Discouraging Life 111
        Disinfection 111
        Sterilization 115
    Encouraging Life 115
    Biological Filtration 115
    Further Reading 116

CHAPTER 10 **Clean Water Hydraulics** **117**
    Introduction 117
    Pump Selection and Specification 118
        Pump Types 118
        Pump Sizing 119
    Open Channel Hydraulics 132
        Channels 132
        Chambers and Weirs 133
        Flow Control in Open Channels 134
    Hydraulic Profiles 134
    Further Reading 135

SECTION 3 MUNICIPAL DIRTY WATER TREATMENT ENGINEERING **137**

CHAPTER 11 **Dirty Water Characterization and Treatment Objectives** **139**
    Wastewater Characteristics 139
        Solids 140
        Biochemical Oxygen Demand 140
        Ammonia 140
    Flowrate and Mass Loading 140
        Selection of Design Flowrates 141
        Selection of Design Mass Loadings 143
    Treatment Objectives 144
        Discharge to Environment 145
        Industrial Reuse 146
        Reuse as Drinking Water 146
    Further Reading 146

CHAPTER 12 Dirty Water Unit Operation Design: Physical Processes **147**

Introduction 147

Flow Measurement 148

Pumping 149

    Airlifts 149

Racks and Screens 151

    Bar Screens 153

    Screens 153

    Comminution 153

Grit Removal 154

    Grit Chambers 154

    Vortex Separators 154

    Suspended Solids Removal 154

    Screens and Filters for Residual Suspended Solids Removal 155

Flow Equalization 155

Primary Sedimentation 156

Final Settlement Tank Design 158

Flotation 160

Filtration 160

Gas Transfer 161

    Stripping 162

Adsorption 162

Other Physical Processes 163

    Physical Disinfection 163

CHAPTER 13 Dirty Water Unit Operation Design: Chemical Processes **165**

Introduction 165

Coagulation 165

Chemical Precipitation 166

Disinfection 166

    Chemical Disinfection 167

    Dechlorination 167

Nutrient Removal 167

    Chemical Removal of Nitrogen 167

    Chemical Removal of Phosphorus 168

Removal of Dissolved Inorganics 168

Other Chemical Processes 169

CHAPTER 14 Dirty Water Unit Operation Design: Biological Processes **171**

An Overview of Biological Processes 172

Aerobic Attached Growth Processes................................172

Aerobic Suspended Growth Processes................................175

Anaerobic Attached Growth Processes...............................176

Anaerobic Suspended Growth Processes 176

Combined Aerobic Treatment Processes 176

Secondary Treatment Plant Design... 177

Suspended Growth Processes 178

Activated Sludge Physical Facility Design and Selection..178

Activated Sludge Process Design.....................................179

Trickling Filters 185

Hydraulic Loading Rate 186

Maximum Organic Loading Rate.....................................187

Rotating Biological Contactors 188

Natural Systems 188

Pond Treatment...188

Constructed Wetlands...189

Aerated Lagoons..190

Stabilization Ponds 191

Tertiary Treatment Plant Design.. 191

The Need for Tertiary Treatment......................................191

Treatment Technologies 191

Tertiary Treatment Plant Design......................................191

Biological Nutrient Removal 192

Novel Processes 195

Bio-Augmentation 195

Deep Shaft Process...196

Pure or Enhanced Oxygen Processes.................................196

Biological Aerated Flooded Filters 196

Sequencing Batch Reactors 197

Membrane Bioreactors 197

Upflow Anaerobic Sludge Blanket Process.......................200

Anaerobic Filter Process 201

Expanded Bed Anaerobic Reactor 201

Two and Three Phase Anaerobic Digestion 201

Further Reading 202

CHAPTER 15 Dirty Water Hydraulics **203**
    Introduction.. 203
    Minimum Velocities 204
    Open Channels.. 204
        Straight Channel Headloss 204
        Circular Channel Headloss 206
        Shock Losses in Channels 206
        Peripheral Channels 208
    Weirs.. 211
        Thin Weirs 211
        Broad Weirs 212
    Screens 213
        Bar Screens 213
    Advanced Open Channel Hydraulics 214
        Critical and Supercritical Flow 214
        Hydraulic Jumps 214
        Choking...215
        Inlet Works 215
        Control Structures...215
        Sump Volumes...216
        Sump Flow Presentation and Baffles 217
    Plant Layout Hydraulics 218
    Further Reading 218

SECTION 4 INDUSTRIAL EFFLUENT TREATMENT ENGINEERING **219**
CHAPTER 16 Industrial Effluent Characterization and Treatment Objectives **221**
    General... 221
    Calculating Costs of Industrial Effluent Treatment.................. 222
    Industrial Wastewater Composition 222
    Industrial Effluent Case Studies................................. 223
        Case Study: Paper Mill Effluents.......................223
        Case Study: Plating Effluents..............................224
        Case Study: Petrochemical Facility Wastewaters.................225
        Case Study: High Sulfate Effluents 225
        Case Study: Vegetable Processing Effluents 227
    Problems of Industrial Effluent Treatment 228
        Batching 228
        Toxic Shocks 228

Nutrient Balance 228
Sludge Consistency..229
Changes in Main Process 229
Further Reading 229

**CHAPTER 17 Industrial Effluent Treatment Unit
Operation Design: Physical Processes** **231**
Introduction... 231
Gravity Oil/Water Separators... 231
Coalescing Media 233
Hydrophobic Media: Packed Beds......................................233
Hydrophilic Media: Nutshell Filters 233
Removal of Toxic and Refractory Compounds 234
Adsorption... 234
Membrane Technologies: Oily Water................................... 235
Membrane Technologies: Removal of Dissolved Inorganics... 236
Further Reading 236

**CHAPTER 18 Industrial Effluent Treatment Unit Operation
Design: Chemical Processes**..................................... **237**
Introduction... 237
Neutralization.. 238
Metals Removal 239
Precipitation as Hydroxides...239
Precipitation as Sulfides 239
Precipitation by Reduction 240
Sulfate Removal 240
Cyanide Removal 240
Organics Removal 241
Other Chemical Processes 241
Oxidants...241
Disinfection.. 242
Chemical Disinfection 242
Dechlorination 242
Further Reading 242

**CHAPTER 19 Industrial Effluent Treatment Unit Operation
Design: Biological Processes**.................................... **243**
Introduction... 243
Inhibitory Chemicals 244
Aerobic Inhibitors..244
Anaerobic Inhibitors 244
Prediction of Heavy Metal Inhibition of Digestion Process 244

Nutrient Requirements 246
Microbiological Requirements 246
Aerobic Treatment Design Parameters 247
Anaerobic Treatment Design Parameters 247
Further Reading 248

CHAPTER 20 Industrial Effluent Treatment Hydraulics 249
Introduction 249
Hydraulics and Layout 249
Hydraulic Design 250
Flow Balancing 250
Sludge Handling 251

SECTION 5 SLUDGE TREATMENT ENGINEERING 253

CHAPTER 21 Sludge Characterization and Treatment Objectives 255
Treatment Objectives 255
Volume Reduction 255
Resource Recovery 256
Other Objectives 256
Sludge Characteristics 257
Wastewater Sludges 257
Potable Water Sludges 259
Disposal of Sludge 261
Beneficial Uses of Sludge 262
Solids Destruction 262
Further Reading 263

CHAPTER 22 Sludge Treatment Unit Operation Design: Physical Processes 265
Sludge and Scum Pumping 265
Grinding 266
Degritting 266
Blending 266
Thermal Treatments 267
Volume Reduction Processes 267
Thickening 267
Dewatering 271
Novel Processes 272
Further Reading 273

CHAPTER 23 Sludge Treatment Unit Operation Design:
Chemical Processes 275
   Introduction.. 275
   Stabilization 275
   Conditioning 276

CHAPTER 24 Sludge Treatment Unit Operation Design:
Biological Processes 277
   Anaerobic Sludge Digestion.................................... 277
   Types of Anaerobic Digestion 278
      Factors Affecting Anaerobic Digestion 278
      Anaerobic Digester Design Criteria 279
   Novel Types Of Anaerobic Reactor.......................... 282
      Anaerobic Filter Reactors.................................. 282
      Anaerobic Contact Reactor 283
      Upflow Anaerobic Sludge Blanket 284
      Fluidized (Expanded) Bed Reactor 285
   Aerobic Sludge Digestion 286
      Composting... 287
      Disinfection.. 287
   Reference 287

CHAPTER 25 Sludge Treatment Hydraulics 289
   Introduction.. 289
   Rigorous Analysis.. 290
   Quick and Dirty Approaches.................................... 290
      Sludge Volume Estimation.................................. 290
      Headlosses for Sludges..................................... 291
   References.. 291
   Further Reading 291

SECTION 6 MISCELLANY 293

CHAPTER 26 Ancillary Processes 295
   Introduction.. 295
   Noise Control.. 295
   Volatile Organic Compounds and Odor Control 296
      Measurement and Setting of Acceptable Concentrations..... 297
      Covers and Ventilation....................................... 297
      Removal of Volatile Organic Compounds and Odors......... 297
   Fly Control... 298
   Reference 299

CHAPTER 27 Water and Wastewater Treatment Plant Layout **301**

   Introduction 301

   General Principles 302

   Factors Affecting Layout 303

      Site Selection 305

      Plant Layout and Safety 307

      Plant Layout and Cost 308

      Plant Layout and Aesthetics 309

   Matching Design Rigor with Stage of Design 310

      Conceptual Layout 310

   Combined Application of Methods: Base Case 312

      Conceptual/FEED Layout Methodology 312

      Detailed Layout Methodology 313

      "For Construction" Layout Methodology 314

   Further Reading 315

CHAPTER 28 Classic Mistakes in Water and Effluent Treatment Plant Design and Operation **317**

   Academic Approaches 317

   Separation of Design and Costing 317

   Following the HiPPO 318

   Running from the Tiger 319

   Believing Salespeople 319

   The Plant Works ... Most of the Time 320

   Water Treatment Plant Design is Easy 320

   Let the Newbie Design It 321

   Why Call in a Specialist? 321

   Who Needs a Process Engineer? 322

   Ye cannae change the laws of physics, Jim 323

   Further Reading 323

CHAPTER 29 Troubleshooting Wastewater Treatment Plant **325**

   Introduction 325

   Sick Process Syndrome 326

   Data Analysis 327

      Too Little Data 327

      Too Much Data 328

   Appropriate Statistics 328

   The Rare Ideal 329

   Site Visits 329

      Using All Your Senses 330

      Interviewing Operators 330

The Operating and Maintenance Manual 331

Maintenance Logs 331

Operator Training 331

Design Verification 332

Putting It All Together 333

Further Reading 334

CHAPTER 30 Water Feature Design **335**

Introduction 335

Tender Stage Design 337

Water Quality 339

Chemical Composition 339

Clarity 339

Biological Quality 340

Detailed Hydraulic Design 342

Nozzles 342

Weirs 342

Water Quality 343

Appendices 345

Appendix 1: General Process Plant Design 347

Appendix 2: The Water Treatment Plant Design Process 359

Appendix 3: Design Deliverables 369

Appendix 4: Selection and Sizing of Unit Operations 381

Appendix 5: Costing Water and Wastewater Treatment Plant 399

Appendix 6: Water Aggressiveness Indexes 405

Appendix 7: Clean Water Treatment by Numbers 411

Appendix 8: Sewage Treatment by Numbers 413

Appendix 9: Industrial Effluent Treatment by Numbers 415

Appendix 10: Useful Information 417

Appendix 11: Worked Example for Chemical Dosing Estimation 421

Index 425

Preface

INTRODUCTION

What qualifies me to write yet another book on water process plant design, in a market seemingly saturated with tomes on the subject?

Well, firstly, there are my qualifications and experience as an engineer. I graduated in 1991 with a Master's degree in Biochemical Engineering and found myself specializing almost exclusively in water and effluent treatment plant.

I worked for several specialist design and build contracting companies before becoming an independent consultant in 1996. Since then, my focus has been on water treatment plant design, commissioning, troubleshooting, and forensic engineering.

This last activity—acting as an expert witness in many legal disputes and several court cases where plants do not work—has opened my eyes to many of the common mistakes made by plant designers, perhaps even more so than the plant commissioning process itself.

These activities have given me a great insight into how not to go about designing water and effluent treatment plant. What is most surprising is how many of the errors I find as a forensic engineer in the design of water treatment plants appear repeatedly. Much of engineering knowledge is knowing what not to do, and that is the subject of this book: engineering know-how (and "know-how-not").

There are several ways in which process design can go wrong. The key way, in my experience, to guarantee a poorly operating plant is to dispense with the process engineer completely. Other ways include:

- Assuming that anyone with a degree in engineering is an engineer; fresh engineering graduates are not engineers yet and therefore require close supervision by a professional engineer
- Lack of clarity on who is responsible for process design
- Failure to manage the design process; active management and formal quality control are required to produce a good design
- Failure to manage the construction and commissioning process; active management and formal quality control are also required to build a good plant
- Failure to manage operation and maintenance properly

The main source of my professional knowledge—other than learning from others' mistakes—has been access to the design manuals held by the various companies I have worked for. While these are jealously guarded commercial secrets, they share many common features. They all contain a core of information that may not be in the public domain, but is common knowledge amongst experienced professional process design engineers.

The second key qualification I have for writing this book is my experience in teaching and training engineers. I have been training other professional engineers around the world to understand water and effluent treatment plant design for more than 20 years.

As a result, I know how (and how not) to go about designing cost effective, safe and robust water and effluent treatment plants. I also know which parts of the discipline are most challenging for the various types of engineers and scientists in different sectors who need to understand or undertake the design of such plants.

For example, many civil engineers have difficulty with chemistry, chemical engineers commonly flounder with biology, and scientists have trouble understanding the difference between engineering and science. Process engineers from other industry sectors are often unfamiliar with water chemistry as well as open-channel hydraulics, far more important in water treatment than in other areas of process plant design.

I have been teaching realistic process plant design in academia for some time, and most of my design examples have been water and effluent treatment plants. I know what beginners to process plant design find difficult; a subject I covered at a generic level in my first book, *An Applied Guide to Process and Plant Design*. Absolute beginners to water and effluent treatment plant design might find it useful to read both books, as I do not repeat very much of what is there in this book.

This book is not intended to cover every aspect of the discipline. Instead, I have focused on those aspects that I had to learn from more experienced engineers in order to design a working water treatment plant. There is not much in here that you can simply Google.

As with both of my previous books I will not be following academic conventions of referencing. The text is based on my personal experience and that of my collaborators, on published codes and standards, and on those parts of in-house design manuals, which are too commonly replicated for anyone to reasonably consider them company-specific know-how. My sources for most of this material themselves came without formal attribution. I asked a more experienced engineer how to solve a problem, and they told me, or gave me a many times photocopied piece of paper with a graph on it. Such is the nature of know how.

I have not, therefore, reproduced very much in the parts of this book (beyond my explanation of engineering science), which is readily available in the public domain. I do, however, suggest texts and web resources in which such material can be found. This book is intended, like my first one, as a substitute for a knowledgeable mentor, willing to freely disclose the core knowledge of the discipline that so many of us like to keep to ourselves. It is written in an informal style appropriate to interactions between mentor and mentee, but the informality does not imply lack of rigor. As with all my books the content has been validated by review by other experienced engineers.

In short, this is the book I wished I could have bought when I started in the sector.

STRUCTURE

The first section of the book is an overview of practical process plant design, based largely on the consensus professional approach I have set out in my two previous books. This is followed by a section giving an overview of relevant engineering science. The book ends with sections on classic mistakes to avoid, troubleshooting and the specialist area of large-scale water feature design.

The main body of the book, between these sections, is split into water and sludge treatment. Water treatment is further split into clean, dirty, and industrial effluent treatment.

I have divided the unit operation design part of these chapters into physical, chemical, and biological processes. I could, alternatively, have split the processes by stage of treatment, but this involves as many compromises as the structure I ultimately chose.

The absence of any perfect way to structure the material generated several orphan chapters, offered in a "Miscellany" section at the end.

Finally, appendixes are provided to cover detailed consideration of some issues in engineering science, general process design, and specific design procedures which, though important and poorly covered in the existing literature, would interrupt the flow if placed in the main body of the book.

Seán Moran
Derbyshire, United Kingdom
2017

Acknowledgments

I would like to thank the many contributors who have helped with the production and planning of this book. Firstly, sincere thanks to the editorial and production team at Elsevier, particularly Fiona Geraghty, Kostas Marinakis, Edward Payne, and Leticia Lima who have all supported me throughout my book-writing ventures and steered this particular project from inception through to publication.

Rosie Fernyhough and Alun Rees both assisted with the tedious but crucial task of proofreading and also made helpful suggestions during the writing of this book.

Many thanks to all those individuals and companies who have kindly supplied images and information, including Tim Allen at Durapipe, Adriaan van der Beek at JWC Environmental, Denise Bennett and Jeremy Dudley at WRc plc, Piotr Brozda at PCI Membranes, Linda Dingley at Grundfos, Ken Edwards at LMNO Engineering, Nazir Haji at EEMUA, Professor Simon Judd, Razib Khan, Tosh Singh at Lutz-Jesco (GB) Ltd, Vicky West at CIWEM, Kim Woochan at ROplant.org and Murphy Yuan at the Shanghai Xunhui Environment Technology Co., Ltd.

Special thanks to Geoffrey Blumber at Hexagon PPM for supplying another striking front cover image from one of their client projects.

Finally, thank you to my wife Annemarie for all her work on preparing and checking the manuscript and managing the project.

Introduction: the nature of water and effluent treatment plant design

1

CHAPTER OUTLINE

What is this Book About?...1
A Brief History of Water Treatment Plant Design 2
The Literature of Water Treatment Plant Design 3
Interaction with Other Engineering Disciplines...3
Hydraulic Calculations 4
Water Chemistry 4
Water Biology..4
Economics...4
Materials Selection..5
The Importance of Statistics in Water and Effluent Treatment 5
    Statistics in Sewage Treatment Plant Design and Performance 6
    Statistics and Discharge Consents 7
Contractual Arrangements in Water and Effluent Treatment...8
Misconceptions in Water and Effluent Treatment Plant Design 10
    Chemical/Process Engineering Misconceptions..10
    Academic Misconceptions...10
    Sales Misconceptions 11
Further Reading 12

WHAT IS THIS BOOK ABOUT?

This is a book about designing water and effluent treatment plants. These plants are groups of machines (plus ancillary machines) which process relatively dirty water into relatively clean water over a series of linked stages.

In professional practice, the design of water and effluent treatment plants is largely based on longstanding heuristics, supported to some extent nowadays by engineering science and mathematics. This book is based upon some of the most elusive parts of the body of design knowledge which I have amassed over my 25 years of professional practice; the sort of knowledge which rarely makes it into the public domain, which I would pass on to someone I was mentoring.

FIGURE 1.1

Bronze Roman hydraulic pump.[1]

Courtesy: Lalupa.

I have gone to great pains to make sure that this book is as close as possible to current consensus best practice. It is however general in nature (though I have offered limits on applicability where I am aware of them), and only offered for guidance.

A BRIEF HISTORY OF WATER TREATMENT PLANT DESIGN

Water engineering has a long pedigree. Fig. 1.1 shows a Bronze Roman hydraulic pump and adjustable nozzle dating from 1 to 200AD, found in a mine in Huelva province, Spain.

While the Romans were impressive water engineers, the modern forms of both dirty and clean water engineering were pioneered in the United Kingdom.

Historically, civil engineers tended to be responsible for the design of municipal water treatment plants, with assistance from chemists for process development research. Civil engineers specializing in this area were known as "environmental" or "sanitary" engineers.

This was the model as far back as the development of the activated sludge process for sewage treatment by Ardern and Lockett in 1915, and the first use of chlorine for drinking water disinfection around the same time.

Since the days of Ardern and Lockett, we have moved from design by a combination of chemists and civil engineers toward integrated process design by chemical engineers. A colleague at a university where I teach water treatment plant design to civil engineers confided recently that the dirty secret of environmental engineering is that everyone who teaches it nowadays is a chemical engineer.

THE LITERATURE OF WATER TREATMENT PLANT DESIGN

There are many substantial and longstanding textbooks on water treatment plant design. Prime examples include *Metcalf and Eddy* for wastewater and *Twort* for clean water supply (see Further Reading).

These books do however tend to have been written both by and for environmental engineers, whose focus is generally not on detailed process design. They can consequently lack much of the information which chemical engineers need to follow their rather different approach to process design.

INTERACTION WITH OTHER ENGINEERING DISCIPLINES

Interactions between chemical process engineers and other engineering disciplines can be slightly complicated because we have traditionally not had a role in water treatment. Nevertheless, there are reasonably clear expectations on the documents which process engineers produce for other disciplines, most notably electrical, civil, and software engineers.

The electrical engineer will need process engineers to produce schedules of drives, actuated valves and instruments, together with a layout drawing showing where drives and instruments are located.

The software engineer will need the functional design specification and the piping and instrumentation diagram from the process engineer to allow them to price the control software and hardware.

The civil engineer will need layout drawings, marked with the weights of any major items to be placed on the slab. If concrete water-retaining structures are being used, they will need to know their internal dimensions and what they are to contain. They may also be relying on the process engineer to comment on the chemical nature of fluids in contact with concrete from the point of view of concrete specifications.

HYDRAULIC CALCULATIONS

Water treatment, especially dirty water treatment, is unusual in making extensive use of open channels, weirs, flumes, and open-topped vessels. This requires "open channel" hydraulic calculations which may be unfamiliar from academic fluid mechanics courses, or professional practice in other sectors.

WATER CHEMISTRY

As a schoolboy I always imagined water chemistry would be the least exciting kind, mainly because of the low potential for explosions, but it is actually an extensive field, about which I am still learning new things after 25 years of practice.

Water chemistry tends to be rather neglected nowadays, both at school/college and university level, though it is of great importance in biotechnology and food processing as well as in water treatment. Other engineers will however usually look to the chemical engineer in the water sector for a detailed understanding of water chemistry.

WATER BIOLOGY

My own background is in applied biology, so I am lucky to have a good grounding in the biological underpinnings of water treatment. It is common for both chemical and civil engineers to struggle with the design of biological processes for water treatment plants.

Biochemical engineering allows a rigorous mathematical and chemical analysis of a monoculture of organisms held under controlled conditions and fed with pure chemicals. However, such approaches are of little use in dirty water treatment. Effluent treatment plants contain thousands of types of organisms in variable proportions under uncontrolled conditions being fed with thousands of different compounds in variable proportions. A balance must be therefore struck between, on the one hand, coarse rules of thumb applied without much understanding and, on the other, an unrealistically precise calculation.

ECONOMICS

In water treatment, it is critical to identify the cheapest way to meet the specification safely and robustly, to the extent that robustness and even safety may be underemphasized by some designers. I was amused to read an academic research paper recently claiming that all the problems of total closed loop water and nutrient recycling had been solved, except for the economics. Economics are, in most cases, the *only* significant problem in water treatment. Even the dirtiest, most polluted water could be made into the cleanest, if cost were not an issue. (Persuading people to drink the product, however, may still be a problem, for a variety of reasons.)

MATERIALS SELECTION

Water treatment engineering usually operates at close to ambient temperature, and at low pressures. Aqueous solutions are often the most aggressive chemical used. Water engineers must almost always be looking for the lowest capital cost solution. This has a strong effect on our material selection choices.

While the bare carbon steel favored in hydrocarbon-based chemical engineering may be relatively cheap, it is readily corroded by all but the most highly treated water. The only time I have seen it used in water treatment plants is when a nonspecialist has been allowed to design them.

If water specialists use ferrous metals, they tend to be either stainless steel or, far more commonly, the much cheaper cast/ductile iron which is significantly more resistant to corrosion than carbon steel. If we use mild steel, it tends to be coated with something such as zinc (galvanized mild steel), glass (enameled, which we call "glass-on-steel"), epoxy, and chlorinated rubber.

Specialists make significant use of thermoplastics, such as polyvinyl chloride (PVC) and acrylonitrile butadiene styrene (ABS), especially for piping systems below 300 mm nominal bore. Plastics are strong, light, cheap, and do not corrode. Some need to be protected from light and heat, but this is straightforward and inexpensive.

We also use copper, brass, and bronzes far more frequently than "traditional" process engineering sectors.

THE IMPORTANCE OF STATISTICS IN WATER AND EFFLUENT TREATMENT

Forensic engineering or expert witness work of the kind I often undertake frequently involves an element of process troubleshooting. There can be no proper process troubleshooting without statistics. Laplace transforms may not be of much use to practicing process engineers, but statistics are essential. It is a pity that university curricula often tend to view things the opposite way around.

I have never had to carry out, say, a three-way analysis of variance with interactions or use Duncan's multiple range test professionally, but if you have tens of thousands of data points, plotting them on a scatter diagram will not be particularly revealing. That said, this exercise tends to have been attempted with every data set I receive from a client, usually with no particular aim in mind.

At a minimum, summary statistics will be needed to make some sense of the data. Using Microsoft (MS) Excel to give the mean, minimum, maximum, and the upper limits of the 95% confidence interval of any data set is a good place to start. It may also help to generate correlation coefficients for relationships between parameters.

I could teach primary school children how to make MS Excel produce summary statistics in a few minutes. What an expert offers, however, is not the knowledge of how to do this, but an understanding of what the results mean and what they do not mean. It is commonplace, even among supposed experts, to misunderstand what we are 95% confident about within the 95% confidence interval, never to check whether the assumptions underlying the valid use of parametric statistics are true, and to be unaware of when and how to apply nonparametric statistics.

If you were to hire a statistician, they would hopefully understand all the above better than me. They would not however understand what the outputs mean, in engineering terms, because these are not just numbers. These are clues as to the state of a process.

Chemical/process engineers understand processes. Mathematics is just a tool to help us to cut through complexity, to see more clearly, and inform our answers to the questions being asked. These questions usually add up to the same thing, namely: "How well does this process work?" (The answer to this question is most often, in my professional practice, "It doesn't work", but that's the nature of expert witness work.)

STATISTICS IN SEWAGE TREATMENT PLANT DESIGN AND PERFORMANCE

Sewage varies continuously over time with respect to both quantity (flows) and composition (loads), and the equipment which makes up a sewage treatment works has to be sized based upon the flows and loads to be treated. Statistics must be used to make sense of these continuously varying flows and loads. In addition, statistics are used to define the limits on the performance of the plant.

A designer of a municipal sewage treatment works needs to have data from both the wet parts of the year, when flows are high and concentrations of pollutants are consequently low; and the dry parts of the year when the reverse situation happens. The designer needs data on the variations which happen on a scale of a year, month, or week in both domestic and industrial wastewater production, as well as those which happen on the scale of an individual day.

The key statistics from the point of view of sewage treatment plant design and performance are as follows:

- The "average," where all the numbers are added together, and divided by the number of figures added. This is known in statistics as the "mean," and there are other kinds of average

- The "standard deviation," calculated from the difference between each individual figure and the mean. The greater the standard deviation, the more variable the data are, and often the less we have of it
- The "95% confidence interval," which is the range of numbers between (mean + 1.96 times the standard deviation) of a set of data and (mean−1.96 times the standard deviation). Correctly, it is the range where we are 95% certain that the true mean (which would be calculated from all theoretically possible samplings) lies. Practically, it represents the uncertainty associated with our data
- "Statistically representative data," from a practical water engineering point of view, is obtained by analyzing many random samples over several years to capture all the variability in flows and loads. It does not matter from an engineer's practical point of view if there are some gaps in the data, as long as these gaps are distributed randomly, and there is a large enough amount of data

So, anyone tasked with writing the specification of a performance trial needs to be very clear about the answer to the question "What do I mean by *works*?" For example:

- If we want something to have a composition which is "25% X," do we mean "no less than 25%X," "25%(± 1%) X," or "on average 25%X"
- If we use the term average, which kind of average do we mean?
- Do we want a statistically significant result? How significant does it have to be? What statistical test must we apply?
- When we use the term "25%," we are implying that there are only two significant figures available. If there were three, it would be "25.0%." As 24.5% is 25% to the implied two significant figures, anything over 24.5% is arguably a pass
- Under what conditions is the trial to be conducted? Who is to operate it? What will be done in the reasonably common situation where the client cannot provide the plant with feedstock to the promised specification at the time of the trial?

From a practical point of view, if the answers to such questions are not decided in advance, the trial becomes a hostage to fortune. From a strictly scientific point of view, the lack of declarations on these issues prior to test commencement invalidates the test, for technical reasons to do with the theory of design of experiments.

STATISTICS AND DISCHARGE CONSENTS

Consents to discharge of treated trade effluent are absolute in the United Kingdom, which means that a single discharge of effluent above consented concentrations or flows is a criminal offence. If an industrial effluent treatment plant breaches consent just once, this is a breach of the law. It is imperative to understand why the breach occurred, in order to avoid it happening again.

I commonly see situations in which breaches of consent have become commonplace. In everyday life, things not working occasionally might be OK, but occasional failures in an industrial effluent treatment point to a problem with the main process which the effluent plant serves, or a problem with the plant itself.

Furthermore, I frequently encounter situations in troubleshooting and expert witness cases where no performance trial has ever been carried out. The plant may consequently have never been capable of reliably meeting consent and therefore could have been running in a noncompliant fashion for years. In such situations, it can be very difficult to persuade clients to understand that discharge consents are absolute, and nonnegotiable, and that every failure is a breach of the law. Sometimes it takes more than one prosecution before they fully understand their position. Environmental regulators tend to exercise discretion and work with the owners of inadequate effluent treatment plants, rather than applying the letter of the law, in a way which has in my experience allowed confusion on the part of treatment plant owners.

The design of industrial effluent treatment plant is intimately linked with reliability. As such, a plant treats a highly variable flow of effluent with a wide variation in contaminants and its designer will have based its design on several scenarios, each with associated probabilities.

So, though discharge consents are absolute, plant design is probabilistic. Designing a plant to handle the 99% confidence interval conditions costs a great deal more than designing one to the 95% confidence interval conditions.

It is possible for plants which meet a minimum specification or standard (such as, for example, EN12566-3 2005 for domestic package plants) to have wildly varying capacities and performance, depending on the cautiousness of their designers.

Any failure to meet consent in industrial or domestic effluent treatment plants should be a wakeup call. More than one failure, and it is time to call in an expert, by which I mean a professional engineer.

CONTRACTUAL ARRANGEMENTS IN WATER AND EFFLUENT TREATMENT

When a business (usually an operating company, and sometimes called "the ultimate client" in the supply chain) wishes to procure a water treatment plant, it will normally start by drawing up a performance specification. This performance specification should detail, among other things, exactly what quantity, and quality of inputs ("feed") the plant will receive, and exactly what quality of product it must make.

The performance specification forms part of an Invitation to Tender which is then published. Contracting companies will be the usual responders to Invitations to Tender.

In response to the ultimate client's Invitation to Tender (containing the performance specification), the contractor then submits a formal Tender document to the ultimate client. The Tender contains information including several key engineering drawings, details of proposed subsystems, and a detailed itemized costing. To draft the Tender, the contractor in turn issues either formal Invitations to Tender or enquiry documents to equipment and services suppliers, as potential subcontractors/suppliers. This is done in order to establish the cost to the contractor of the main plant items and other equipment and services, which will inform the pricing of the Tender response to the ultimate client.

So, the contracting company must effectively complete a detailed design of the plant before they can even issue their own Invitations to Tender to suitable potential suppliers. Every plant is bespoke, and its key subcomponents are also usually customized. Even if they are standard items, they are taken from a range of possible unit sizes and specifications. No supplier sells a single product of a single size.

Both suppliers and purchasers tend to have their standard terms and conditions of contract. In addition, engineering institutions and other professional organizations publish standard terms. Whether third party, supplier or purchaser's standard terms are ultimately adopted is a matter of negotiation.

In any case, equipment is normally bought outright, and payment terms requiring upfront payment of 100% of the purchase price are extremely unusual. Stage payments are the norm, both from contractor to supplier and from ultimate client to contractor. The equipment is often installed by specialist subcontractors, who are again paid by stages.

The contractor's own highly expert staff tend to carry out process commissioning in which the plant is put to work, and to supervise performance trials.

The plant is handed over after a successful performance trial to the ultimate client, whose staff are trained to run it during or after the commissioning and performance trials period.

Plants are, in summary, usually designed and built by specialist contractors for well-financed ultimate clients who purchase them outright, and run them using their own workforce.

There are variations on this process such as the Build Own Operate contract, in which a contractor will offer not just to build a plant but to operate it for a cost per unit of feed treated. The only difference between this situation and the one described earlier is that the plant is not handed over after performance trials, but remains the property of the contractor, who operates it for the client, and charges per unit of feed treated. There is also a BOOT variant in which the plant is eventually transferred to the client. These are not commonly used models in developed countries, because they are designed to accommodate clients with limited capital reserves, technical skills, or skilled manpower at the expense of what are usually far higher prices per unit treated.

MISCONCEPTIONS IN WATER AND EFFLUENT TREATMENT PLANT DESIGN

CHEMICAL/PROCESS ENGINEERING MISCONCEPTIONS

Process engineers from other sectors can find it difficult to assimilate our often nebulous design specifications and the vague and sometimes rather flexible nature of the "substances" we work with. Many of these "substances" are more accurately viewed as test results than chemicals.

"Suspended solids" can (unlike matter) be created and destroyed in an effluent treatment plant. Chemicals can also switch from suspended to dissolved forms if the water chemistry changes, with no loss of matter.

Another example is biochemical oxygen demand (BOD). Although specialists discuss it between themselves as if it were a chemical, BOD is not really a chemical substance at all, but the amount of oxygen a complex mixture of substances will absorb under given physical, chemical, and biological conditions in a given time.

Also, buffering effects in natural waters make prediction of the amount of acid required to give a specified pH change something of a black art. See discussion of this at Chapter 2, Water Chemistry and Appendix 6.

Many water and effluent treatment processes are simply too complex for them to be sufficiently well characterized to allow a traditional chemical engineering approach.

ACADEMIC MISCONCEPTIONS

Much academic research in water and effluent treatment is not merely highly speculative, but is based on a misunderstanding of the basic constraints of water and effluent treatment.

It may, for example, be the case in future that higher regulatory costs to discharge nutrients to environment make the economics of recovery more viable. It will, however, never be economically viable to operate 100% recovery of all components of an aqueous waste stream, as each successive increment in recovery toward 100% costs more than the previous one. Engineers must always consider the balance of costs and benefits.

By this logic, any research based upon a supposed need, say, to remove amounts of trace organics only detectable by means of the most sensitive laboratory tests which have not been shown to have any effects on humans or environment is probably entirely spurious and a waste of public funds.

On the other hand, nothing less than an entirely robust technique will ever be suitable in a water treatment environment, though the reasons for this differ. Clean water treatment must be an entirely reliable source of safe drinking water, even when operated nonideally.

Sewage and effluent treatment plant designers may in future be given slightly more flexibility with respect to absolute reliability, but the industry is incredibly price sensitive and conservative. Industrial effluent treatment plants are frequently operated by the cheapest staff available, and/or ignored until they break down entirely. Effluent also contains a variable mixture of grit, grease, harsh chemicals, textiles, and other components which block, blind, abrade, and corrode anything they meet.

Equipment and processes which need treating with great care are fundamentally unsuitable for water treatment plants, a fact which is ignored by most researchers. I can illustrate this point with two anecdotes from personal experience.

In recent years, I was an advisor to a UK government program promoting the use of novel anaerobic treatment designs in industrial effluent treatment. The promotional materials I was issued with have been proven over time to be wildly optimistic, and since then I have been professionally instructed on more than occasion to report on why these technologies subsequently proved not to work as well in practice as they had in the laboratory.

My second anecdote is based on a conversation I had with a researcher about a new nanopore diffuser he had designed for aerating sewage. His tests were proceeding well with simulated sewage, as would be expected: the smaller the bubbles, the better the mass transport. Of course, as soon as the nanopores met real sewage, the diffuser did not diffuse any more. That was not the only problem— the tiny bubbles did not rise as quickly as the larger ones normally used, so they carried over into the next stage, and did not stir the aeration basin to keep the solids in suspension. Nano-stuff was, however, the flavor of the moment for funding bodies. As a result, an obviously unsuitable but fashionable technology was investigated purely to attract research funding, rather than because it had a realistic chance of success.

SALES MISCONCEPTIONS

To call some of the following "sales misconceptions" is arguably overly generous. However, technical salespeople in engineering tend not to lie outright. Salespeople tend to rely on the materials they are given, just as I did myself in the anecdote I gave in the previous section.

The same promises have been being made and broken repeatedly over the time I have worked in water engineering. Any of the following must be treated with extreme caution:

- Zero solids yield biological treatment processes
- Biological treatment processes which are radically smaller than an equivalent conventional plant
- Magnetic water treatment (and any other treatment process with an unexplained, secret, unknown, or unproven mechanism, supported largely by anecdotes)
- Novel processes involving filtration in place of settlement for solids removal
- Comminution in place of screening for gross solids removal
- Special bacterial cultures, nutrients, or mixtures of these which are claimed to enhance biological treatment

FIGURE 1.2

Advertisement for a "marketing breakthrough."

Source: RPM Ltd.

- Any process which does not have at least three installations which treat the same water type to the same standard at the same size as you are considering or larger, at least 5 years old, whose owners are willing to speak candidly to you (see, for example, electrodialysis)
- Plant owners who are commercially interested in further sales in any way

Finally, it is worth noting that it is not just academia which has fashions. Although there is virtually nothing new in water treatment, old ideas get rebranded and recycled every few decades, as a "marketing breakthrough" (see Fig. 1.2 for an amusing example of this). Despite the salesmen's claims, they have the same advantages and disadvantages as they always did.

FURTHER READING

Brandt, M. J., et al. (2016). *Twort's water supply* (7th ed.). Oxford: Butterworth-Heinemann.

Metcalf & Eddy Inc., et al. (2014). *Wastewater engineering: Treatment and resource recovery* (5th ed.). New York, NY: McGraw-Hill.

Water engineering science

INTRODUCTION

Despite the fact that most chemical engineering syllabi contain only limited water chemistry, process engineers in the water sector are far more commonly expected to perform their own "chemistry" than in other sectors. Indeed, many civil engineers still think of us more as process chemists than as engineers for historical reasons.

Similarly, we will be expected to understand enough biology and biochemistry to go beyond simple rules of thumb if we are involved in secondary treatment plant design, despite most engineers knowing little about biology.

In this section, I cover the absolute minimum knowledge of water chemistry and biology to get by, as well as a few important areas, which are comparatively poorly understood even by some experienced engineers.

Water chemistry

2

CHAPTER OUTLINE

Hardness and Alkalinity 15
Acids and Bases 15
    Buffering 17
    Scaling and Aggressiveness 19
    Oxidation and Reduction 21
    Dechlorination..23
    Catalysis..23
    Osmosis 23
Further Reading 24

HARDNESS AND ALKALINITY

Even relatively experienced engineers sometimes confuse hardness and alkalinity, perhaps because of how water engineers standardize their units.

Alkalinity is determined by titration against acid, and is generally expressed as an equivalent amount in mg/L of $CaCO_3$, though the titration result will be affected by other bases such as hydroxide.

Hardness is the amount of dissolved calcium and magnesium in water. Calcium hardness is the hardness based solely on the calcium ion, and it is often also expressed as mg/L $CaCO_3$.

Expressing both calcium hardness and alkalinity as mg/L $CaCO_3$ is confusing, but if you remember that hardness is the Ca^{2+} part, and alkalinity the CO_3^{2-} part, this should help to avoid confusion.

We also have to be familiar with many other ways to express these two parameters, such as milliequivalents, English (Clark), German, and French degrees. These can be converted to standard units as given in Table 2.1.

ACIDS AND BASES

Although water based chemistry is often poorly covered in chemical engineering courses, the concepts of dissociation, pH, and acid/base reactions are hopefully familiar to the reader.

Table 2.1 Conversion Table for Hardness and Alkalinity Parameters

Water Hardness Classification	American Degrees (mg/L CaCO$_3$)	German Degrees (dGH/°dH)	English/Clark Degrees (°Clark, °e or e)	French Degrees (°fH or °f)	American gpg	Milliequivalents/L (meq/L mval/L)
Soft	0–60	0–3.4	0–4.2	0–6	0–3.5	0–3
Moderately hard	61–120	3.4–6.7	4.2–8.4	6.1–12	3.6–7.0	3–6
Hard	121–180	6.7–10.1	8.4–12.6	12.1–18	7.0–10.5	6–9
Very hard	≥181	≥10.1	≥12.6	≥18	≥10.5	>9

Acids are—as far as water engineers are concerned—substances which form low pH solutions and react with bases (which have high pH solutions) to form salts which tend toward neutral pH. If both acid and alkali are highly dissociated ("strong"), their salts have a neutral pH. The salt of a strong acid and a weak base (such as ferric chloride) tends to form a low pH solution, and the salt of a weak acid and a strong base (such as sodium acetate), a high pH solution. The pH of a solution of a weak acid and a weak base depends upon the relative strengths of acid and base.

Things can get very complex and hard to engineer when there are (as is often the case) many weak acids or bases in a water to be treated. Therefore, we tend to use strong mineral acids such as hydrochloric or sulfuric and strong bases such as sodium hydroxide to control pH, to minimize the number of possible variables. The reaction will yield either neutral or acid salts.

We are not however able to size pumps used to dose sufficient acid or alkali to obtain a chosen pH by simple stoichiometry, due to the phenomenon of buffering.

BUFFERING

Buffering reduces the effect on pH of added acid or alkali. Buffer solutions achieve their resistance to pH change because of the presence of an equilibrium between a weak acid and the "conjugate base" left behind when a proton (H + ion) is separated from it by dissociation.

When a strong acid is added to an equilibrium mixture of a weak acid and its conjugate base, the protons from the highly dissociated strong acid shift this equilibrium toward the undissociated form of the acid (by Le Chatelier's principle) and the hydrogen ion concentration consequently increases by less than the amount expected for the quantity of strong acid added. Similarly, if strong alkali is added to the mixture, the hydrogen ion concentration decreases by less than the amount expected for the quantity of alkali added.

Buffer capacity

Buffer capacity, β (Fig. 2.1), is a quantitative measure of the resistance of a buffer solution to pH change on addition of hydroxide ions. As process designers, we do not need to worry about calculating it precisely. In practice, we only need to be able to estimate the chemical addition which will overcome the buffering capacity of our specific source water. We should however understand that there are three regions of high buffer capacity:

- At very low pH, β increases proportional to hydrogen ion concentration. This is independent of the presence or absence of buffering agents
- In the region pH = $pK_a \pm 2$, β rises to a maximum at pH = pK_a. Buffer capacity is proportional to the concentration of the buffering agent, C_A, so dilute solutions have little buffer capacity. It is also proportional to the acid

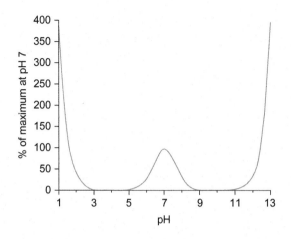

FIGURE 2.1

Buffer capacity for a 0.1M solution of an acid with pK_a of 7.

dissociation constant, pK_a (rather than K_a); thus, the weaker the acid the greater its buffering capacity

- At very high pH, β increases proportional to hydroxide ion concentration, due to the self-ionization of water and is independent of the presence or absence of buffering agents

Carbonate buffering is likely to be the most important kind in the case of natural waters. Even confining ourselves to just buffering by carbonates, there are many variables to consider in order to understand how pH, and various forms of unbound, partially bound and bound carbon dioxide are related. The relevant various forms of carbon dioxide are known to water engineers as "free CO_2," hydrogen carbonate, and carbonate.

To explain how pH might not change as expected when acid or alkali are added to a carbonate-buffered system, it may help to think of free CO_2 acting like a weak acid, and hydrogen carbonate like a weak alkali, keeping pH stable whether small amounts of acid or alkali are added. The higher the concentration of the buffering substances, the stronger the effect (in the region where we are usually working). To give a broad idea of what that means in practice, Table 2.2 might help.

A more sophisticated analysis may be enlightening. Much of what practicing engineers rely on in this area is ultimately based on the 1912 work of Tillmann (in German), and several types of terminology are used in discussing these relationships. A few definitions are therefore needed to clarify matters:

- "Combined" or "firmly bound" CO_2 is the CO_2 bound in carbonates: it is absent below pH 8.3
- "Semicombined" or "half-bound" CO_2 is the CO_2 bound in hydrogen carbonates (also known as bicarbonates)

Table 2.2 Relationship Between P and M Alkalinity and Alkaline Ion Concentrations

P Alk =	Hydroxide (OH⁻) Alkalinity	Carbonate (CO_3^{2-}) Alkalinity	Bicarbonate (HCO_3-) Alkalinity
0	0	0	M
$<0.5 \times$ M Alk	0	2P	M − 2P
$0.5 \times$ M Alk	0	M	0
$>0.5 \times$ M Alk	2 P − M	2 (M − P)	0
M Alk	M	0	0

M Alk, methyl orange alkalinity (or simply "alkalinity" or "total alkalinity") is alkalinity measured at pH 4.2−4.5.
P Alk, phenolphthalein alkalinity (or "carbonate alkalinity" or "hydroxide alkalinity") is alkalinity measured at pH 8.2−8.3.
Courtesy: Watermaths by S. Judd (publishers: Judd & Judd Ltd).

• "Free CO_2" is undissociated carbonic acid.

Free CO_2 concentration can be calculated by Tillmann's formula:

"[Free CO_2]''(including aggressive CO_2) = antilog 10(6.3 + log 10[alk] − pH).

Conversely, pH can be calculated as = log10 [alk] × 0.203 × 10⁷/[CO_2].

"Equilibrating CO_2" is the part of the "free CO_2" required to keep hydrogen carbonates in solution.

"Aggressive CO_2" is "free CO_2" in excess of the "equilibrating CO_2," and its presence is considered to be a marker of the corrosion potential, or aggressiveness of a water.

Fig. 2.2 shows how the proportions of the different forms of carbon dioxide vary with pH in otherwise pure water. Left to right, these are "free CO_2," hydrogen carbonate, and carbonate.

However, if the water is not pure (as is the case for all real water and effluent sources), the curves fall in different places, as shown in a simplified form in Fig. 2.3.

The graph shows how the ratio of free CO_2 to hydrogen carbonate is different if there are other dissolved solids, and that it matters what those dissolved solids are. Hardness and alkalinity are themselves both types of dissolved solids. In addition, the ratios will also vary based on temperature.

So, even leaving aside similar interactions between other types of dissolved solids, and considering only "carbonate buffering," there are still many variables which determine the buffering capacity of the water.

SCALING AND AGGRESSIVENESS

"Aggressive CO_2" and other measures of aggressivity, such as the Langelier Saturation Index (LSI) and Aggressive Index (AI), are rough heuristics. They make use of some of the variables discussed earlier to attempt to say something

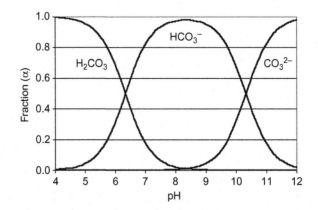

FIGURE 2.2

Relationship between pH and proportion of free CO_2/hydrogen carbonate and carbonate in pure water.

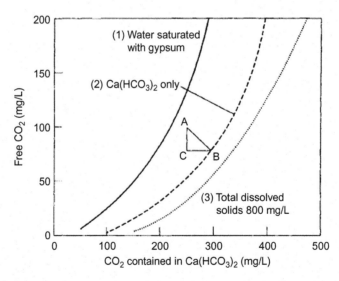

FIGURE 2.3

Relationship between pH and proportion of free CO_2/hydrogen carbonate and carbonate in water containing dissolved solids.

Reprinted from Eglinton, M. Resistance of concrete to destructive agencies in Hewlett, P.C. (ed.) (1998).
Lea's chemistry of cement and concrete, 4th ed., Copyright 1998 with permission from Elsevier.
Courtesy: Elsevier Press.

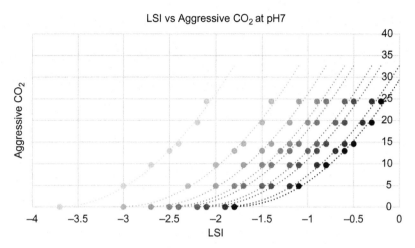

FIGURE 2.4

Relationship between LSI and aggressive CO_2 at pH7.

useful about the water for practical purposes such as specifying construction materials, or sizing acid or alkali dosing pumps.

These rough heuristics are related to one another, but their interrelationships are complex, because each heuristic emphasizes different factors. For example, I produced a graph of LSI versus Aggressive CO_2 (Fig. 2.4) for a site I worked on. Each curve represents the relationship between the two heuristics at a different level of hardness, and the whole graph only applies at a single pH value.

In summary, the real-world situation is irreducibly complex, but engineers deal with this by using rough measures, which they understand to be useful approximations with limited applicability. This description applies to all calculated values of "aggressive CO_2," LSI, AI, and so on.

The British Standard "aggressive CO_2" test (see Further Reading) gets around this from the point of view of the concrete designer by simply measuring how much lime the water dissolves.

OXIDATION AND REDUCTION

Redox reactions are key to several water treatment processes.

In borehole drinking water treatment, reduced manganese and iron salts may be oxidized by chlorine or by oxygen in biological filters, and biological oxidation is the key to secondary treatment of both municipal and industrial effluents.

In industrial effluent treatment, sulfite is used to reduce soluble Cr^{6+} into insoluble Cr^{4+}, and cyanides are oxidized by chlorine under alkaline conditions.

Oxidizing Disinfectants

Chlorine

Chlorine is the most widely used water biocide. The amount which needs to be added to bring about disinfection is determined by chlorine demand, contact time, water pH, and temperature.

When chlorine gas is mixed with water, it forms hypochlorous and hydrochlorous acid.

$$Cl_2 + H_2O \rightarrow HOCl + HCl$$

Hydrochlorous acid diffuses through cell walls and oxidizes the cytoplasm of microorganisms, disturbing the production of adenosine triphosphate, which is essential for respiration, thus inactivating more or less any living organisms present.

The dose of chlorine required is often determined based upon the amount of chlorine remaining after reaction time, known as the free chlorine residual. "Free chorine" is just our name for chlorine itself, differentiating it from bound chlorine (an assortment of chlorine compounds).

As a rule of thumb, assuming a residence time of 30 minutes, and pH values below 9, 0.4 ppm of free chlorine residual is required for effective chlorination. With pH in the range of 9−10, 0.8 ppm of free chlorine residual is required.

Chlorine dioxide

Chlorine dioxide is an oxidizing biocide which does not form hydrochlorous acids in water, but instead exists as dissolved chlorine dioxide, which is a stronger biocide at high pH.

Chlorine dioxide is explosive above 5% w/w in air, and therefore it must be produced or generated on site, by means of the following reactions:

$$Cl_2 + 2\,NaClO_2 \rightarrow 2\,NaCl + 2\,ClO_2$$

or

$$2\,HCl + 3\,NaOCl + NaClO_2 \rightarrow 2\,ClO_2 + 4\,NaCl + H_2O$$

ClO_2 is an effective biocide over a wide pH range at concentrations as low as 0.1 ppm, and is used principally in potable water treatment for surface waters with odor and taste problems. It is also claimed to have superior disinfectant ability against cryptosporidium oocysts.

Hypochlorite

Hypochlorites are salts of hypochlorous acid, usually supplied as sodium hypochlorite (NaOCl) solution or (less popularly) as solid calcium hypochlorite (Ca(OCl)$_2$). Both function in much the same way as chlorine, but are slightly less effective. Hypochlorination is less used for drinking water disinfection than before it was found that its use can produce excessive (toxic) bromate concentrations.

Ozone

Ozone, O_3, is a very strong oxidizing agent with a short life span. It can act as a biocide in much the same way as chlorine, as well as oxidizing things which chlorine and oxygen cannot.

Ozone readily splits into O_2 and an oxygen free radical, which rapidly reacts with bacterial cell contents:

$$O_3 \rightarrow O_2 + (O \cdot)$$

The usual dose in drinking water applications is 0.5 ppm of ozone, but the required dose for oxidation varies with pH, temperature, the concentration, and type of oxidable matter present, and the concentration of reaction products.

DECHLORINATION

If we wish to remove an excess of chlorine, we can dose sulfur dioxide or sodium metabisulfite.

From sodium metabisulfite:

$$S_2O_5^{2-} + H_2O = 2HSO_3^{2-}$$

From sulfur dioxide:

$$SO_2 + H_2O = H_2SO_3 = H^+ + HSO_3^{2-}$$
$$HSO_3^{2-} + HOCl = H_2SO_4^- + OCl^-$$
$$HSO_3^{2-} + Cl_2 + H_2O = HSO_{4^-} + 2\,HCl$$

CATALYSIS

Unlike the kind of process engineering taught in universities, physical/chemical catalysis is far less important in water treatment engineering than biological catalysis.

Greensand filters are used to catalyze the conversion of reduced manganese to MnO_2. That is more or less it, but see Chapter 8, Clean Water Unit Operation Design: Chemical Processes, for further information.

OSMOSIS

Osmosis is the process by which water moves across a semipermeable membrane from the side of the membrane with a lower concentration of dissolved substances to the side with a high concentration until the two sides are at the same concentration. Although diffusion drives the process, it exerts a measurable pressure across the membrane.

Reverse osmosis reverses the process by exerting hydrostatic pressure on the high concentration side of the membrane, effectively pushing water against the concentration gradient.

FURTHER READING

Judd, S. (2013). *Watermaths: Process fundamentals for the design and operation of water and wastewater treatment technologies.* UK: Judd & Judd Ltd.

Biology

3

CHAPTER OUTLINE

Mind the (Biology/Biochemistry) Gap..25
Important Microorganisms 27
    Bacteria...27
    Protozoa and Rotifers...28
    Algae..28
    Fungi 30
Bacterial Growth 30
Biochemistry..31
    Proteins..32
    Fats 32
    Carbohydrates 33
    Nucleic Acids..33
    Metabolism...34
The Biology of Aerobic Water Treatment..35
The Biology of Anaerobic Water Treatment 36
    Stage 1: Hydrolysis...36
    Stage 2: Volatile Fatty Acid Production..36
    Stage 3: Conversion of Higher Fatty Acids to Acetic Acid by Acetogenic
    Microorganisms...36
    Stage 4: Conversion of Acetic Acid to Methane and Carbon Dioxide by
    Methanogenic Bacteria 36

MIND THE (BIOLOGY/BIOCHEMISTRY) GAP

I have trained many chemical engineers, from first year undergraduates to experienced professionals, and biology is almost always their weakest science. This is ultimately to do with the way we select entrants to chemical engineering courses. Based on the distances between subject areas in Fig. 3.1, chemical engineers are most like mathematicians, and they are also a lot more like chemists and physicists than biologists and architects. This tends to be reflected in what they have studied before going to university. Math, more math, physics, and chemistry tends to be what they bring with them, and that is also a lot of what is taught in university.

An Applied Guide to Water and Effluent Treatment Plant Design. DOI: https://doi.org/10.1016/B978-0-12-811309-7.00003-5

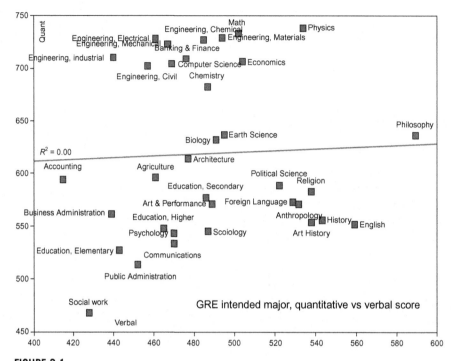

FIGURE 3.1

Quantitative and verbal reasoning skills in entrants to US academic programs.

Courtesy: Razib Khan.

That is a pity, because biology has a lot to show us about how complex, self-regulating systems can be made to work. It is also a handicap from the point of view of the budding water process engineer, because engineered biological systems are used in both clean and dirty water treatment.

You cannot just think of a biological organism as a catalyst. It is alive, and if it dies it will stop working. If you treat it nicely, it will replicate itself so efficiently that you will be presented with the problem of what to do with the surplus. It also tends to be a challenge from the point of view of handling. If you are too rough with it, it might die, and it often forms non-Newtonian fluids.

The organisms we use in water treatment include many kinds of bacteria, fungi, protozoa, algae, and also some higher organisms. The relative importance of these groups varies between types of treatment processes, but it is nearly always bacteria that are responsible for most of the work.

All bacteria belong to the group known as prokaryotes, the organisms without a cell nucleus.

Eukaryotic organisms have a nucleus in the cell. Representatives of this group (colloquially known as higher organisms) important in wastewater treatment include protozoa and rotifers, algae, fungi, worms, snails, and insect larvae in some circumstances.

IMPORTANT MICROORGANISMS

BACTERIA

Bacteria are the most common organisms in wastewater treatment processes. Most of the treatment processes we use operate under aerobic conditions meaning, colloquially, in the presence of air.

Technically, aerobic environments contain molecular oxygen (O_2). Anoxic environments contain no oxygen (<0.5 mg/L), but contain electron acceptors such as nitrate and sulfate. Anaerobic environments have no such electron acceptors. There are organisms which can thrive in all three conditions, and reliably engineering process conditions to culture such organisms is important in water and wastewater treatment.

The clear majority of the bacteria in aerobic treatment systems are of the more modern type known as Eubacteria. Archaebacteria are of greater importance in anaerobic treatment, oxygen-rich atmospheres being a relatively recent phenomenon.

Bacteria tend to be the smallest of the organisms doing useful work in the wastewater treatment process (you can see a scale bar on Fig. 3.2). In common

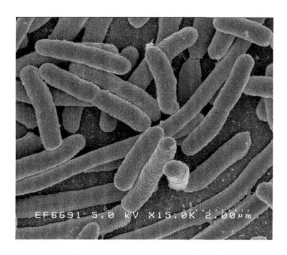

FIGURE 3.2

Scanning electron micrograph of *Escherichia coli* bacteria.

Courtesy: NIH.

with all life, bacteria comprise a cell membrane containing water, proteins, nucleic acids, fats, carbohydrates, and other compounds. They lack, however, certain distinctive structures that arrange these components in eukaryotic organisms. Their nucleic acids also have several differences in composition and associated substances from eukaryotic organisms.

They may have a feature lacking in higher organisms that is very important in wastewater treatment, namely a capsule, often of polymeric materials. These slimy polymers have an important role in bacteria's attachment to each other and to surfaces, as well as capturing materials from suspension. Slime is not amorphous. It is a network of sticky strands which bind bacteria to surfaces, control mass transfer of oxygen, nutrient and waste products, and trap passing particulate matter, giving the biofilm the appearance in dirty water of an inorganic collection of mud and debris, and providing an environment which is ideal for growth of bacteria important in attached growth biological processes (and in microbial corrosion in the case of sulfate reducing bacteria).

As well as having requirements for energy and chemical building materials, bacteria have preferences with respect to the temperature, pH, and osmotic conditions of their surroundings. Individual types of bacteria will have different preferences, varying quite widely for temperature. Bacteria that grow best under cold conditions are called cryophilic, those preferring elevated temperatures, thermophilic, and those preferring intermediate temperatures, mesophilic. With a few exceptions, the optimal pH for bacterial growth lies close to pH 7.

Bacteria vary widely in shape and size. They may be spherical, rod shaped, filamentous or helical, and vary in size from 0.1 to 100 μm in length.

PROTOZOA AND ROTIFERS

Protozoa such as that illustrated at Fig. 3.3 are very simple animals, usually comprising a single cell. The majority require oxygen, and obtain their carbon and energy from long-chain organic molecules. The *Amoeba*, *Paramecium*, and other common pond life fall within this category. Their function from an engineering point of view is to consume the bacteria and particulate organic matter within the system.

Rotifers (see Fig 3.4) have two sexes. They are all oxygen requiring, and obtain energy and carbon from long-chain organic molecules. They perform a similar role to the protozoa, at a slightly larger scale.

ALGAE

Algae are plants (though so-called "blue green algae" are not, they are bacteria). Algae are a varied group, with small unicellular types, very large multicellular types, and a wide range in between. They are of greatest water engineering

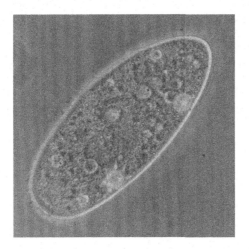

FIGURE 3.3

Optical microscope image of *Paramecium aurelia*.

Courtesy: Barfooz. Licensed under CC BY-SA 3.0: https://creativecommons.org/licenses/by-sa/3.0/deed.en

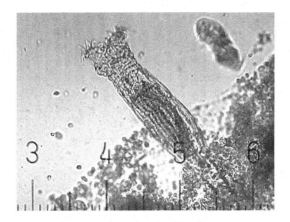

FIGURE 3.4

Rotifer.

Courtesy: Bob Blaylock. Licensed under CC BY-SA 3.0: https://creativecommons.org/licenses/by-sa/3.0/deed.en

importance as suppliers of oxygen through photosynthesis in oxidation ponds. They obtain carbon from CO_2, and energy from light.

They can be nuisance organisms in both clean and dirty water treatment systems if they grow where we do not want them to. Their requirement for light is the key to preventing nuisance growth.

FUNGI

Fungi are neither plants nor animals. In sewage treatment they have a useful ability not possessed by any other group, namely the ability to digest cellulose (which is what toilet paper is mostly made of). They are all oxygen requiring, and have unicellular, multicellular, and syncytial (intermediate between uni- and multicellular) types. They have a nitrogen requirement half that of bacteria, and are more tolerant of low pH than most bacteria.

They are therefore important contributors the overall community of degradative organisms, and especially important where the ability to withstand low pH and nitrogen levels is important, e.g., in the treatment of paper mill wastes.

Fungi can also be important in corrosion. Certain fungi are capable of producing organic acids and have been blamed for corrosion of steel and aluminum, as in some highly publicized corrosion failures of aluminum aircraft fuel tanks. In addition, fungi may produce anaerobic sites for sulfate reducing bacteria and can produce metabolic by-products that are useful to various bacteria.

BACTERIAL GROWTH

The most important consideration for understanding bacterial growth is that reproduction is usually by binary fission. Once a bacterial cell gets big enough, it splits in two. The time taken for this can vary from a few minutes to days. In pure culture, this leads to enormous rates of growth. In the absence of predation by other organisms, limiting nutrients, and slugs of toxic materials we would see these rates of growth in wastewater treatment plant (Fig. 3.5).

In a pure batch culture, the growth curve for bacteria is conventionally thought of as having four distinct phases. Plotting numbers of cells against time, there is an initial period of slow growth, as the bacteria acclimatize to their environment, perhaps activating the production of enzymes necessary for living on the nutrients available. This is known as the Lag Phase. Next, the acclimatized bacteria approach theoretical minimum generation times, producing a logarithmic/exponential growth in cell numbers known as the Log (or exponential) Phase, until either some nutrient becomes limiting, or the rate of cell death equals the rate of cell growth. This condition leads to a state where viable cell numbers become steady, known as the Stationary Phase. The exhaustion of system resources during this period leads to a Death Phase, where cell numbers may drop as quickly as they increased during the Log Phase.

This system may be looked at in a different way, plotting total microorganism mass against time. This is an important way of considering the system much used in determining sludge yields in wastewater treatment plant. The four phases are now seen as acclimatization, rapid growth in microorganism mass, slower growth

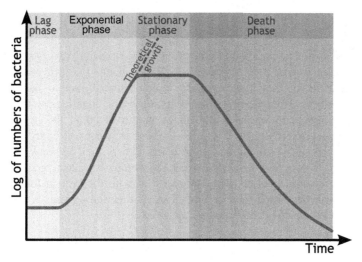

FIGURE 3.5

Bacterial growth curve.

in mass, and declining mass. These are known as the Lag, Log, Declining growth, and Endogenous Phases. The decline in system mass during the last phase is of most interest, as this results from bacteria metabolizing their own mass, and lysing (breaking open) the other organisms present to use their biomass. The implications of this for reduced sludge yields are clear.

In the treatment facility, the above systems are of interest only as theoretical limits to system performance, and for input into models attempting to predict the behavior of the mixed, usually continuous system. The performance of the complex ecosystem of a real works will bear little resemblance to these curves. The actual performance will be a product of the curves of all the organisms involved, as well as complex interactions between them in terms of temperature, pH, nutrients, oxygen availability, and so on.

BIOCHEMISTRY

Living things are mostly made of proteins, fats and carbohydrates: substances we all have a passing familiarity with, as we are made of them, and we eat them as our source of energy and nutrients. Chemical engineers are often somewhat sketchy when asked what these substances are chemically, so let us start with them.

PROTEINS

Proteins are mixed polymers, whose monomers are amino acids. Amino acids have an amino and a carboxylic acid group as their name suggests, and these groups are used to link them together via peptide bonds into polypeptides (proteins). Only 20 or so of the 500 + possible amino acids are used to make proteins.

Though proteins are unbranched polymers, there is a high degree of interaction between the sidechains of the amino acids not involved in the peptide bond. Some of these interactions may be quite strong, as when sulfhydryl bonds are formed between sulfur containing amino acids. Others may be weak, relying on *Van der Waals* forces, but these weak interactions may be crucial to protein function. This is because the sum of these forces leads to the protein folding into a characteristic shape, which may be crucial to its function, especially in the case of those proteins which act as catalysts: enzymes.

Enzymes

Enzyme activity is usually the objective of growing bugs in a plant. Enzymes are biological catalysts. Just like other catalysts, they reduce activation energy, allowing reactions to proceed in a desired direction. Because their activity is dependent on the folding of a protein, they are usually sensitive to the temperature, pH, and salt concentration of the environment they work in. (Thus, pH, salt, and temperature are effective in preserving foods). There are exceptions to this, such as the enzymes put in washing powder (which come from bacteria evolved to survive in hot springs), but the general rule is that small deviations from their tolerable ranges can (reversibly) reduce activity. Larger excursions may denature them irreversibly.

As engineers, we wish to control reactions. If we want high enzyme activity, we need to control temperature, pH, and salt concentration within a range which encourages life. If we want to kill enzyme activity, extremes of pH, temperature and (to a lesser degree), salt concentrations are used to disinfect or sterilize equipment.

FATS

Fats are esters of a trihydric alcohol (glycerol) and three "fatty acid" sidechains ("fatty acid" being another name for carboxylic acid). Humans tend to find the smell of fatty acids repugnant, perhaps because they are released when bacteria degrade fats. Table 3.1 gives my homologous series of stench.

Short-chain fatty acids in combination can produce some truly memorable smells. I once had a job collecting samples from fat traps in fast food restaurants. From personal experience, I know that opening the lid of a chamber full of partially hydrolyzed fats can empty the restaurant of customers in under a minute.

Table 3.1 Homologous Series of Stench

1	Formic acid	Methanoic acid	HCOOH	Acrid/vinegar
2	Acetic acid	Ethanoic acid	CH_3COOH	Acrid/vinegar
3	Propionic acid	Propanoic acid	CH_3CH_2COOH	Body odor
4	Butyric acid	Butanoic acid	$CH_3(CH_2)_2COOH$	Vomit
5	Valeric acid	Pentanoic acid	$CH_3(CH_2)_3COOH$	Sweaty socks
6	Caproic acid	Hexanoic acid	$CH_3(CH_2)_4COOH$	Smelly goats
7	Enanthic acid	Heptanoic acid	$CH_3(CH_2)_5COOH$	Rancid
8	Caprylic acid	Octanoic acid	$CH_3(CH_2)_6COOH$	Rancid goats
9	Pelargonic acid	Nonanoic acid	$CH_3(CH_2)_7COOH$	Rancid
10	Capric acid	Decanoic acid	$CH_3(CH_2)_8COOH$	Sweaty goats
11	Undecylic acid	Undecanoic acid	$CH_3(CH_2)_9COOH$	Waxy
12	Lauric acid	Dodecanoic acid	$CH_3(CH_2)_{10}COOH$	Waxy/soapy
13	Tridecylic acid	Tridecanoic acid	$CH_3(CH_2)_{11}COOH$	Waxy
14	Myristic acid	Tetradecanoic acid	$CH_3(CH_2)_{12}COOH$	Waxy
15	Pentadecylic acid	Pentadecanoic acid	$CH_3(CH_2)_{13}COOH$	Waxy
16	Palmitic acid	Hexadecanoic acid	$CH_3(CH_2)_{14}COOH$	Old people's homes

CARBOHYDRATES

Carbohydrates are (broadly) sugars or sugar polymers or "polysaccharides" (Fig. 3.6).

The most interesting aspect of carbohydrates to a water engineer is the fact that bacterial slime and cellulose are polysaccharide in nature.

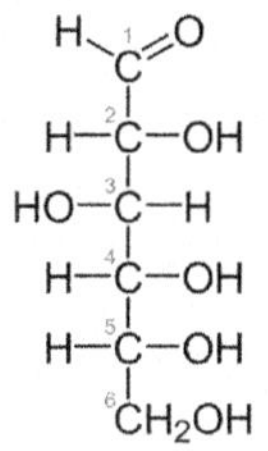

FIGURE 3.6

Glucose, a sugar.

NUCLEIC ACIDS

So now we have looked at the three food groups, what else is life made of? Most importantly nucleic acids (Fig. 3.7). Deoxyribose nucleic acid (DNA) is a long chain polymer made of deoxyribose (a sugar), purine and pyrimidine bases, and phosphate. RNA differs in its sugar component (ribose) being different, and using uracil instead of thymine as one of its four bases.

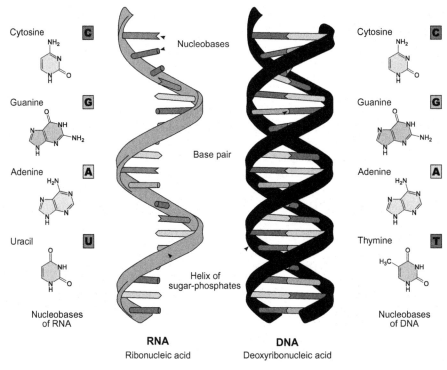

FIGURE 3.7

Nucleic acids.

Courtesy: Roland1952. Licensed under CC BY-SA 3.0: https://creativecommons.org/licenses/by-sa/3.0/deed.en

The arrangement of the bases on the sugar/phosphate backbone carries information on how to produce all the components of a complete new organism, as well as the full range of biochemical capabilities of the organism. Broadly, DNA is used to make RNA which is used to make proteins.

From an engineer's point of view, the dependence on phosphate for nucleic acid production of all life is what can make phosphate a limiting nutrient.

METABOLISM

All living things need a source of energy, and a source of carbon, nitrogen, sulfur, phosphorus, potassium, and other lesser nutrients.

Higher organisms (both plants and animals) need free oxygen. Almost all animals get their energy and nutrients by ingesting and breaking down (metabolizing) complex molecules ready-made by other organisms. Fungi are more like animals than plants in this respect. Almost all plants get their nutrients from the environment as simple molecules, which they build into more complex molecules using energy from light.

Bacteria are more varied in their requirements for growth than higher organisms. They have types which can survive both with and without oxygen, types which obtain their carbon from CO_2 and build up longer molecules by photosynthetic or chemical energy sources, and types which break down long chain molecules to obtain energy. There are also types that combine these abilities.

The presence of nutrients such as nitrogen and phosphorus may be the limiting factor in certain circumstances. This is the reason for installing nutrient removal stages at wastewater treatment works. Discharge of effluent rich in these nutrients encourages runaway growth of microorganisms (especially algae and cyanobacteria) in the receiving body of water.

Some of these organisms also require or can use ready-made organic building blocks, such as amino acids (used to make proteins), the bases used to construct DNA and RNA, and vitamins. Some of these vitamins may however differ from those human beings use.

THE BIOLOGY OF AEROBIC WATER TREATMENT

Aerobic water treatment is a great deal more complex from a biological point of view than anaerobic treatment, due to the far higher number of organisms which can live in the presence of oxygen. The general idea is that complex molecules are broken down into simple ones which are either oxidized for energy, turned into more microorganisms ("bugs"), or released to the environment. The reactions involved do not necessarily happen in a single organism.

Bugs which get their energy from complex reduced organic molecules are (broadly) called heterotrophs (as well as other, longer names). Autotrophs which fix carbon from carbon dioxide using either light or chemical energy are the environment's ultimate producers of fixed carbon, and heterotrophs the ultimate consumers. The sources of chemical energy some of them can use are quite surprising.

The nitrifying bacteria (*Nitrosomonas*) which turn ammonia into NO_2 and *Nitrobacter*, which turns NO_2 into NO_3 are an important representative of this class from a water engineer's point of view. *Thiobacillus* and *Sulfolobus* which turn H_2S or Sulfur into SO_4 will also be mentioned later in the book, as will the iron bacteria (*Gallionella*, *Thiobacillus*) which convert Fe^{2+} into Fe^{3+}. There are also those who think that the methanogens discussed in the next section can be classed as "chemoautotrophs"—the name given to those bugs which fix carbon using chemical energy.

Organisms may compete with each other for the same nutrients, eat each other, or one organism's waste may be another's food (syntrophy) as with the nitrifiers discussed above.

THE BIOLOGY OF ANAEROBIC WATER TREATMENT

Anaerobic treatment is a lot simpler in terms of both biochemistry, and the number of different types of bugs involved. The total number of organisms will be around 10^6 to 10^8 organisms/g volatile solids. Anaerobic treatment follows four stages are described in following sections.

STAGE 1: HYDROLYSIS

Large molecules/polymers like cellulose and proteins are broken down by hydrolytic microorganisms to monomers like glucose and amino acids. These heterotrophic organisms can be "facultative anaerobes," capable of tolerating anaerobic conditions, even if they do not prefer them.

$$(C_6H_{10}O_5)n \quad + \quad H_2O \quad \rightarrow \quad nC_6H_{12}O_6$$
$$\text{cellulose} \quad + \quad \text{water} \quad \rightarrow \quad \text{glucose}$$

STAGE 2: VOLATILE FATTY ACID PRODUCTION

The monomers are converted to a mixture of volatile fatty acids (VFAs) by fermentation, again involving facultative anaerobes. Typical acids produced are acetic, propionic, butyric, and valeric.

$$C_6H_{12}O_6 \quad + \quad 2H_2O \quad \rightarrow \quad C_3H_6O_2 \quad + \quad 3CO_2 \quad + \quad 5H_2O$$
$$\text{glucose} \quad + \quad \text{water} \quad \rightarrow \quad \text{propionic acid} \quad + \quad \text{carbon dioxide} \quad + \quad \text{water}$$

STAGE 3: CONVERSION OF HIGHER FATTY ACIDS TO ACETIC ACID BY ACETOGENIC MICROORGANISMS

Three groups of more specialized types of facultative anaerobic bacteria turn the mixture of fatty acids into acetate, carbon dioxide, and hydrogen. One of these types of bacteria also incidentally reduces sulfate to hydrogen sulfide.

$$C_3H_6O_2 \quad + \quad 2H_2O \quad \rightarrow \quad C_2H_4O_2 \quad + \quad CO_2 \quad + \quad 3H_2$$
$$\text{propionic acid} \quad + \quad \text{water} \quad \rightarrow \quad \text{acetic acid} \quad + \quad \text{carbon dioxide} \quad + \quad \text{hydrogen}$$

STAGE 4: CONVERSION OF ACETIC ACID TO METHANE AND CARBON DIOXIDE BY METHANOGENIC BACTERIA

COD is removed from the wastewater as gaseous products are formed by bacteria which are "obligate anaerobes"—oxygen poisons them. Methanogenesis is rate limiting for Stage 3, bacteria are present at 1000th the concentration of Stage 1. The bacteria are often archaea, not strictly bacteria at all, survivors from a time before photoautotrophic microorganisms (and their successors, plants) made the atmosphere oxygen rich as a result of photosynthesis.

$$C_2H_4O_2 \rightarrow CO_2 + CH_4$$
$$\text{acetic acid} \rightarrow \text{carbon dioxide} + \text{methane}$$

Engineering science of water treatment unit operations

4

CHAPTER OUTLINE

Introduction 39
Coagulation/Flocculation 40
    Coagulants...40
    Flocculants..40
Sedimentation/Flotation 40
    Sedimentation..40
    Flotation...42
Filtration 43
    Membrane Filtration/Screening 44
    Depth Filtration 45
Mixing...46
    Combining Substances...46
    Promotion of Flocculation 46
    Heat and Mass Transfer...47
    Suspension of Solids 47
    Gas Transfer..47
Adsorption 48
Disinfection and Sterilization...49
Dechlorination 51
Further Reading 51

INTRODUCTION

Heuristic unit operation design in water treatment long predates modern science, but science has been applied to our traditional processes after the fact, and newer processes have been developed alongside scientific analyses for a long time now.

Most professional water process plant design practice still involves extensive use of heuristics, but scientific approaches can aid the designer where they provide insight into what is happening in a unit operation. This is especially true for the equipment designer, rather than the process designer.

In this chapter, I will look at the most common application of this principle in my experience.

An Applied Guide to Water and Effluent Treatment Plant Design. DOI: https://doi.org/10.1016/B978-0-12-811309-7.00004-7

COAGULATION/FLOCCULATION
COAGULANTS

The most commonly used coagulants are metallic ions with a high charge density, though there are also now some organic coagulants, similar to flocculants but with lower molecular weight.

Aluminum and iron are the most popular ions used: aluminum as compounds like $Al_2(SO_4)_3$ (alum) and iron as compounds like $FeCl_3$ or $Fe_2(SO_4)_3$.

It has become commonplace in recent years to use ferric sulfate or, to a lesser degree ferric chloride solution, as an iron based coagulant. Ferric salts are much better flocculants than ferrous salts, but copperas ($FeSO_4$), which is oxidized to $Fe_2(SO_4)_3$ during aeration or by chlorination may be used, allowing ferric chloride to be produced cheaply by chlorinating a solution of copperas. Iron doses usually vary between 10 and 90 mg Fe^{3+}/L.

$$6FeSO_4 - 7H_2O + 3Cl_2 = 2Fe_2(SO_4)_3 + 2FeCl_3 + 42H_2O$$

There are also more complex versions of these chemicals such as polyaluminium silicate sulfate (PASS), with advantages under certain circumstances, but a higher price tag.

Whichever coagulant is used, coagulation efficiency is dependent on the dose of coagulant, pH and colloid concentrations, and intensity of agitation. Lime may be used both to adjust pH levels, and as coflocculent.

FLOCCULANTS

Water soluble organic polymer coagulants with a molecular weight in the range 10^5 and 10^6 are used to promote the formation of flocs from suspended solids in water.

Flocculants can be anionic (−ve), cationic (+ve) or ampholytic (mixed +ve and −ve). Cationic polymers are usually based on nitrogen, anionic polymers on carboxylate.

SEDIMENTATION/FLOTATION
SEDIMENTATION

Sedimentation (also called settlement or sometimes clarification) is the process by which heavier particles settle from suspension under gravity, and continue settling to reduce the water content of the settled mass. When a suspension of particles settles under gravity, the relative importance of several physical forces changes as the concentration of solids increases with tank depth.

Type 1 or "Discrete" Settling occurs when discrete particles settle freely without interacting with each other significantly. This process is important in grit removal. The settling rate is independent of solids concentration, but the surface loading rate is critical. Stokes' Law (see also Chapter 5, Fluid Mechanics, and Eq. 5.6) is used to analyze type 1 settling (Fig. 4.1).

Type 2 or "Flocculent" Settling occurs when the suspension is a little more concentrated, and inter-particle events become more important. The processes of coagulation and flocculation occur as faster settling particles impinge upon slower ones, resulting in particles settling faster (Fig. 4.2). These are the important processes in primary settlement. Surface loading and HRT are critical in type 2 settling.

Type 3 or "Hindered" Settling (Fig. 4.3), occurs when the concentration of particles has reached the point where inter-particle forces predominate. The whole settled mass forms a discrete interface between solids and liquid, and the contents below the interface settle as a mass rather than as individual particles. Agitation can be of benefit in type 3 settling. HRT and solids loading are critical design parameters.

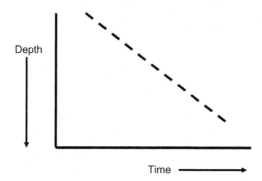

FIGURE 4.1

Relationship between time and depth in type 1, "discrete," settling.

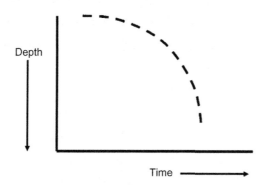

FIGURE 4.2

Relationship between time and depth in type 2, "flocculant," settling.

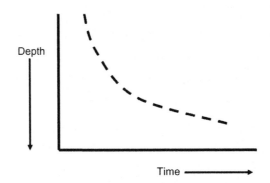

FIGURE 4.3

Relationship between time and depth in type 3, "hindered," settling.

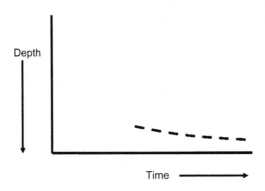

FIGURE 4.4

Relationship between time and depth in type 4, "compressive," settling.

Type 4 or "Compressive" Settling (Fig. 4.4) is the process by which the upper layers of a sludge mass compress the lowest layer, forcing further water out of the mass. This is the predominant process in sludge consolidation and thickening processes.

FLOTATION

Gravity flotation

Just as heavier than water particles sink naturally in water, lighter than water particles float. Where these particles do not interact, a context specific application of Stokes' law (Eq. 4.1) from API 421 can be used to predict their rise rate.

Applied Stokes's law:

$$V = 0.545(SG(\text{oil}) - 1) \times d^2 \times 60/\text{liquid viscosity (cP)} \qquad (4.1)$$

where

  V = rise rate mm/minute at 60°F
  SG = specific gravity relative to water
  d = droplet diameter

  This is the basis of the API 421 oil water separator design standard.

Dissolved air flotation

Heavier than water particles can also be made to float. Dissolved air flotation, the most common approach, works by attaching small bubbles of air to suspended solids. The bubbles are generated by saturating a recycled stream of water with air under pressure, then releasing the pressure rapidly to produce clouds of micro-bubbles. Attaching the bubbles to the solids requires a reduction in charge of the particles and the production of hydrophobic spots on the surface of the solids via chemical/physical pretreatment.

A first-principles design would involve predicting how much air per kg of water could be dissolved in water at a given temperature and pressure using Henry's law, then working out the required recycle water flowrate based upon an amount of air per incoming solids load. It is this second factor which cannot be determined from first principles. It can be measured experimentally, or more commonly estimated based on experience.

FILTRATION

From a theoretical point of view, there are two kinds of filtration. "Surface" or "absolute" filtration passes the filtered fluid (known as filtrate or permeate once it has passed through the filter) through a relatively thin barrier, such as a screen or membrane, and "depth" (or as I call it "probabilistic" filtration) through a deep bed of particles.

Absolute filters have holes of a controlled range of sizes which filtrate must pass through. Particles larger than the largest holes in the filter are excluded by straining, some hit the filter medium and stick to it (impingement), or are trapped in the matrix of the filter (entanglement). There are also cases where particles are attracted to the filter medium by electrostatic and other forces.

It is generally the case that the finer the filter, the less important simple straining will be. Simple straining is however the designer's guarantee that oversize particles are absolutely excluded from the filtrate.

Depth filters such as rapid gravity sand filters generally mainly remove particles by impingement, so the size of the retained solids is far smaller than the media grain size, but only a percentage of incoming solids are filtered, as it is possible for any given particle by chance to escape impingement (Fig. 4.5).

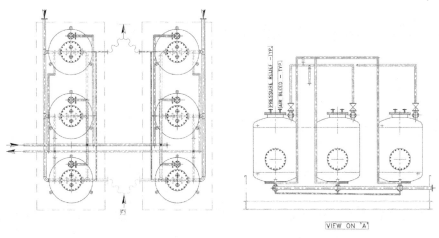

FIGURE 4.5

Pressure filters in plan and elevation view.

Courtesy: Expertise Limited.

Chemical engineers analyze both kinds of filtration with our standard approach: as a flux of something over a boundary, with driving and opposing forces. The flux (flow of water per unit filter area) declines over the time a filter is in service. With depth filters, it is usually the case that the original flux rate is entirely recoverable by cleaning, but with membrane filters, getting back to the original flux by cleaning becomes increasingly difficult, and the membrane cartridges consequently need replacement after a time.

In the usual case of semibatch water filtration, we are therefore most often interested in the inlet pressure required to generate the required flow of water per unit of filter area at the point where a filter must be taken off-line for cleaning, as this is the condition which is likely to have the largest effect on pump specification.

MEMBRANE FILTRATION/SCREENING

Screens are the oldest engineered examples of absolute filtration. They have been used to exclude oversize particles from sewage and drinking water treatment for many decades, though the hole sizes have become much finer in recent years, especially when used upstream of membrane processes. They are engineered products, so process plant designers need not concern themselves too much about how to design them, though I include some design heuristics at Chapter 7, Clean Water Unit Operation Design: Physical Processes, which can be used to predict their performance.

Theoretical analysis of membrane filtration has two main aspects. There is the analysis of the efficiency of separation and buildup of headloss through the filter itself, and the extent to which a layer of filtered material sometimes known as a cake builds up on the membrane. This may be deliberate, as in the filter press

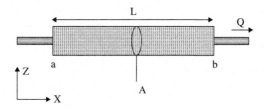

FIGURE 4.6

Darcy's equation.

Courtesy: Peter Capitola. Licensed under CC BY-SA 2.5: https://creativecommons.org/licenses/by-sa/2.5/

used for solids recovery, or problematic as in the case of the "concentration polarization" which reduces flowrate at a given pressure through ultrafiltration membranes. Filtration through a cake is depth filtration, rather than membrane filtration.

When dealing with the finest filters, such as those used for reverse osmosis, it can be a layer of ions which builds up on the membrane surface, to the point where it exceeds the solubility product of sparingly soluble ions. This requires the exclusion of certain species such as boron and silica from RO feed water, as well as prevention of scaling conditions being developed based on LSI analysis (see Appendix 6). Membranes are always used in practice as engineered modules, whose solids removal performance, driving head and application-specific pretreatment requirements will need to be discussed with manufacturers.

Darcy's equation (Eq. 4.2 and Fig. 4.6) provides the most commonly used mathematical model for surface filtration.

Darcy's equation:

$$Q = -\frac{\kappa A(p_b - p_a)}{\mu L} \tag{4.2}$$

where

Q = volumetric flowrate (m³/s)
κ = cross-sectional area for flow (m²)
$p_b - p_a$ = pressure drop (Pa)
μ = viscosity (Pa·s)
L = length over which the pressure drop is taking place (m)

DEPTH FILTRATION

Although depth filtration is so long established that a process designer will be unlikely to even need to so do, it can be modeled mathematically by means of the Poiseullie or Kozeny-Carman equations.

The most complex analysis I have ever had to undertake as a process designer is the empirically derived method of Amirtharajah and Cleasby (see Further Reading).

FIGURE 4.7

Mixers (A) static; (B) dynamic.

MIXING

Mixing in wastewater treatment is important in combining one substance with another, the promotion of flocculation, heat, and mass transfer, and retaining solids in suspension.

There are a variety of dynamic mixing processes, where the mixing element is powered in some way, as well as a variety of static mixers, where dissipation of previously input pumping energy in the body of the mixer provides the mixing force (Fig. 4.7).

COMBINING SUBSTANCES

We usually want to carry out this process as rapidly as possible, using as high a degree of shear as is practical. Some of the materials found in biological stages of treatment such as Return Activated Sludge are shear-sensitive, and mixing must take account of the requirement to minimize damage to the material in the mixing operation. Analysis of this process will in the first instance center on the energy input per unit volume as a measure of mixing effectiveness.

PROMOTION OF FLOCCULATION

Once we have mixed a substance with a coagulating agent, we will often wish to increase the size of the flocculated particles to be able to separate them from the surrounding fluid by means of settlement, flotation, or filtration.

The mixing processes used for this application are far gentler than those used for combining substances as above, and the geometry of the tank and mixer may be selected with a mind to the size of eddies formed as much as the quantity of energy input. The minimum size of such eddies is predictive of the size of the particles formed. The Kolmogorov microscale of turbulence can be used to predict the size of these eddies.

HEAT AND MASS TRANSFER

Heat transfer is not as important in wastewater treatment as in many other branches of process engineering, as most of the processes are carried out under ambient conditions, but is an important consideration in anaerobic digesters, as well as some other sludge processing activities. Sludge is usually therefore the medium being heated, and mixing is necessary to overcome the problems of high viscosity, poor heat transfer characteristics, and lack of tolerance of over-temperatures that sludge often exhibits. The major problem of anaerobic treatment plant is obtaining quite a narrow range of elevated temperature in a very large volume of sludge. The reactor must be mixed to maintain the acceptable temperature range, in the face of heat transfer to the environment, local high temperatures from any heated recycle, local low temperatures from any colder feed, and so on. This mixing is often carried out by means of sparging of biogas recycle.

The key mass transfer processes of importance in wastewater treatment include, most importantly, gas and nutrient transfer in biological treatment. These are once again usually effected by means of gas bubbling, although there are a variety of mechanical aerators, and mixing devices used to carry out these processes.

SUSPENSION OF SOLIDS

The intensity of mixing used in retaining solids in suspension tends to be intermediate between the vigorous mixing used to blend substances, and the gentle mixing used to promote flocculation. Analysis of requirements will be likely to center on simple velocity effects, aiming to obtain an upflow or side velocity sufficient to overcome the tendency of the heaviest particles to settle. What may be thought of as an exception to this is where these velocities are selected so as to allow certain classes of materials to settle preferentially, for example in an aerated grit channel. This process is termed sedimentation or settlement, a very important process in wastewater treatment.

GAS TRANSFER

This is of course a special case of mass transfer, and is bound by all the usual rules of that field. Since wastewater treatment has evolved to some extent alongside general process engineering, there are several commonly used empirical relationships used in the analysis of gas transfer, especially oxygen transfer in wastewater treatment.

It is generally accepted however, that mass transfer of gases follows the usual form of rate equation in process engineering, namely that the rate of transfer is proportional to the relationship between the driving force and the resistance to transfer. The resistance to transfer is in turn inversely proportional to the area over which transfer is occurring, and proportional to the coefficient of gas transfer.

Due to a complication of attempts to analyze the situation in water—lack of knowledge of the true coefficient of gas transfer, and the lack of knowledge about the true area of gas transfer—these two terms are in practice rolled up into a single term KLa, the overall mass transfer coefficient, with units of s^{-1}.

KLa may be determined empirically using a clean solution of sodium sulfite, and the results of this test adjusted to account for such factors as the differential solubility of oxygen in wastewater, temperature, system geometry, and wastewater impurities. This is covered in practical terms in Chapter 14, Dirty Water Unit Operation Design: Biological Processes.

ADSORPTION

Adsorption is a surface process, in which hydrophobic substances are removed from solution by attraction to a hydrophobic surface. Though this differs from a similar process which happens at a larger scale when plastic packing is used to aggregate small impinging oil droplets from water, adsorption media such as granular activated carbon (GAC) (Fig. 4.8) are very effective at oil removal, leading to premature exhaustion as the droplets block the medium's larger internal pores, blocking access to the smaller pores where adsorption mostly occurs.

The nature of the process means that adsorption performance can be predicted to some extent from a substance's Octanol/Water Partition coefficient (K_{ow}).

FIGURE 4.8

GAC filters.

Courtesy: Expertise Ltd.

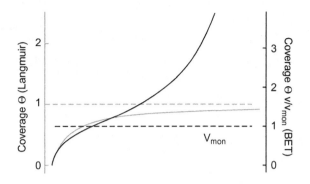

FIGURE 4.9

Langmuir (blue (gray in print version)) and BET (red (black in print version)) isotherms.

Courtesy: Toshiyouri. Licensed under CC BY-SA 4.0: https://creativecommons.org/licenses/by-sa/4.0/deed.en

More formal analyses rely upon the availability of specific laboratory results for a given substance summarized as isotherms such as the Freundlich, BET, or Langmuir (Fig. 4.9).

Even in a simple lab experiment with a single pure substance in water, the amount of a substance adsorbed by a given commercial grade of GAC varies according to incoming concentration, and temperature. GAC's adsorption capacity can also vary greatly if even one other substance capable of adsorption is present.

In the real world, where temperature, flowrates, and the concentration of multiple (possibly interacting) contaminants vary continuously, and the media become progressively more contaminated over time, the process designer cannot be overly concerned with theoretical matters. Empty Bed Contact Time (EBCT) is the key design parameter, though a familiarity with relevant isotherms can inform professional judgement.

DISINFECTION AND STERILIZATION

Three different processes (cleaning, disinfection, and sterilization) are commonly referred to as "disinfection," so it would be useful to start by defining our terms. Cleaning simply reduces the number of contaminants present and, in doing so, removes a proportion of organisms present. Disinfection removes most pathogenic organisms. Sterilization is the killing or removal of all organisms.

Drinking water treatment only involves cleaning and disinfection, but the production of water for higher grade uses, such as the water for injection (WFI) used in pharmaceutical formulation requires sterilization.

The analysis of these processes from the point of view of approaching sterilization often involves the use of kill curves, which display a logarithmic

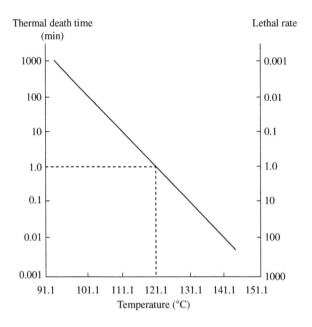

FIGURE 4.10

Example of a kill curve.

Courtesy: Food & Agriculture Organization of the United Nations. Excerpted from Canning Principles,
available online at http://www.fao.org/docrep/003/R6918E/R6918E02.HTM (accessed 2 November 2017).

relationship between the "dose" of sterilizing agent applied and the percentage of
the initial number of organism present (Fig. 4.10).

This illustration tells us that three things matter in the removal of all life from
a system: how much life you started with, plus both the intensity and duration of
your killing agent. This holds true whether we are using electromagnetic radiation, heat, or chemicals.

Therefore, we tend to analyze all three types of process with respect to log
reductions, whether that be in organisms in general or pathogens in particular.
We frequently use "indicator organisms" such as "colony forming units" or
"coliforms" to stand in for the harder-to-test-for organisms which also tend to be
markers of fecal contamination.

Consequently, regulators tend to specify a certain number of colony forming
units (CFUs) and coliforms per unit volume of water. We do not know which
microorganisms these proxies represent, but we assume that if these easy to culture organisms made it through our process, many pathogens did too.

Conventional drinking water treatment processes prior to the disinfection stage
give a 4-log reduction in CFUs, a 2-log reduction in coliforms, a 2-log reduction
in (far smaller) *Cryptosporidium* and *Giardia* cysts and a 2-log reduction in
(smaller still) viruses.

Conventional disinfection such as chlorination after such treatment give a 2-log reduction in CFU and coliforms, in viruses, but no reduction in (far smaller) *Cryptosporidium* cysts. This last is a problem because at least a 3-log reduction in *Cryptosporidium* is required in many regulatory regimes. Conventional treatment and chlorination may not be enough to ensure safety.

For comparison, sterilization processes are normally specified as providing at least a 12-log reduction in all types of organisms.

DECHLORINATION

We may add chlorine in excess as our disinfecting agent, or as an oxidant. We may then need to reduce or "trim" the levels of chlorine to specification for supply, or be required to dechlorinate entirely before discharge to environment. Sulfur dioxide or sodium metabisulfite are most commonly used for this duty. Unusually in water engineering, the required dose can be determined by straight reaction stoichiometry (see Chapter 2, Water Chemistry).

FURTHER READING

Amirtharajah, A., & Cleasby, J. L. (1972). Predicting expansion of filters during backwash. *Journal AWWA, 64*(1), 52−59.

Metcalf., & Eddy Inc., et al. (2014). *Wastewater engineering: Treatment and resource recovery.* (5th ed). New york, NY: McGraw-Hill.

Fluid mechanics

5

CHAPTER OUTLINE

Introduction 53
The Key Concepts 53
    Pressure and Pressure Drop, Head and Headloss 54
    Rheology 54
    Bernoulli's Equations 56
    Darcy-Wiesbach Equation 56
    Less Accurate Explicit Equation 57
    Stokes Law 57
Further Reading 58

INTRODUCTION

Since all civil, mechanical, and chemical engineering students study fluid mechanics, I have assumed that readers will be familiar with the derivation of Navier-Stokes' and Bernoulli's equations, and once had to pass an exam in which they derived them from first principles.

If you need a refresher on this topic, I would recommend reading Coulson and Richardson Vol. 1 (see Further Reading). However, fluid mechanics are in my opinion of limited value in professional practice where their more practical relative, hydraulics, tends to be of far greater utility.

THE KEY CONCEPTS

You may recall from university that the equations used in fluid mechanics are generated from assumptions of conservation of mass, energy, and momentum, together with the continuum assumption (that fluids are not particulate at the scale relevant to fluid mechanics).

These assumptions do not always hold in practice—energy can be lost from systems by hydraulic jumps, real fluids contain particles at all scales and, at very small scales, the presence of molecules starts to affect fluid flow.

An Applied Guide to Water and Effluent Treatment Plant Design. DOI: https://doi.org/10.1016/B978-0-12-811309-7.00005-9

PRESSURE AND PRESSURE DROP, HEAD AND HEADLOSS

Does this seem too basic a starting point? Quite possibly not, for it is apparent to me from talking to new graduates that it is possible to graduate with a good grade in chemical engineering whilst not really understanding what pressure is, let alone pressure drop. Many graduates do not really understand that water can flow downhill without being pumped, and that this can be used to design a controlled gravity flow system.

Pressure is a continuous force applied perpendicular to the surface of an object. As is it a force applied to a surface, its SI units are N/m², also known as Pascals (Pa). Since a Newton is about the amount of force exerted by a mass of 100 g under gravity, Pascals are relatively small, so engineers commonly work in kilopascals. As is implied above, pressure can be produced by a mass accelerating under gravity.

A column of water 1 m high exerts a hydrostatic pressure at its base which can be calculated using Eq. (5.1).

Hydrostatic pressure at the base of a column of fluid:

$$p = \rho g h \tag{5.1}$$

where

ρ (rho) = density of the fluid (around 1000 kg/m³ for water)
g = acceleration due to gravity (around 9.81 m/s² on Earth)
h = height of the fluid column (in meter)

Water engineers tend to refer to this pressure, when exerted by a 1 m column of water as 1 m water gage (WG). We also commonly call pressure a "head" of water, and use the term head interchangeably with pressure.

Pressure can also be produced by means other than gravity, in pumps, compressors and so on. The pressure produced by such devices is of course the same thing as the pressure produced by gravity.

Pressure drop, or headloss, is the reduction in the pressure of a fluid through a system. Headloss is caused by losses of energy through friction with pipe walls and obstructions, when changing direction in bends and so on.

RHEOLOGY

Rheology is the science of viscosity. From a practical point of view, we may not need to concern ourselves with rheology far beyond the point at which we can differentiate between Newtonian and non-Newtonian fluids, and know the viscosity or range of viscosities we need to assign to our fluid in our hydraulic calculations.

A Newtonian fluid is a fluid whose shear stress is in linear proportion to the velocity gradient in the direction perpendicular to the plane of shear. As an equation this is set out at Eq. (5.2).

For Newtonian fluids:

$$\tau = -\mu \frac{dv}{dy}$$

(5.2)

where

τ = shear stress (force per unit area)
μ = dynamic viscosity
dv/dy = velocity gradient perpendicular to the direction of shear

Newtonian fluids have a constant dynamic viscosity (at a given temperature) because, for such fluids in laminar flow, shear stress is proportional to strain rate (deformation with respect to time), and the factor of proportionality for the fluid is the dynamic viscosity.

The key message therefore is that Newtonian fluids have constant viscosity.

Non-Newtonian fluids conversely do not have constant viscosity. Fig. 5.1 shows how non-Newtonian fluids, whose viscosity changes with shear, differ from a Newtonian Fluid.

There are also non-Newtonian fluids whose viscosity changes with duration of stress. *Rheopectic* fluids have a viscosity which increases with duration of stress, while *thixotropic* fluids have a viscosity which decreases with duration of stress.

Non-Newtonian properties are common in the sludges, slurries, flocs, and slimes especially common in water treatment.

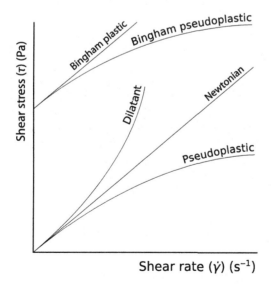

FIGURE 5.1

Rheology of time-independent fluids.

Courtesy: g-sec. Licensed under CC BY-SA 3.0: https://creativecommons.org/licenses/by-sa/3.0/deed.en

BERNOULLI'S EQUATIONS

Bernoulli was a mathematician, rather than an engineer. His equations are widely taught in academic courses on fluid mechanics, because they can be derived directly from Newton's laws, allowing a standard exam question requiring regurgitation of the derivation to be set.

The underlying principle of Bernoulli's equations is that an increase in the speed of a fluid is associated with a decrease in its pressure or potential energy. I know from teaching the subject that this can seem counterintuitive to beginners, but that's how it is.

The equation for incompressible flow is at Eq. (5.3).

Bernoulli's equation for incompressible flow:

$$\frac{v^2}{2} + gz + \frac{p}{\rho} = \text{constant} \qquad (5.3)$$

where

v = velocity of the fluid
g = acceleration due to gravity
z = elevation in the direction opposite to the gravitational acceleration
p = pressure of the fluid
ρ = density of the fluid

Bernoulli's equation is not used in professional practice, but the underlying principle should nevertheless be grasped.

DARCY-WIESBACH EQUATION

You might have been taught at university that this equation (used to determine practical headlosses in pipes), descended from Bernoulli, but in fact it did not. It is a development by two engineers of an earlier empirical equation which was authored by one Monsieur Prony. (Engineer-in-Chief of the École Nationale des Ponts et Chaussées in France, and one of the 72 names inscribed on the Eiffel Tower.)

The equation expressed in terms of headloss, Δh, is set out at Eq. (5.4).

Darcy-Wiesbach equation for headloss:

$$\Delta h = f_D \times \frac{1}{2g} \times \frac{v^2}{D} \qquad (5.4)$$

where

g = acceleration due to gravity
D = hydraulic diameter of the pipe
v = velocity of the fluid
f_D = Darcy friction factor

The difficulty with the equation historically was that its solution always involved iteration, though MS Excel's "What If Analysis" functions make this straightforward nowadays.

LESS ACCURATE EXPLICIT EQUATION

You can use an explicit equation based on that in Coulson and Richardson Vol. 6 for a quick and dirty answer avoiding the hassles of iterative solutions (Eq. 5.5). It is only good enough for initial pipe size selection, being slightly more accurate than superficial velocity, but inferior to nomograms. It does, however, lend itself to use in spreadsheets for sensitivity analysis. The equation is based on turbulent flow in clean commercial steel pipes (close enough hydraulically to the kinds of pipes water engineers use, at least in the new condition).

Explicit equation for headloss:

$$\Delta P = 0.125 G^{1.84} \mu^{0.16} \rho^{-1} d^{-4.84} \tag{5.5}$$

where

ΔP = pressure drop (Pa)
G = mass flow (kg/s)
μ = viscosity (N/m^{2}s)
ρ = density (kg/m^3)
d = pipe internal diameter (mm)

STOKES LAW

Stokes Law is the only other element of fluid mechanics which I have found of common practical use during my career.

It assumes laminar flow in homogeneous fluids, and that the particle is smooth, spherical and not interacting with other particles. As a result, the equation (Eq. 5.6) often yields imprecise results, but it nevertheless has many uses. For example, we can use it to determine how the balance of forces on a very small particle falling (or rising) through a fluid affect its terminal velocity.

Stokes law:

$$V = \frac{2}{9} \frac{(\rho_p - \rho_f)}{\mu} g R^2 \tag{5.6}$$

where

V = terminal velocity
R = particle radius
g = acceleration due to gravity
ρ_p = particle density
ρ_f = fluid density
μ = fluid dynamic viscosity

FURTHER READING

Coulson, J. M., Richardson, J. F., Backhurst, J. R., & Harker, J. H. (1999). *Coulson & Richardson's Chemical engineering, Vol. 1: Fluid flow, heat transfer and mass transfer* (6th ed.). Oxford, UK: Butterworth-Heinemann.

Green, D. W., & Perry, R. H. (2007). *Perry's Chemical engineers' handbook* (8th ed). New York, NY: McGraw-Hill.

Sinnot, R., & Towler, G. (2009). *Chemical engineering design: SI edition. Chemical engineering series* (5th ed). Oxford, UK: Butterworth-Heinemann.

Clean water treatment engineering

2

INTRODUCTION

Having largely dealt with engineering science in Section 1, Water Engineering Science, we can now move on to the design of clean water treatment plant. Clean water treatment engineering involves more engineering science than sewage treatment engineering, though it still has as much art as science.

Ignoring the preliminary treatment of "raw" water supplied for onsite use or further treatment by customers, the minimum standard of clean water treatment is the potable (drinking) water standard. Potable water is often used as a feed to produce the higher grades of water used for cooling and boiler feed duties, and the still-higher grades of water used in production of pharmaceuticals and electronic components.

The production of safe and palatable potable water from naturally occurring waters has a very long history, and the design methods for the core conventional technologies used are therefore as well known as their dirty water counterparts. Even the more novel processes now in common use for drinking water production, such as dissolved air flotation, have very long track records and are well characterized.

Designers of potable water treatment plants are, if anything, more conservative than their dirty water counterparts. Killing one person through dirty drinking water from a badly designed or operated plant is generally held to be far more serious than the loss of a whole river's worth of fish.

Great care is thus taken in selection of the source of water used to produce potable water. Not all bodies of water are suitable as a feedstock to a plant which must produce safe water under all foreseeable circumstances. Even the occasional presence of certain chemical or biological contaminants usually rules out the use of some potential sources of water, though there are places where scarcity means that drinking water is now made directly from sewage or industrial effluent.

Clean water characterization and treatment objectives

CHAPTER OUTLINE

Impurities Commonly Found in Water 61
Characterization 62
   Microbiological Quality 62
   Color 62
   Suspended Solids...63
   Alkalinity..63
   Iron..63
   Manganese 64
   Total Dissolved Solids 64
Treatment Objectives..64
   Potable Water 65
   Cooling Water...66
   Ultrapure Water 66
Further Reading 67

IMPURITIES COMMONLY FOUND IN WATER

Water normally contains many impurities taken from the materials it has come into contact with. Inland water can contain:

- dissolved inorganic compounds, e.g., compounds of sodium, calcium, potassium, and magnesium
- suspended inorganic compounds, e.g., clays, silt, sand, and metal oxides
- dissolved organic compounds, e.g., humic acid, fulvic acid, and tannins
- suspended organic compounds, e.g., leaves and humus
- dissolved gases, e.g., oxygen, nitrogen and carbon dioxide
- microorganisms, e.g., bacteria, algae, and fungi

The composition of a given water depends on its local environment and history. For example, in the granite regions of Scotland, calcium levels below 10 mg/L are common. Conversely, the calcite rocks of say Doha (Qatar) and Derbyshire (England) are associated with waters containing 350 mg/L of calcium carbonate.

An Applied Guide to Water and Effluent Treatment Plant Design. DOI: https://doi.org/10.1016/B978-0-12-811309-7.00006-0

The type and quantity of impurities present determines the quality of the water, the problems that can arise from its use, and the types of treatment required to make it suitable for use.

CHARACTERIZATION

Internationally, the "bible" of water testing procedure is the American Water Works Association's "Standard Methods for the Examination of Water and Wastewater."

In the United Kingdom, we have the Standing Committee of Analysts (SCA) blue books, available free of charge (see Further Reading).

Knowing which sampling and testing protocols were used to measure a water quality parameter can be very important, though it is beyond the scope of this book to examine individual methods.

MICROBIOLOGICAL QUALITY

This is the key characteristic of water because it is frequently the most important from the point of view of human health. However, in practice, it is not a key design parameter, as sources with high levels of microbiological contamination are considered unsuitable for producing drinking water. Microbiological quality is assessed by passing a water sample through a sterile fine filter, which retains any microorganisms present, then placing the filter on a growth medium, and incubating. A viable organism on the filter may proliferate to produce a colony of bacteria, visible to the naked eye.

A growth medium with no inhibitors of bacterial growth gives a total bacterial count. The inclusion of soap in the growth medium will kill any bugs which do not come from animal digestive systems. These gut bacteria are known as coliforms, and are taken as a sign of fecal contamination. Some of these coliforms are of the species known as *Escherichia coli* or *E. coli* for short.

There are almost universally regulatory limits on Total Viable Count, coliforms, and *E. coli* in drinking water. The coliform organisms which have been cultured are not necessarily themselves harmful to health. They are "indicator organisms," proxies for harmful organisms, found in association with them in water contaminated by feces. Fig. 6.1 shows how the relationship between indicator organisms and pathogens is used to keep drinking water safe.

COLOR

The measurement of color involves measuring the degree of absorption of a given set of wavelengths of light. The test protocol needs to define whether the sample has been filtered prior to testing.

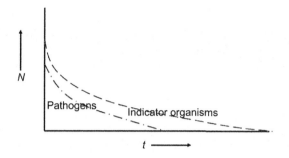

FIGURE 6.1

Relationship between indicator organisms and pathogens during disinfection. N, number of organisms, t, time.

There are several different protocols and several different units of measurement. Degrees Hazen (also known as APHA color or the Platinum Cobalt (Pt/Co) scale) is the most commonly used unit in water treatment and is standardized in ASTM D1209. This is a "yellowness scale" commonly used as a proxy measure of organic content in drinking water treatment, as well as degrees of contamination in effluent treatment.

SUSPENDED SOLIDS

Suspended solids are properly measured by filtration and drying at 105°C, which drives off all water without oxidizing too much of the organic matter present. Volatile suspended solids (VS) are measured as the loss of weight after the dry suspended solids have been ignited.

ALKALINITY

As discussed at Chapter 2, Water Chemistry, there are many kinds of alkalinity, and it is important to know which of these a given measurement refers to. Most crucially, what was the pH endpoint of the determination, and what units is the given measurement in? Sampling and sample handling protocols can sometimes make a large difference to alkalinity measurements.

IRON

Iron in water can be determined by atomic absorption spectrometry (AAS, detection limit 1 μg/L) or by colorimetric methods (detection limit 5 μg/L). Colorimetric methods are appropriate to drinking water treatment, as they are simpler and cheaper than AAS, and the taste threshold for iron is 40 μg/L, even in distilled water. Colorimetry is easily accurate enough for the water engineer's needs.

MANGANESE

A concentration of 0.1 ppm of manganese ion tastes unpleasant, and stains laundry. Even 0.02 mg/L of manganese can form black coatings on water pipes that may later slough off. Several countries have set standards for manganese of 0.05 mg/L, easily measured by ISO standard colorimetric methods, which have detection limits of about 10 μg/L.

TOTAL DISSOLVED SOLIDS

Total dissolved solids (TDS) is usually low for freshwater sources, at less than 500 ppm. Seawater and brackish (mixed fresh and sweater) water contain 500–30,000 and 30–40,000 ppm TDS, respectively. TDS is most accurately measured by weighing a filtered sample, and drying at 105°C until no further mass is lost. A more rapid method is to use conductivity, which is proportional to TDS. This is particularly important in the case of very high purity water, whose quality is often specified as a conductivity in microsiemens per centimeter. Deionized water has a very low conductivity, around of 0.06 μS/cm.

TREATMENT OBJECTIVES

As with every process engineering product, drinking water is made by the cheapest robust and safe method allowable by law. There is a large body of law and regulation concerning drinking water quality but, unlike sewage treatment, water quality standards are not constantly getting tighter (although there are perhaps those who would like them to, irrespective of any rational basis for this).

We consequently still use venerable techniques such as settlement tanks and sand filters wherever we can. The only genuine recent novelty has been the advent of membrane treatment, and even this is many decades old now. Clean water treatment plants (Fig. 6.2) tend to be somewhat less robust than effluent treatment plants, slightly more closely controlled, and slightly more expensive per volume of water treated.

There is far more high-accuracy chemical dosing in clean water treatment than in dirty water plants. Dosing acid or alkali for pH control is the most common requirement, but we also commonly dose chemicals to coagulate colloids, flocculate solids, and disinfect treated water, among other things. Clean water treatment design consequently tends to involve the process engineer in a lot more chemistry than the design of effluent treatment plants.

The minimum potable water standard is the World Health Organization (WHO) standard, followed by national standards for potable water. For non-potable applications, we have manufacturers' boiler/process water specifications, standards for "Water For Injection" (WFI) set by pharmaceutical regulators, and standards for ultrapure water determined by process quality assurance requirements.

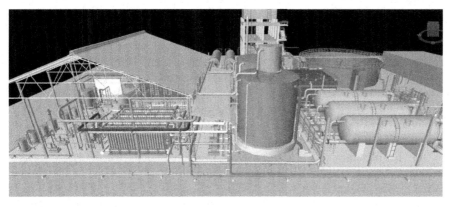

FIGURE 6.2

Membrane clean water treatment plant.

Courtesy: Hexagon PPM.

POTABLE WATER

The quality of potable water must be reliably controlled such that it is safe and pleasant to drink, suitable for use for other household purposes such as personal hygiene and laundry, and harmless to plumbing.

WHO offers guidelines on this, but it does not prescribe international standards for drinking water, and instead advocates a country by country, risk based local approach. WHO states, in its guidelines for drinking water quality: *"The judgement of safety – or what is a tolerable risk in particular circumstances – is a matter in which society as a whole has a role to play. The final judgement as to whether the benefit resulting from the adoption of any of the health-based targets justifies the cost is for each country to decide."*

I am, however, personally of the opinion that professional engineers, when designing plant to produce drinking water, should not work to lower water quality standards than those set by WHO on ethical grounds. The guidelines state that: *"The primary purpose of the Guidelines for Drinking-water Quality is the protection of public health. Water is essential to sustain life, and a satisfactory (adequate, safe and accessible) supply must be available to all. Improving access to safe drinking-water can result in tangible benefits to health. Every effort should be made to achieve a drinking-water quality as safe as practicable. Safe drinking-water, as defined by the Guidelines, does not represent any significant risk to health over a lifetime of consumption, including different sensitivities that may occur between life stages. Those at greatest risk of waterborne disease are infants and young children, people who are debilitated or living under unsanitary conditions and the elderly."* This seems to me to represent the minimum ethical standard.

The guidelines provide (among many other things) a defensible minimum standard of drinking water treatment, as well as reminders for the unwary to check all aspects of water quality. I was approached a few years ago by a group of technically naïve students hoping to offer a cheap fix to make safe groundwater contaminated by farmers in a less developed country. The real fix here is to get the farmers to stop polluting the groundwater but, in any case, they were proposing to put locally produced granular activated carbon into discarded soft drinks bottles to make a water free from solids and pesticide residues. This would not, however, have treated the arsenic or microorganisms present in the water, and by producing a more palatable, but nonetheless still nonpotable water, would have actively encouraged the use of these polluted sources of water.

The list of things which WHO suggest should be controlled in drinking water is long, but most of them will be effectively controlled by making an appropriate initial choice of source water.

For a suitable freshwater source, the engineer will be mainly concerned with the characteristic range of suspended solids, iron, manganese, color, alkalinity, pH, and temperature of the water. These are the parameters which their treatment processes influence, and are influenced by.

For brackish or saline waters to be desalinated by reverse osmosis for potable use, the engineer will be concerned in addition with several ions which interfere with membrane operation, such as boron, strontium, and silicon, as well as the potential presence of things such as small polysaccharide particles.

COOLING WATER

There are three primary concerns in cooling water treatment: scale and deposit formation; corrosion and biological growth.

To reduce the risk of scaling or corrosion in cooling water systems, hardness, alkalinity, and pH of supplied water must be controlled in accordance with the Langelier (or similar) index. The TDS must also be controlled. Uncontrolled biological growth can produce thick layers of bacteria, algae, and fungi as well as larger organisms such as mussels.

ULTRAPURE WATER

This term is used for a wide range of different water specifications in different industries. There are three main types: water used to make steam, various pharmaceutical water grades, and semiconductor production washing waters.

Boiler feed water

There are three primary concerns in boiler water treatment: prevention of boiler water carryover into the steam; prevention of deposits and/or scale forming in the boiler, and prevention of corrosion in the boiler. These are controlled by modifying water chemistry broadly as described for cooling water above, but to a higher standard. Steam and condensate have higher temperatures which make microbiological growth less of a concern.

Pharmaceutical water

Examples of waters used in the pharmaceutical industry include the following:

- Deionized/demineralized/softened water: Water treated by distillation, ion exchange or reverse osmosis to have reduced levels of TDS or hardness
- Purified water: Water which has been softened and UV disinfected
- Water for injection: Purified water which has been certified free of pyrogens (substances capable of causing fever). From a practical water engineering point of view, this means that WFI must be produced by either reverse osmosis or distillation

Semiconductor processing water

The highest purity grades of water are required when manufacturing silicon chips. These waters must not deposit anything onto nor dissolve anything from the surfaces they wash. The two most common standards used are ASTM D5127 "Standard Guide for Ultra-Pure Water Used in the Electronics and Semiconductor Industries" and SEMI F63 "Guide for ultrapure water used in semiconductor processing."

FURTHER READING

American Society for Testing and Materials (ASTM). (2013). *ASTM D5127-13 standard guide for ultra-pure water used in the electronics and semiconductor industries.* West Conshohocken, PA: Author.

Moran, J., Jardine, N., & Davies, C. (2003). Design of utilities and services. In B. Bennett, & G. Cole (Eds.), *Pharmaceutical production: An engineering guide.* Rugby, UK: IChemE.

Semiconductor Equipment and Materials International (SEMI). (2016). *SEMI Standard F63-1016 guide for ultrapure water used in semiconductor processing.* Milpitas, CA: Author.

UK Environment Agency. (2014) *Standing Committee of Analysts (SCA).* Blue Books. Retrieved from https://www.gov.uk/government/publications/standing-committee-of-analysts-sca-blue-books

World Health Organization. (2006) *Guidelines for drinking-water quality. First addendum to third edition, Volume 1: Recommendations.* Retrieved from http://www.who.int/water_sanitation_health/dwq/gdwq0506.pdf

Clean water unit operation design: physical processes

7

CHAPTER OUTLINE

Introduction 69
Mixing 70
   High Shear 70
   Low Shear 70
Flocculation 70
Settlement 71
   Clarifier Design 71
Flotation 73
Filtration 74
   Coarse Surface Filtration 75
   Fine Depth Filtration 76
   Fine Surface Filtration: Membranes 87
Distillation 96
   Pretreatment 96
   Types of Distillation Process 96
Gas Transfer 97
Stripping 98
Physical Disinfection 98
   UV Irradiation 98
Novel Processes 99
   Electrodialysis 99
   Membrane Distillation 99
   Forward Osmosis 100
Further Reading 100

INTRODUCTION

Physical processes tend to be cheaper, simpler, and more robust than chemical and biological processes, and water engineers do like it cheap, simple, and robust. Where it is possible to achieve something using a physical process, we will tend to favor it.

Physical processes tend to be used at earlier stages of the process, reserving the more complex and expensive chemical and biological processes for later stages.

An Applied Guide to Water and Effluent Treatment Plant Design. DOI: https://doi.org/10.1016/B978-0-12-811309-7.00007-2

This chapter is arranged in the approximate order of where these unit operations are likely to be used in the process.

MIXING

Mixers are used for several purposes in drinking water treatment. We can usefully split the applications into high-shear and low-shear types as follows.

HIGH SHEAR

High-shear or high-intensity mixing of short duration (a few seconds) is used to mix dosed chemicals with the water being treated. The mixers used can be "static" or "dynamic." Static mixers are short baffled channels or (more commonly) pipes in which mixing takes place without moving parts by one of several mechanisms. Flash mixers are small mixed tanks of few seconds' hydraulic retention time (HRT) in which powered mixers blend water and dosed chemicals. It is commonplace to see far larger tanks used in error for this purpose in industrial effluent treatment and sometimes in drinking water treatment. Their heterogeneity and high HRT makes process control very poor.

LOW SHEAR

The main application for low-shear mixing in drinking water treatment is flocculation (see "Flocculation" section) though the principles of controlled shear need to be applied to all handling of coagulant flocs. As with high-intensity mixers, flocculators may be plug-flow or stirred tanks. The lower mixing intensity means that they have higher residence times than high-intensity mixers, between 5 and 15 min. Traditionally, picket fence mixing elements were used, but propeller mixers can give similar performance in many applications with lower capital and running costs.

FLOCCULATION

Flocculation is commonly confused with coagulation, and the terms are consequently used to some extent interchangeably but they are very different processes. Flocculation is a physical process in which the fine solids produced by coagulation are brought together into larger particles using controlled low levels of shear. This may be done in either static (Fig. 7.1) or dynamic mixers.

Although chemical flocculants are used to promote flocculation in effluent treatment, in drinking water treatment flocculation is usually effected solely using

FIGURE 7.1

Static flocculator.

Courtesy: FRC Systems, a JWC brand.

shear. Certain organic flocculants with low monomer residuals may however be used in drinking water sludge treatment.

SETTLEMENT

Settlement tanks that are very like their dirty water equivalents were historically commonly used as the first stage in drinking water treatment, especially with lowland surface water as feedstock. The main difference between the clean and dirty water versions of this process lies in the chemical dosing and flocculation stages which precede them. Settlement tanks are however rarely used on new designs though many historical units are still in use (see Fig 7.2).

To obtain 95% solids removal, HRTs of at least 0.75 h and surface loadings of 1 m/h are common starting points for conceptual designers.

Tank sizes above 30 m diameter are more troublesome to design well, as are rectangular tanks. Personally, I avoid both as there can be distribution problems with large circular tanks, and the scrapers used for sludge collection on rectangular tanks are prone to mechanical failures.

For circular tanks, half-bridge scrapers are used on tanks up to 30 m diameter, and (more expensive) full bridge scrapers on tanks 30–50 m diameter. There is also such a thing as a three-quarter bridge scraper used for intermediate size tanks, though I have never used one.

CLARIFIER DESIGN

Several types of settlement tank may be used for clarification; many of these being offered as proprietary systems.

FIGURE 7.2

Potable water treatment clarifiers (rectangular, center of picture), Waikato Water Treatment Plant, New Zealand.

Courtesy Google Earth 2017.

Table 7.1 Basic Design Parameters for Settlement Tanks

Surface loading low acidic water	1−2 m/h
Surface loading alkaline water	2−5 m/h
Side wall depth	3−5 m
HRT	1−3 h

Sludge blanket clarification

After the coagulation and flocculation stages, flow is introduced upward through a "sludge blanket" in the lower part of a gravity settlement tank. Though provision of a delay time between coagulant and polymer addition is essential, this is sometimes omitted in error. Good flow distribution across the tank bottom is key. The tank may be rectangular or circular, inclined or hopper bottomed.

Settling sludge particles are captured in the bottom of the tank by the sludge blanket, and clarified water is drawn from weirs at the top of the tank. The height of the sludge blanket is controlled at a design level by removal of sludge via weirs, bellmouths, hoppers, sludge cones, or pumping. Basic design parameters for settlement are given in Table 7.1.

Inclined plate settlement tanks

Inclined plates at about 300 mm centers, usually 60° to the horizontal, positioned above and within the sludge blanket can be used to enable a higher surface loading for a given treated water quality. Inclined tubes may be used instead of plates,

Table 7.2 Basic Design Parameters for Horizontal Flow Tanks

Surface loading	1 m/h
Aspect ratio rectangular tanks	3:1 minimum
Velocity at inlet baffle	<0.3 m/s
Outlet weir length	0.15 m weir/m^3/h
Scraper speed	0.3 m/min
Side wall depth	2–5 m
Floor slope	7.5°

in which case there is no sludge blanket. There are other variants (see also Chapter 17, Industrial Effluent Treatment Unit Operation Design: Physical Processes, for discussion of parallel plate separators).

The increase in effective surface area is proportional to the number of plates used, multiplied by the cosine of their horizontal angle, with a reduction to allow for the reduced plan area available for the plates and inefficient distribution of flows at the edges.

Sludge dry solids production is often about equal to the mass of the coagulant added though more detailed methods for calculation can be found at Chapter 21, Sludge Characterization and Treatment Objectives. The sludge blanket concentration will typically be about 0.1% dry solids, but this may be increased using a sludge concentrator cone to 0.5% dry solids, dependent on raw water type and use of polymers.

Horizontal or radial flow clarification

Horizontal flow tanks are generally scraped and require a sludge hopper. They may be either circular or rectangular. Incoming flow is introduced after coagulation and flocculation at one end of a rectangular tank or in the center of a radial tank allowing horizontal flow to launders or weirs at the top water level of opposite end or periphery of the tank. Basic design parameters are given in Table 7.2.

Proprietary systems

Equipment suppliers and EPC companies design proprietary clarification systems based on the above principles, as well as in-house experience. Expert purchasers may wish to request the design calculations from such suppliers, so that they can evaluate for themselves the sufficiency of the design.

FLOTATION

Dissolved air is the most common type of flotation gas used in potable water treatment. The dissolved air flotation (DAF) process mixes a clarified stream from the outlet of the unit with air at 3–9 bar, to produce a supersaturated (compared with saturation at atmospheric pressure) solution of air in water. This is rapidly depressurized at the inlet of the unit to produce a mass of microbubbles which attach to the solids present, floating them to the surface.

Its 15-min HRT and 15 m/h surface loading to give 95% solids removal makes DAF a compact alternative to settlement tanks for drinking water treatment, favored by many designers since the 1980s for treatment of upland waters and more recently for algae removal. Generally the higher the feed solids, the higher the %removal as the effluent quality is substantially constant.

More recently, high-intensity DAF has been used to pretreat seawater to protect against algal blooms prior to its desalination for drinking or industrial water. In this application, surface loadings of 50 m/h are common as the algal cells have densities close to or lower than seawater.

An additional advantage of DAF is that it yields sludge at maybe 5% dry solids content as opposed to approximately 1% dry solids from a settlement tank, which means that sludge pumping and dewatering are cheaper. 5% sludge may however be hard to remove and operators may feel they need hosepipes running to wash it away! Consequently, for new build it is sometimes thought better to use hydraulic desludging (also known as flooding) and use a separate thickener, which also overcomes some problems with level control. It does however remove a key benefit of DAF.

Designers should bear in mind that it is important to mix the DAF sludge well with any thinner sludges which are being cotreated prior to subsequent sludge treatment. A suitable mixed buffer tank is recommended to avoid problems caused by variable sludge solids content on downstream processes and for degassing DAF sludge.

The main disadvantages of DAF over simple settlement tanks are its far greater mechanical complexity and power costs.

In low-rate drinking water treatment duties, around 9 g air per m^3 of recirculated water are required. A rough rule of thumb is that the air compressor should deliver a volumetric flow of air (measured at atmospheric conditions) equivalent to 25% of recycle water flow rate. The recycle water flow rate should be at least 10% of the unit throughput. An HRT of at least 15 min is required, and a surface loading of 15 m/h.

For high-rate DAF, sludge treatment, and industrial applications, required recycle rates may be several hundred percent of the throughput. Air solubility in water is temperature-dependent, which may also be a factor in selecting a DAF process, since sludges and industrial effluents may be warm, as may seawater used by those countries which desalinate for drinking water treatment.

FILTRATION

We can separate filtration into coarse and fine filtration, as well as surface and depth filtration classes. Coarse surface filtration (screening) is used at the front of the plant to exclude gross solids. Fine depth filtration or fine surface (membrane) filtration is then used to remove finer solids. Very fine surface filtration can be used to remove dissolved solids.

COARSE SURFACE FILTRATION

Screens are proprietary items, designed by equipment suppliers rather than process plant designers, so I will cover them here only in outline as process designers specify rather than design them.

Trash racks

Trash racks are very coarse bar screens used to remove the gross solids for feed water, especially in lowland water supply and desalination applications.

Band screens

Band screens are usually installed in a channel, with flow driven by gravity head across the screen. They use a traveling arrangement of finely perforated plates to collect gross solids from the flow passing through them. There are several flow patterns used. Fig. 7.3 shows the two most popular patterns, with flow (in gray) passing left to right through (black) screen elements.

The (top) dual flow or "out to in" flow pattern is the most suitable flow pattern for most water intakes, avoiding the debris carryover to the clean side which the through flow (bottom) can be prone to.

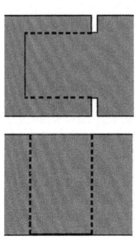

FIGURE 7.3

Typical flow patterns through band screens.

FIGURE 7.4

Slow sand filter with *schmutzdecke* layer.[1]

Courtesy: Olov Eriksson.

FINE DEPTH FILTRATION

Sand filtration is the most common solids removal process used in water treatment. Slow gravity sand filtration used to be commonplace, but most plants nowadays use either "rapid gravity" or pressure filters.

Slow sand filters

A slow sand filter is a shallow, low-surface-loading bed of fine sand irrigated with the water to be treated. After a maturation period of several weeks a gelatinous layer of bacteria, algae, fungi, and higher organisms which feed on them known as a *schmutzdecke* (German for "dirty skin") grows over the filter (Fig. 7.4).

The sand in a slow sand filter is just a support medium for this layer. It is the schmutzdecke which forms the filtration medium, and where biological activity removes the solids which build up. Either small areas of schmutzdecke and support sand are removed sequentially all over the filter surface by a traveling device, and the sand is returned after being washed free of the solids which have

---

[1]Licensed under CC BY-SA 4.0: https://creativecommons.org/licenses/by-sa/4.0/deed.en.

FIGURE 7.5

Microstrainer.

built up; or more commonly, whole filters are taken offline for manual or mechanical skimming and then brought back on line after rematuration.

Though they are simple and cheap to build and run, slow sand filters take up a lot of land. They have been declining in popularity for many years, but they have made somewhat of a comeback in recent years as they can be installed on top of a bed of granular activated carbon, to provide combined solids and dissolved organics removal. Thames Water (United Kingdom) developed the approach to keep their existing slow sand filters operational and avoid adding further processes, e.g., GAC contactors. They also did work on uprating the loading rates to cope with increased demand.

Slow sand filters are normally fed with untreated raw water, giving additional savings in terms of cost and complexity over other filter types although high turbidity feed water will require pretreatment with rapid gravity filters (RGFs) (see later) or microstrainers (Fig. 7.5).

Slow sand filters can reduce water turbidity to less than 1 NTU, total organic content by 10%, remove 95% of coliform bacteria, improve a water's color, taste, and odor, and do not result in significant sludge production. They are also good for ammonia and manganese removal but are not reliably good for color removal. They have been used with ozone dosing upstream to break the color down but plants that did this have now been decommissioned as the THM reduction downstream was not good enough.

Slow sand filter design

Basic design parameters are given in Table 7.3.

Table 7.3 Basic Design Parameters for Slow Sand Filters

Depth of water over sand bed	1–2 m
Depth of sand bed	0.5–1.5 m
Sand size	0.15–0.35 mm
Gravel layer thickness	300 mm of 10–20 mm gravel
Filtration rate	0.1–0.3 m/h
Filter run time	50 days

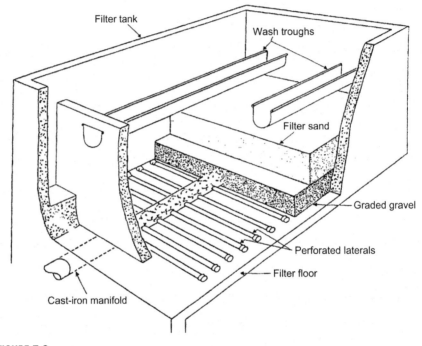

FIGURE 7.6

Rapid gravity filter schematic.

Courtesy: US EPA.

Rapid gravity filters

The single medium RGF (Figs. 7.6 and 7.7) has a bed of up to 2 m depth of coarser sand. This provides a different solids removal mechanism to the slow sand filter and takes up a lot less plan area. It is however taller than the slow sand filter, to accommodate the vertical extension of the wall for 2 m above the sand bed as well as, in many cases, the false floor used to distribute flow evenly.

There are many solids removal processes in RGFs, of which the key one is impingement: particles hit sand grains and stick to them. Trapping solids between

FIGURE 7.7

Rapid gravity filter.

particles, which many people believe to be the mechanism of action, only occurs if too fine a grade of sand has been selected. In this case, headloss through the filter builds up rapidly as the top layer of the filter blocks with solids.

The correct size of sand will distribute solids fairly evenly throughout the sand bed. The Rose equation may be used to find a quick value for clean bed headloss as shown in Metcalf and Eddy. Table 7.4 shows the equations required to construct a MS Excel spreadsheet allowing you to calculate fluidizing velocity and expanded bed height by their method.

Filter run time can then be calculated using the Rose equation (Eq. 7.1) as follows:

Rose equation and methodology

Initial filter headloss in the clean condition:

$$h_L = \frac{1.067}{\phi} \frac{C_D}{g} D \frac{V_a^2}{\varepsilon^4} \frac{1}{d} \qquad (7.1)$$

where

h_L = headloss
φ = shape factor
C_D = drag coefficient
D = bed depth
ε = resting porosity (called Iε in Table 7.4)
V_a = approach velocity
d = particle diameter

Table 7.4 Minimum Fluidizing Velocity Calculation

Amirtharajah and Cleasby Method	Symbol	Value	Units	Notes
Flow per filter	Q	Designer choice	m³/h	
Filter diameter	D_F	Designer choice	m	
Filtration area	A	Designer choice	m²	
Backwash velocity	V	Designer choice	m/h	
Particle diameter	d	60% finer size	mm	
Fluid specific weight	g_f	1000/16.0185	lb/ft.³	Water = 1000 kg/m³
Particle specific weight	g_x	165.43	lb/ft.³	Sand SG
Fluid viscosity	μ	1.10	cP	Water at 20°C
Minimum fluidizing velocity	V_{mf}	$(0.00381 \times (d^{1.82})) \times ((g_f \times (g_x\text{-}g_f))^{0.94})/(\mu^{0.88})$	gpm/ft.²	Use when $Re_{mf} < 10$
Hypothetical settling velocity	V_s	$8.45 \times V_{mf}$	gpm/ft.²	Use when $Re_{mf} < 10$
Reynolds number	Re_{mf}	$(d \times 0.1/(\mu_x\,0.01)) \times 30.48 \times V_{mf}/(448.8)$	–	
Reynolds number	Re_o	$8.45 \times Re_{mf}$	–	Use when $Re_{mf} < 10$
Correction factor if $Re_{mf} > 10$	k_{mf}	$1.775 \times Re_o{}^{-0.272}$	–	Use when $Re_{mf} > 10$
Corrected, V_{mf}	CV_{mf}	$V_{mf} \times k_{mf}$	gpm/ft.²	Use when $Re_{mf} > 10$
Corrected, V_s	V_s	$8.45 \times CV_{mf}$	gpm/ft.²	Use when $Re_{mf} > 10$
Corrected Reynolds number	CRe_o	$Re_o \times k_{mf}$		
Expansion coefficient	n	$4.35 + (17.5 \times (d/1000/Q) \times Re_o{}^{-0.03})$	–	$0.2 < Re_o < 1$
Expansion coefficient	n	$4.45 + (18 \times (d/1000)/D_F) \times (Re_o{}^{-0.1})$	–	$1 < Re_o < 10$
Expansion coefficient	n	$4.45 + (18 \times (D/1000)/A) \times (Re_{eo}{}^{-0.1})$	–	$10 < Re_o < 200$
Resting porosity	l_ε	Look up for grade selected	–	
Amirtharajah and Cleasby constant	k	V_{mf}/e^n	gpm/ft.²	N.B. use correct value of n and V_{mf}!
Porosity at backwash rate	ε	$0.4089979V/k^{(1/n)}$	–	NB use correct value of n!
Initial bed height	l_o	Designer choice	mm	
Expanded height of bed	l_e	$l_o \times (1 - l_\varepsilon)/1 - \varepsilon$	mm	

As with the Rose equation, the drag coefficient is given by:

$$C_D = \frac{24}{Re} \quad \text{for } Re < 1 \quad Re = \frac{\phi \rho d V_a}{\mu}$$

$$C_D = \frac{24}{Re} + \frac{3}{\sqrt{Re}} + 0.34 \quad \text{for } Re > 1 \text{ but} < 10^4$$

To calculate the build up of headloss over time, first calculate the mass of solids captured per hour by carrying out a mass balance over the filter. With proper selection of sand grade, solids should be deposited throughout the bed, but I make the conservative assumption that they are all deposited in the top 500 mm for the purposes of this exercise. This allows the loading (mass of solids per hour per m^3 of filter) to be calculated.

I then produce a MS Excel spreadsheet (Fig. 7.8) which calculates the headloss after a given number of hours of running for sand with $d = 0.5-1.0$ as below. Each of the "mWG" cells contains the following formula:

= [initial headloss] + (0.0538 × ([hours of running] × [loading])2.2267)

Let us say that we would like at least 12 h (ideally 24 h) of running between backwashing, and to backwash at less than or equal to 10 mWG headloss (for pressure filters). Fig. 7.8 shows an example in which 0.6 mm sand gives a 24 h run time between backwashes.

Sand does not come in single grain sizes at a commercial scale, but is supplied in grades retained by one mesh size and passed by another. Table 7.5 shows the

0.5 mm Sand	h	mWG	0.6 mm	h	mWG	0.7 mm	h	mWG	0.8 mm	h	mWG	0.9 mm	h	mWG	1.0 mm	h	mWG
Headloss after Hour	1	1		1	1		1	1		1	1		1	1		1	1
Headloss after Hour	2	1		2	1		2	1		2	1		2	1		2	1
Headloss after Hour	3	1		3	1		3	1		3	1		3	1		3	1
Headloss after Hour	4	1		4	1		4	1		4	1		4	1		4	1
Headloss after Hour	5	2		5	1		5	1		5	1		5	1		5	1
Headloss after Hour	6	2		6	2		6	1		6	1		6	1		6	1
Headloss after Hour	7	2		7	2		7	2		7	1		7	1		7	1
Headloss after Hour	8	2		8	2		8	2		8	2		8	1		8	1
Headloss after Hour	9	2		9	2		9	2		9	2		9	1		9	1
Headloss after Hour	10	3		10	2		10	2		10	2		10	1		10	1
Headloss after Hour	11	3		11	3		11	2		11	2		11	2		11	2
Headloss after Hour	12	4		12	3		12	3		12	2		12	2		12	2
Headloss after Hour	13	4		13	3		13	3		13	2		13	2		13	2
Headloss after Hour	14	5		14	3		14	3		14	2		14	2		14	2
Headloss after Hour	15	5		15	4		15	3		15	2		15	2		15	2
Headloss after Hour	16	6		16	4		16	4		16	3		16	3		16	2
Headloss after Hour	17	6		17	4		17	4		17	3		17	3		17	2
Headloss after Hour	18	7		18	5		18	4		18	3		18	3		18	2
Headloss after Hour	19	8		19	5		19	4		19	3		19	3		19	2
Headloss after Hour	20	9		20	6		20	5		20	4		20	4		20	2
Headloss after Hour	21	10		21	6		21	5		21	5		21	4		21	3
Headloss after Hour	22	10		22	7		22	5		22	5		22	4		22	3
Headloss after Hour	23	11		23	7		23	6		23	6		23	5		23	3
Headloss after Hour	24	12		24	8		24	6		24	6		24	5		24	3

FIGURE 7.8

MS Excel spreadsheet for headloss calculation.

Courtesy: Expertise Ltd.

Table 7.5 Sand Characteristics

Sand Grade	Lower Size (mm)	Upper Size (mm)	Uniformity	60% Finer Size (mm)	Bed Depth (m)	SG	Bulk Density (t/m³)	Washwater Upflow Rates (m/h)	Notes
22/36	0.4	0.7	–	–	–	2.65	1.5	12@15°C 9@5°C	Pressure filters
16/30	0.5	1	1.26	0.73	–	–	–	–	–
16/30	0.5	1	1.4	0.77	–	–	–	22@15°C 16@5°C	With prior clarification
16/30	0.54	0.71	1.4	–	0.9	–	–	–	–
14/25	0.6	1.18	1.28	0.88	–	–	–	–	–
14/25	0.63	0.85	1.4	–	0.9	–	–	–	–
12/22	0.71	1.4	1.25	1.13	1.1	–	–	–	–
12/18	0.8	1.4	1.25	–	1.1	–	–	32@15°C 24@5°C	No prior clarification
8/16	1	2	1.25	1.35	–	–	–	–	–
8/16	1.05	1.27	1.4	–	0.9	–	–	–	–
8/16	1	2	1.33	1.49	–	–	–	–	–
7/14	1.18	2.36	–	–	–	–	–	–	–

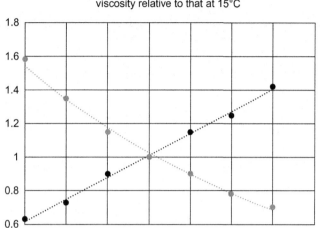

Filter wash rate for given expansion and water
viscosity relative to that at 15°C

● Wash rate ⋯⋯   ● Viscosity ⋯⋯

FIGURE 7.9

Effect of temperature on the viscosity of water and wash rate for equivalent expansion.

characteristics of filter sand supplied in commercial grades in the United Kingdom, together with some notes on usage, and Fig 7.9 shows how the water viscosity and the required wash rate for a given bed expansion changes with temperature.

The grades of sand used for filtration are selected for roundness and a high uniformity of size, shape, and density. Bed fluidization during backwashing selects denser particles to rest at the bottom of the bed. This allows more than one size of media to be used, creating a bed in which coarse particles are at the top of the bed and fine at the bottom, giving successively finer filtration in a single bed. If the coarse particles are less dense than the fine ones, they will rearrange correctly upon backwashing. Multimedia filters of this type give a high degree of solids removal and are well suited to higher solids loads. The use of a multimedia bed will improve filtration performance, enable longer run times for a given loading rate, and possibly help remove *Cryptosporidium*.

The coarsest particles in a three-layer bed are often anthracite or bituminous coal (in Australia); the medium size particles the denser sand; and the finest garnet or ilmenite, both of which are denser than sand (you can get even more layers if you wish using pumice or plastic media). Design data for multimedia filter beds are presented in Table 7.6.

Anthracite and sand layers may usefully be increased in depth from the minimum depth given to accommodate higher solid loadings, but increasing the depth of the garnet layers is unlikely to be beneficial.

Table 7.6 Basic Design Parameters for Multimedia Filter Beds

Layer	Material	Specific Gravity	Grain Size (mm)	Minimum Bed Depth (mm)
Top	Anthracite	1.5	1.5	300
Middle	Sand	2.6	0.6	180
Bottom	Garnet	4.2	0.4	120
Support	Garnet	4.2	1.5	100

Backwash requirements

Backwash water rate	35 m/h
Air scour rate	25 m/h
Head above filter	Up to 2.5 m

The high wall around and above the sand bed in a rapid gravity filter allows for water to build up over the bed, providing a driving force opposing the increase in headloss which happens as the filter collects solids. Unlike the slow sand filter the RGF must be taken offline frequently (at least every 2–3 days for low loadings, and commonly daily) for cleaning.

For conceptual design purposes, surface loadings between 5 and 20 m/h are recommended. Although lower loadings generally make for better performance, loadings below 5 m/h should be avoided as they can result in maldistribution. A further concern is headloss increase and solids breakthrough if flow increases when near full solids loading. Maximum sizes of RGF filter cells are 120–140 m². The limiting factors are distribution of backwash air and water, and size of backwash pumps. Crossflow washing is used on large RGFs.

There may be a range of design flow rates and loadings to accommodate. At the very least, there will be high- and low-flow scenarios.

Filter design run times are normally 12–24 h, and 24-h filter runs usually require the following conditions to be met:

- ≤ 15 ppm incoming SS/ < 2NTU turbidity
- 1–2 kg/m²/day SS
- Normal coagulant dose 8 g alum/m²/h (for very alkaline waters, 0.6 g alum/m²/h)

The sand grades used for RGFs are coarser grades. 16/30 grade sand tends to be used in filters with a prior clarification stage, or underlying anthracite, whilst the coarsest 12/18 grades are used in filters with no prior clarification.

Filter headloss increases should be calculated using the Rose equation and the charts in Metcalf and Eddy (assuming, for example, 500 mm solids penetration).

Suspended solids concentrations (including those resulting from added coagulant) above 15 mg/L should be avoided as they are likely to result in difficulty in maintaining reasonably short intervals between backwashes.

The requirement for backwashing of RGFs means that there must be sufficient standby capacity to give full rated flow with at least one filter out of service for backwashing.

Backwash flow must be sufficient to give at least 10−30 m/h surface loading, and an air scour at 30−50 m/h enhances cleaning. Filters without air scour will mud-ball within 3 years. Air scouring may be done before backwash, or combined in the "continental wash." It should never be done after backwash as air bubbles caught between sand particles can blind the filter.

Filters: air scour distribution

Air scour must be distributed evenly across the base of a filter if it is to work effectively. The mixing of air and water in distributing pipework or suspended floors can result in distribution problems.

Filter nozzles (Fig. 7.10) commonly used to assist with this can be installed through the wall of pipework or through a suspended floor of concrete, plastic, or steel, mounted on dwarf walls or pillars.

In all cases the variation in laid floor level from one side of the filter to the other should be ±3 mm. Nozzles passing through such floors should be installed such that the variation in level measured from the top of two adjacent nozzles is ±2 mm.

Dwarf walls should span the full width of the filter floor at a height varying by ±5 mm. They should be equally spaced with paired 200 × 100 mm horizontal distribution slots located within the expected air cushion.

Nozzle stems should be long enough to extend below the filter floor, permitting an air cushion at least 75 mm deep and no greater than the total available length of air slot or stem extending beneath the floor.

Nozzle stems should have 2 mm wide by 30 mm long slots or an orifice as well as a slot in high flow conditions (such as continental wash) to make them

FIGURE 7.10

Filter nozzles.

less sensitive to level variations. They may be supplied with slots, or they may be drilled on site to match conditions.

Whichever method is used to introduce air into the plenum underneath a nozzle floor, and whatever the shape or material of construction of the floor, it is important that the lowest exit or discharge point of the air must never be below the air slot in the nozzle stem.

Air in the main header from the blower should have a superficial velocity of less than 30 m/s, and velocity through the nozzle orifices should be 20−30 m/s.

Filters with suspended floors may be split into three types: those with dwarf walls and distribution ducts, those with dwarf walls but no distribution ducts, and those with neither.

For the first of these types the distribution duct normally runs the length of the filter underneath the washout channel. The dwarf walls create discrete zones within in the plenum, each fed from the distribution duct by one air and two water orifices.

The air inlet to the main distribution duct enters horizontally, located such that its invert is at least 25 mm above the air/water interface during steady state air scour.

Air distribution orifices from the duct to the underside of the slabs must also be placed no less than 25 mm above the air/water interface.

For the second and third types the main air inlet enters horizontally and generates branches along approximately 80% of the filter length, to suit the number of discrete zones present.

The branches need to be U shaped, connecting the underside of the main air inlet to the plenum chamber, with air outlets above the air/water interface during air scour.

Branches should be on 1−2.5 m centers, typically leading in practice to 2−6 branches per filter.

Pressure filters

Pressure filters are essentially a RGF inside a pressure vessel (Fig. 7.11). Their advantage is that they allow for higher operational pressures, whilst restricting unit height.

The use of a pressure vessel tends to make them circular rather than rectangular in plan view, though the cylindrical vessel can be laid down horizontally to approximate a rectangular plan view.

Pressure filters require finer grades of sand, grade 22/36 being the finest grade used.

The pressure filter height should be less than 5 m.

When used in downflow, design loadings and limitations are the same as RGFs, but the vessel (especially if vertical) should not exceed 3 m in diameter, to avoid flow distribution problems.

FIGURE 7.11

Vertical pressure filters.[2]

Courtesy: Stephen Craven—geograph.org.uk/p/1121363.

Upflow pressure filters

Pressure filters can be run with flow going upwards, rather than downwards, allowing higher solids concentrations to be accommodated. Table 7.7 gives basic design parameters for upflow filters.

Table 7.7 Basic Design Parameters for Upflow Filters

Surface loading	12 m/h
SS removal	90%
Bed depth 1.5 m	1.5 m
Backwash surface loading	30–35 m/h

FINE SURFACE FILTRATION: MEMBRANES

There are several membrane technologies used in water treatment which are named based on the size of particles which they remove. Relatively coarse micro- and ultra-filtration membranes remove fine solid particles, the finest reverse osmosis (RO) membranes allow water and dissolved ions to be separated.

Fig. 7.12 shows the relationship between particle size and membrane separation process nomenclature.

---

[2]Licensed under CC BY-SA 2.0: https://creativecommons.org/licenses/by-sa/2.0/.

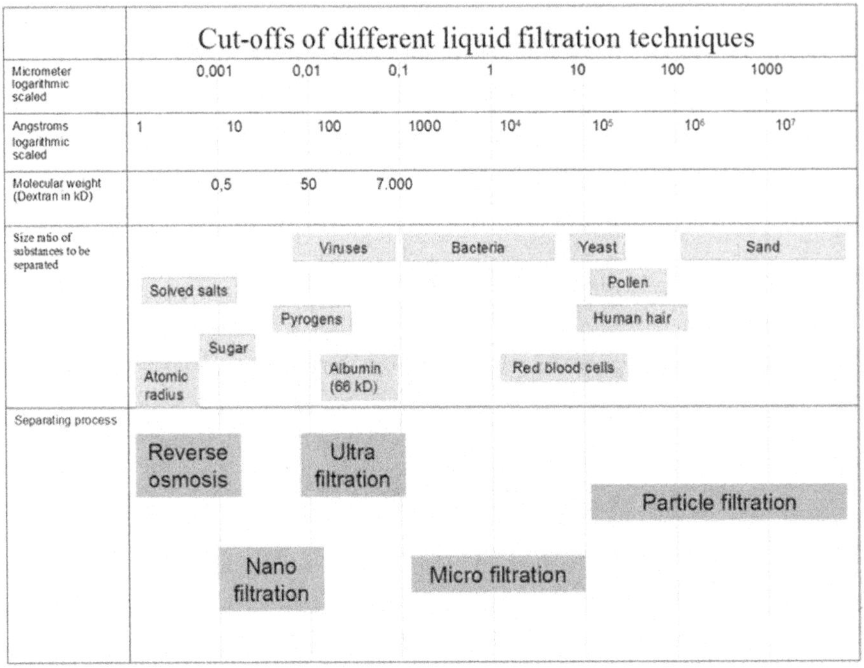

Cut-offs of different liquid filtration techniques							

FIGURE 7.12

Particle size and membrane separation processes.

Ultrafiltration/microfiltration

We would tend to use ultrafiltration (UF) or microfiltration (MF) to give an absolute barrier against the passage of particles smaller than a given size. This can be useful in color removal, removal of pathogens (especially *Cryptosporidium*) and physical disinfection prior to discharge, amongst other things.

UF and MF membranes may be constructed from a wide variety of materials, including sintered materials, cellulosic, polyvinylidene fluoride, polyacrylonitrile, polypropylene, polystyrene, polyethersulfone, or other polymers.

Each of these materials has different properties with respect to surface charge, degree of hydrophobicity, pH and oxidant tolerance, strength, and flexibility.

Nanofiltration/reverse osmosis

Nanofiltration (NF) and RO allow us to selectively remove dissolved ions from a solution. We can use this for producing high-quality water from seawater or industrial effluent. There is no cheap way of doing this, but when energy is cheap, and water is expensive, it can be a very attractive option.

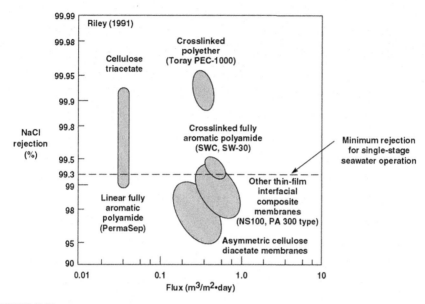

FIGURE 7.13

Performance characteristics of membranes operating on seawater at 56 kg/cm^2 (840 psi) and 25°C.

Courtesy: Baker, R. W., et al., (1991) Membrane separation systems, 1st ed. Recent developments & future direction *(Elsevier Press)*.

There are two commonly used NF/RO membrane materials: polyamide and cellulosic (Fig. 7.13). The polyamide membranes can be further divided into solid aromatic polyamide and thin film composite.

Configurations of reverse osmosis membrane systems

There are four main configurations used for RO membrane systems, namely plate and frame (Fig. 7.14), tubular (Fig. 7.15), spiral wound (Fig. 7.16), and hollow fiber (Fig. 7.17).

Advantages/disadvantages

Each of the various configurations has different advantages and disadvantages. In summary, capital cost is highest for tubular and plate and frame systems, and lowest for hollow fiber and spiral wound systems.

Tubular systems have the highest power costs, which are successively lower for plate and frame, hollow fiber, and spiral wound systems. The process footprint is highest for tubular membrane systems, with plate and frame, spiral and hollow fiber systems being successively smaller. Tubular systems are the most robust, with plate and frame, hollow fiber and spiral systems becoming increasingly prone to irreversible failure.

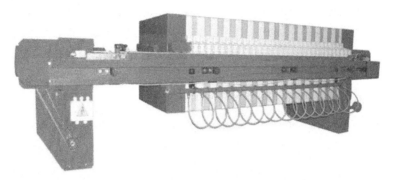

FIGURE 7.14

Plate and frame RO system.

Courtesy: Shanghai Xunhui Environment Technology Co., Ltd.

FIGURE 7.15

Hollow fiber RO system.

Courtesy: ROplant.org.

Some of the disadvantages of RO generally can be offset against potential advantages. For example, we can recover energy from the permeate stream to pre-pressurize the feed stream, creating consequent savings in energy costs. This is far from universally applicable, however, as RO is most often used in countries where energy is cheap.

In addition, the more effort and expense we put into pretreatment of feed water, the lower the capital and running costs of cleaning the systems, and the longer the expected life in service of RO membranes.

Membrane materials selection

Membranes are generally made from either cellulose, polyamide, or thin film composites.

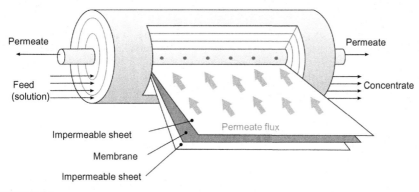

FIGURE 7.16

Spiral RO system.[3]

Courtesy: Daniele Pugliesi.

FIGURE 7.17

Tubular RO system.

Courtesy: PCI Membranes.

Cellulose membranes are prone to biological hydrolysis, which is avoided by adding 1 mg/L chlorine to the feed water. Chemical hydrolysis is avoided by operating at around pH 5. Cellulose membranes may be subject to compaction, so we must avoid excessive feed water temperature and pressure. Cellulose di/tri acetates are less affected by these problems although polyamides are still more tolerant of feed water temperature and pressure variations.

Polyamides resist hydrolysis and can operate from pH 3−11, but feed chlorination will degrade the membrane so, for some applications, it may be necessary to dechlorinate the feed water. Table 7.8 gives materials selection data for RO systems.

Design for fouling

Designers need to consider the possibility of membrane fouling, and specify suitable membranes and cleaning systems to allow a reasonable period of reliable

---

Table 7.8 Membrane Materials Selection

Feature	Cellulosic	Aromatic Polyamide	Thin Film Composite
Rejection of organics	L	M	H
Rejection of low molecular weight organics	M	H	H
Water flux	M	L	H
Acceptable pH range	4–8	4–11	2–11
Upper operating temperature (°C)	35	35	45
Oxidant tolerance	H	L	L
Compaction tendency	H	H	L
Biodegradability	H	L	L
Cost	L	M	H

operation before membranes need replacement. Foulants on RO membrane surface can reduce permeate flux; or increase differential pressure, product water conductivity, and required feed pressure to maintain output or a combination of these effects.

There are several possible types of fouling.

Scaling arises from either high bulk concentrations of an ion, or local high concentrations arising because of concentration polarization. We can reduce scaling problems by operating in *crossflow mode*: with most of the flow passing over the membrane rather than through it, any build up on the membrane tends to be washed away. More generally, we can reduce recovery of the permeate.

We can use lime softening to remove sulfate and silica, and ion exchange to remove Ba/Sr. To help prevent scaling due to hardness, we can use pH adjustment to remove CO_2 prior to filtration.

We can assess the potential for scaling due to hardness using a heuristic technique such as the Langelier Index. However, this is not strictly scientific. There are several of these indices, all of which are at best an indicator of the tendency of a water to scale or be aggressive.

Metal oxide fouling is usually caused by iron or manganese, especially in reduced forms such as Fe^{2+}. To avoid metal oxide fouling, we can remove iron and manganese from the feed. In the case of iron, we want <4 ppm if there is no dissolved oxygen (DO), and <0.05 ppm if DO is present. This is normally achieved for Fe^{3+} by filtering, or by aeration and filtration of Fe^{2+} in groundwater feeds. Alternatively, we can prevent oxidation of the reduced iron and manganese, or use zeolite softening to remove ferrous iron.

Citric acid cleaning at pH 4 will restore reductions in membrane flux caused by metal oxide deposition.

Plugging is caused by the blockage of membrane pores by fine solids. Coarser solids form a cake on the membrane, finer ones enter and block pores completely, and finer ones still enter pores and block them partially, but increasingly over time. This plugging of pores is normally avoided by pretreatment with $5-10\,\mu m$ cartridge filters. This is the standard/minimum pretreatment used in NF and RO applications.

Colloidal fouling is usually caused by clays in the feed. The standard pretreatment processes usually solve this problem, but a coagulation stage prior to filtration will enhance operation by breaking colloids.

Biological fouling due to overgrowth can be prevented by feed water chlorination (for resistant membranes such as cellulosic ones) or, in the case of polyamide membranes, through the use of formaldehyde. It is also good practice to avoid open feed/surge tanks in RO installations, and disinfectants can also be added to feed water to discourage overgrowth. It has been reported that the finer details of filter element design such as spacer dimensions can strongly affect biological fouling.

Solute rejection by reverse osmosis membranes

Predicting how well RO membranes will retain dissolved ions can be very complex, but there are a few rules of thumb which can be used to predict performance.

Generally speaking, multivalent ions are retained better than univalent ones. Traces of univalent ions are particularly poorly rejected, and nitrate rejection is anomalously low for a univalent ion.

The lower the degree of dissociation, the lower the rejection is, which is why silica can be a problem.

Acids and bases are rejected less well than their salts, and co-ions can affect rejection—for example, $NaSO_4$ is rejected better than $NaCl$. Low molecular weight organic acids have poor rejection though their salts are well rejected.

There are several problem salts from the point of view of RO system designers. The most commonly recognized are:

- Barium sulfate (>0.05 ppm Ba^{2+})
- Calcium carbonate/sulfate (>1500 ppm)
- Strontium sulfate (>10 ppm Sr^{2+})
- Reactive (neither polymerized nor colloidal) silica

These and any other salts with low solubility can cause scaling.

Design methodology

Design begins, as ever, with determination of the quantity and quality of feed and product. This will require determination of feed source, and analysis of the quality of that source.

Then the designer must decide on the flow configuration and number of passes, and select the required membrane and element type. Standard elements are 8-in. in diameter and 40-in. long, but smaller elements are used for smaller systems.

Next the designer selects a realistic design flux, ideally based on pilot data, but commonly based on customer experience or typical values. This allows them to calculate the number of elements needed by determining the required permeate flow rate/design flux to get a required surface area, then dividing that by the surface area of an element.

The required number of pressure vessels needed can then be determined from the known number of elements per pressure vessel (six elements per vessel is the usual standard).

Next the configuration of the system is determined. This involves specifying the number of process stages in series and the staging ratio (the ratio of the number of parallel vessels in each stage of the series, also known as the array ratio).

If that seems confusing, each stage usually consists of a number of parallel process vessels. There are multiple (usually two or three) stages in series.

By this point in the design, or ideally earlier, you should be consulting specialist suppliers, who can help you with the last stage, balancing permeate flow rates. Permeate flows drop as you pass through the stages of the system, especially with high feed total dissolved solids (TDS) or high recovery ratios. These need to be balanced by increasing feed pressure (or reducing backpressure) in later stages, or increasing backpressure in earlier stages.

Adsorption

Adsorption onto inorganic coagulant floc particles and subsequent removal by settlement, flotation, or filtration is an important mechanism of color removal, especially in highly colored peaty upland waters. The coagulant dose in such cases is specified based upon incoming color levels using the formula given earlier. Taste and odor compounds may also be removed by this mechanism.

Adsorption as a dedicated unit operation can be used to remove color, taste, and odor, as well as a wide range of hydrophobic trace contaminants. The most common adsorption medium is granular-activated carbon, often contained in pressure filters with around 45 min of "Empty Bed Contact Time" (EBCT) (the HRT of the portion of the vessel's filled with GAC, ignoring the volume of the GAC itself). Alternatively, GAC can also be sandwiched in a slow sand filter as discussed earlier, or used in RGFs as a combined filtration and adsorption medium.

The design of GAC filters from first principles using Freundlich isotherms, as practiced in academia, is impractically complex once more than two components are considered, even if the variation of flow rate, concentration, and temperature in a water treatment plant is ignored.

A combination of experience, EBCT, and isotherm approaches is used to specify a conservative bed volume which gives a sufficient projected life in service. This bed volume is then used as a minimum total volume in service in the design of a filter via the normal process for an RGF, slow sand filter, or pressure filter as appropriate. Adsorption (especially of organic acids) is proportional to pH. An extra 10% EBCT should be allowed for every pH unit away from neutrality of the feed water under operating conditions.

As a general rule, EBCT for GAC is 5–45 min in the absence of sizing criteria, and surface loadings of between 5 and 30 m/h are recommended. While lower loadings generally make for better performance, loadings below 5 m/h should be avoided as they are uneconomic and can tend to result in maldistribution, though it can be argued that poor underfloor design ultimately causes maldistribution.

Ion exchange

Ion exchange was once a popular technique in water treatment but has now been largely superseded in many of its old markets by membrane technologies. It can provide high degrees of softening and demineralizing, and consequently still finds use in industrial applications.

Ion exchange is an exchange of ions between two electrolytes or between an electrolyte solution and a complex. In most cases the term is used to denote the processes of purification, separation, and decontamination of aqueous and other ion-containing solutions with solid polymer or mineral "ion exchangers."

Ion exchangers can be ion exchange resins (functionalized porous or gel polymer), zeolites, montmorillonite, clay, or even soil humus. Ion exchangers are either cation exchangers that exchange positively charged ions (cations) or anion exchangers that exchange negatively charged ions (anions). There are also amphoteric exchangers that can exchange both cations and anions simultaneously. However, the simultaneous exchange of cations and anions can be more efficiently performed in mixed beds that contain a mixture of anion and cation exchange resins, or passing the treated solution through several different ion exchange materials.

Ion exchangers can be unselective or have binding preferences for certain ions or classes of ions, depending on their chemical structure. This can be dependent on the size of the ions, their charge, or their structure. Typical examples of ions that can bind to ion exchangers are:

- H^+ (proton) and OH^- (hydroxide)
- Single-charged monatomic ions like Na^+, K^+, and Cl^-
- Double-charged monatomic ions like Ca^{2+} and Mg^{2+}
- Polyatomic inorganic ions like SO_4^{2-} and PO_4^{3-}
- Organic bases, usually molecules containing the amino functional group $-NR_2H^+$
- Organic acids, often molecules containing $-COO^-$ (carboxylic acid) functional groups
- Biomolecules that can be ionized, such as amino acids, peptides, proteins, etc

Along with absorption and adsorption, ion exchange is a form of sorption.

Ion exchange is a reversible process and the ion exchanger can be regenerated and loaded with desirable ions when exhausted by washing with an excess of the desirable ions.

DISTILLATION

My own professional experience does not extend to the design of distillation plants, so this section is a very brief overview based on external sources. Distillation is the oldest method of desalination and has been in use since the 1920s. It used to be the standard approach in the Middle East, where energy costs are low (and integration with power plants is a possibility), but even there, it is increasingly being supplanted by RO.

The Source Book of Alternative Technologies for Freshwater Augmentation in Latin America and the Caribbean (see Further Reading) states:

> *Distillation is a phase separation method whereby saline water is heated to produce water vapor, which is then condensed to produce freshwater. The various distillation processes used to produce potable water, including MSF, MED, VC, and waste-heat evaporators, all generally operate on the principle of reducing the vapor pressure of water within the unit to permit boiling to occur at lower temperatures, without the use of additional heat. Distillation units routinely use designs that conserve as much thermal energy as possible by interchanging the heat of condensation and heat of vaporization within the units. The major energy requirement in the distillation process thus becomes providing the heat for vaporization to the feedwater.*

PRETREATMENT

Distillation plants generally have lower pretreatment requirements than RO plants, especially with respect to suspended solids and disinfection control. They may however require greater posttreatment as they produce an output at around 100 ppm TDS vs RO's 400 ppm (it should be noted that variations in feed salinity do not affect distillation plant output quality as they do RO).

We can divide pretreatment for distillation into two classes: treatments like those on any water treatment plant (covered in the preceding filtration section), and treatments like those on boiler feed water plants, used for corrosion and foam control.

TYPES OF DISTILLATION PROCESS

There are three types of distillation plant used for desalination:

- Multistage flash (MSF)
- Multiple effect distillation (MED)
- MED—thermal vapor compression (MED-TVC) processes

Of these the MSF process is by far the most commonly used.

GAS TRANSFER

The most important application of gas transfer in clean water treatment is probably the dissolution of chlorine or ozone in water to facilitate disinfection (and to a lesser extent the oxidation of iron, manganese, ammonia, taste, and odor compounds). This is normally achieved via diffusers, venturi, or static mixers. Sulfur dioxide dissolution to facilitate dechlorination and reduction uses the same equipment, which has a long pedigree in water treatment (see Fig. 7.18).

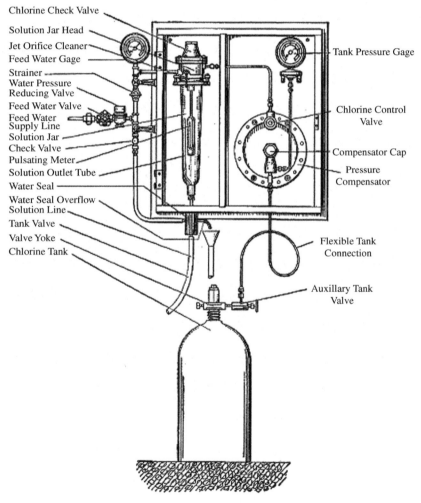

FIGURE 7.18

Water chlorinator from 1918.

Chlorination of water by Joseph Race, 1918.

There are less common gas transfer applications, such as the use of carbon dioxide for pH and alkalinity control. Traditionally carbon dioxide was bubbled directly through the water being treated in contactors with a HRT of around 30 min, but carrier water systems, or direct gas injection into a static mixer as used for other gases is now becoming the norm.

STRIPPING

Applications of gas stripping in clean water treatment are rare though it has been used for removal of THMs and chlorinated solvents. In ultraclean water the removal of carbon dioxide between stages in ion exchange is important. TR131 (see Further Reading) reviews the literature on CO_2 stripping though this is quite an old reference now.

PHYSICAL DISINFECTION

Physical treatments tend to be energy intensive, and leave no residual. This is an advantage when used for treating water to be discharged to environment, or ultra-pure water, but not for drinking water treatment. Heat has been used at small scale, but UV light is popular for small to medium scale potable and ultrapure applications. Other types of electromagnetic radiation, including gamma radiation, have been used. However, gamma radiation is unpopular, despite its effectiveness, due to the added complexities of handling radioactive materials.

UV IRRADIATION

Ultraviolet (UV) light can be used as a physical disinfectant. A UV disinfection system exposes water to a specific intensity and wavelength of UV for a specific time. The optimum wavelength for water disinfection is about 254 nm.

The UV light is usually produced by mercury vapor lamps housed in quartz sleeves, to allow the lamps to be placed near the water to increase efficiency.

UV transmission is reduced by absorbance by colored compounds such as dissolved iron and manganese as well as scattering by turbidity. Broadly speaking, killing 99% of the bacteria present will require twice as high a dose as that required to kill 90%.

Maximum contaminants for effective UV disinfection are set out in Table 7.9.

The system should have the capacity to produce a minimum dose of 25 mWs/cm^2, providing a better than 99.999% inactivation of *Escherichia coli*. For drinking water the United Kingdom's DWI guidelines require the water to be like that dosed with chlorine, e.g., <1 NTU, and validation is a necessity. A dose

Table 7.9 Maximum Contaminants for Effective UV Disinfection

Description	Allowable Maximum
Transmittance (254 nm, 5 cm pathlength)	80%
Turbidity	15 NTU
Color	10° hazen
Iron	5 ppm
Manganese	8 ppm
Calcium	110 ppm
Total organic carbon	4 ppm

(or fluence) of $40 \, \text{mJ/cm}^2$ is common for disinfection but lower values are accepted for *Cryptosporidium* inactivation.

NOVEL PROCESSES

ELECTRODIALYSIS

Electrodialysis systems use a selectively permeable membrane to move ions from one side to the other under the influence of an electric potential. In almost all practical electrodialysis processes, multiple electrodialysis cells are arranged into a configuration called an electrodialysis stack, with alternating anion and cation exchange membranes forming the multiple electrodialysis cells.

In normal potable water production without the requirement of high recoveries, RO is generally believed to be more cost-effective when TDSs are 3000 parts per million (ppm) or greater. Electrodialysis is more cost-effective for TDS feed concentrations of less than 3000 ppm or when high recoveries of the feed are required when it is properly developed.

It is however not a robust technology. In my own professional experience, many owners of electrodialysis systems have reported problems with operation and reliability.

The process requires extensive pretreatment, including removal of all particles bigger than $10 \, \mu\text{m}$, as well as hardness, large organic anions, colloidal matter, iron, and manganese oxides.

MEMBRANE DISTILLATION

Membrane distillation is essentially distillation through a membrane. A hydrophobic membrane holds back liquid water but allows water vapor pass through the membrane's pores. A partial vapor pressure difference gives the driving force of the process commonly triggered by a temperature difference.

When it is better developed, it may well provide an alternative to ion exchange, distillation, or RO to produce deionized water. Although currently very popular with researchers, it is not yet market ready as a number of technical problems such as nontraditional fouling still remain unsolved.

FORWARD OSMOSIS

"Forward osmosis" uses a high osmotic pressure "draw" solution to move water across a semipermeable membrane by osmosis, leaving behind dissolved solutes.

There are some small commercial plants using the technology as a first stage, prior to RO. If it can be made to work reliably and economically at scale, it might greatly reduce the fouling and energy demands problems of membrane desalination of seawater.

Practical full-scale applications are presently limited to a small number of plants desalinating seawater by FO, particularly in Gibraltar and Oman. When it is developed, it might provide an alternative to ion exchange, distillation, or RO for production of deionized water.

FURTHER READING

Amirtharajah, A., & Cleasby, J. L. (1972). Predicting expansion of filters during backwash. *Journal-American Water Works Association, 64*(1), 52–59.

Gauntlett, R. B. (1980). *Water research centre technical report: A literature review of CO_2 removal by aeration, TR131*. Marlow, UK: Water Research Centre. Available from: <https://www.ircwash.org/sites/default/files/251-80LI.pdf> Accessed 21.11.17.

Haarhoff, J., & Van Vuuren, L. (1993). *A South African design guide for dissolved air flotation: Report for the Water Research Commission*. Pretoria, South Africa: Water Research Commission.

United Nations Environment Programme. (2017). Sourcebook of alternative technologies for freshwater augumentation in Latin America and The Caribbean. Available from: <http://www.unep.or.jp/ietc/Publications/TechPublications/TechPub-8c/index.asp#1> Accessed 9.10.17.

Clean water unit operation design: chemical processes

CHAPTER OUTLINE

Introduction 101
Drinking Water Treatment...102
    pH and Aggressiveness Correction...102
    Coagulation...104
    Precipitation 105
    Softening...105
    Iron and Manganese Removal by Oxidation 107
Boiler and Cooling Water Treatment..108
    Cooling Water Treatment 108
    Boiler Water Treatment 109

INTRODUCTION

Chemicals are dosed into drinking water for a variety of reasons. Dosing may be required to:

- correct pH
- oxidize or reduce undesirable substances
- form insoluble compounds with undesirable substances
- coagulate colloidal suspensions
- adsorb undesirable substances
- disinfect
- medicate (in the case of fluoride addition)

The general idea is to remove substances which are present at levels above that allowed by the specification and add substances which are not present at levels required by the specification. In the case of drinking water treatment the specification will be set by a national regulator, or in the absence of one, by WHO standards.

The standards apply to water at the point of delivery. It may sometimes be necessary to exceed the standards at the works to achieve compliance at the tap. For example, if supply to homes is via legacy lead piping, great care must be taken to supply water that does not dissolve lead by controlling aggressiveness (see Langelier Index) and adding phosphate.

An Applied Guide to Water and Effluent Treatment Plant Design. DOI: https://doi.org/10.1016/B978-0-12-811309-7.00008-4

Clean water also encompasses cooling water, boiler feed water, and ultraclean grades of water. As these are not used for drinking, additional chemicals may be added, which are not suitable for use in potable supplies.

Carrier water systems may be used to add (often recirculated) water to the main stream, improving mixing effectiveness, and sometimes allowing dosing of hard to handle chemicals.

DRINKING WATER TREATMENT

PH AND AGGRESSIVENESS CORRECTION

Current specifications for drinking water tend to require a pH of between 6.5 and 9.5. Water for process duties may have a far tighter specification. There may also be a requirement that the water is neither scaling (more common in process water applications) nor aggressive (a general requirement).

Correction of pH may be achieved by adding acid or alkali as required. Reducing aggressiveness may require the addition of alkalinity as well as an increase in pH. Lime is the only dosed chemical which simultaneously reduces acidity, and adds both alkalinity and hardness.

All pumps sufficiently accurate for dosing pH correction chemicals will be of the positive displacement type. Piston diaphragm pumps are most commonly used for acids or alkali solutions, and peristaltic or progressing cavity pumps for lime suspensions that can block the valves in piston diaphragm pumps.

To obtain consistent mixing, the pulsating flow provided by piston diaphragm type positive displacement pumps needs to be smoothed, especially if delivering to a static mixer. An excellent way of doing this is to insert a loading valve in the discharge side downstream of a pulsation damper as shown in Fig. 8.1. With this design, it should be possible to control pH to within 0.1 pH units.

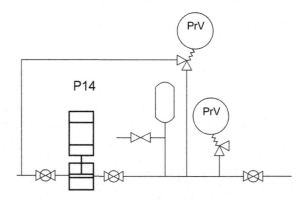

FIGURE 8.1

Use of loading valve and pulsation damper to control pulsation from positive displacement pump.

The control of pH requires signals from both flow and pH meters to be used. Piston diaphragm pumps are normally controlled by motor speed being controlled by the flow signal and stroke length being controlled based on deviation from the pH set point. Modern digital pumps driven by stepper motors are arguably making this approach obsolete, though they still work from flow and pH probe signals.

It is in my experience common to see a failure to control pulsations in dosing lines, and consequent cyclical variability in effluent quality on effluent treatment plants, but potable water designers are more careful (though installers and operators may not be so conscientious!).

Directions on sizing acid and alkali pumps for pH control are, as far as I am aware, not in the public domain. I was given an empirical method by a more experienced engineer in a fax back in 1991. I found out years later that the method was based on Tillmann's formula (Eq. 8.1), which is found in a paper written in German in 1912.

Tillmann's formula

$$pH = \log \frac{ALK \times 0.203 \times 10^7}{[CO_2]} \tag{8.1}$$

The basic idea is that we can view all the various carbon dioxide derived species in the system as "potential alkalinity" (Fig. 8.2).

1. Use the nomogram to find the ratio of methyl orange alkalinity and "free CO_2" at the worst case initial pH
2. Use the known initial alkalinity and this ratio to work out what the free CO_2 is
3. The potential alkalinity of water is initial alkalinity + [Free CO_2] × 50/44
4. Use the nomogram to find MOA/Free CO_2 at the desired pH
5. Reallocate potential alkalinity in accordance with the ratio
6. The difference in alkalinity between the initial and desired states can be used to calculate how much acid or alkali is needed to give pH change

The method is most valid in the range pH 5−8, and particularly close to neutral pH.

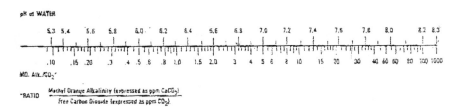

FIGURE 8.2

Nomogram based on the Tillman formula.

It should also be noted that inorganic coagulants are acidic. The pH drop associated with adding them thus needs to be taken into consideration. This is especially true with low alkalinity waters.

I have included a worked example at Appendix 11 of how to use this approach.

COAGULATION

To obtain sufficient removal of solids (as well as other things such as color), it is usual to add a chemical that "breaks" colloidal suspensions. The two most commonly used types of chemicals are aluminum and iron salts, and the most popular types of these are strongly acidic.

Precipitation of as much of the added metal as possible is desirable, both to comply with water quality parameters, and to enhance the removal of color and organic compounds by adsorption onto flocs. This is achieved by operating as close as possible to the pH giving minimum solubility of aluminum or iron hydroxide as appropriate.

The two pH changes produced by first adding a strongly acidic coagulant and then a strongly basic pH correction chemical may enhance coagulation (especially if they are large), but broadly, the main effect of coagulant addition is a result of the high charge density of the Al^{3+} and Fe^{3+} ions.

There are also organic coagulants, though they tend to be more often used for industrial effluent treatment than for drinking water treatment. Organic coagulants are not generally approved by regulatory authorities for potable water use, due to concerns over potential carcinogenic effects associated with byproducts.

The design of ideal coagulation systems involves creating a high shear environment of low residence time, with accurate, smooth dosing of coagulant chemical followed by accurate, smooth dosing of the correct amount of acid or alkali to obtain the correct pH. The optimum relationship between these processes is best determined by experimentation at commissioning stage.

The high shear environment can be a static mixer, or an open topped tank with a high intensity powered mixer, and a hydraulic retention time (HRT) measured in seconds known as a flash mixer.

I personally favor the static mixer, following the "KISS" principle, but its disadvantage is that as it is a plug flow reactor, it only mixes radially, so ensuring a smooth flow of dosed chemical is more important than with a flash mixer, which is a low residence time continuous stirred tank reactor (CSTR). However, mixing may often be incomplete within a static mixer and it may therefore be necessary to move sampling points some distance downstream.

Dosing pumps as described in the last section are required. Automatic control of coagulant dose is usually proportional only to flowrate, though companies increasingly use automatic coagulation control based on algorithms, variants on DOC monitors, streaming current, etc.

Aluminum coagulants work in the pH range 5–7, and are best between 5.5 and 6.3. Iron coagulants work in the pH range 5–8, best between 4.5 and 5.5, and are highly dependent on water type.

The required dose of alum can be estimated from an empirical formula (given in TR189) as follows:

$$\text{Alum dose} = 17.8 + 0.295 \times \text{color in}^{\circ}H / 5.41 \times (\text{pH-7}) \text{mg}/L \ Al_2(SO_4)_3 \cdot 18H_2O$$

This implies that alum doses are seldom lower than 18 mg/L, with an average of around 29 mg/L. US practice is to sometimes overdose coagulant in a process known as "enhanced coagulation."

Most UK companies follow a WRc lead and always express coagulants in terms of milligrams per liter as metal. Roughly, doses of aluminum are equimolar to iron doses. This means that required alum doses are half iron doses (and likewise for residuals), so an iron dose of 6 mg/L as Fe would be equivalent to an alum dose of 3 mg/L Al. Paradoxically, the average iron dose is however around 6mg/L as Fe. This may be explained by the fact that the two coagulants are used with different types of water.

PRECIPITATION

Precipitation is the creation of a solid from a solution. It is therefore distinct from coagulation, which starts from a colloid rather than a solution. In water engineering, this is done to facilitate removal of an undesirable substance.

SOFTENING

Hardness in natural water results almost entirely from the presence of calcium and magnesium salts. Temporary hardness (which is removable by boiling) comes from their bicarbonates, and permanent hardness is caused by their sulfates and chlorides. (Note that I have included some physical processes in this section to keep all of the softening techniques together.)

Calgon dosing may be employed, mainly in nonpotable applications, to inhibit deposition by complexing any calcium and magnesium present, or hardness may be removed from water by softening using one of the following methods:

- Base exchange
- Ion exchange
- Lime, soda ash, or combined lime/soda processes
- Membrane, by nanofiltration

Pellet reactors are also available, but are not recommended. These are a subset of lime and soda processes.

Softening processes rely upon chemical precipitation, ion exchange, or ion exclusion. Chemical precipitation is more suitable for surface waters, especially where there are existing clarifiers. Base exchange is likely to be economic only for small groundwater sources.

Softening costs money, so only a proportion of water may be softened and blended with unsoftened water to give a total hardness concentration not exceeding 400 mg/L as calcium carbonate. Designers should note that water softening and blending may result in variable or aggressive water, requiring pH correction before distribution. It should also be noted that softening is not required on health grounds, therefore there is no regulatory impetus to do so.

Lime/Soda softening

Adding lime (calcium hydroxide) or soda ash (sodium carbonate) precipitates out hardness as virtually insoluble calcium carbonate and magnesium hydroxide, readily removable by sedimentation and filtration. Temporary hardness is removed by the addition of lime and permanent hardness by the addition of soda ash.

The process produces large quantities of sludge, whose treatment and disposal should be considered during preliminary design.

Lime is usually supplied as a dry powder, and it can be added dry or as a slurry, made up on site from the powder. Soda ash is also usually supplied dry, and is made up on site into a solution for dosing.

The required lime dose if total hardness is equal to or less than total alkalinity can be estimated as follows:

$$\text{Lime dosage} = 1.2 \times ([CO_2] + [\text{Total hardness}] + [Mg])$$

Base exchange softening

In the base exchange process, calcium and magnesium ions are exchanged for sodium by passing water through a cationic resin loaded with sodium. Resins are regenerated using brine to displace the calcium and magnesium and replace the lost sodium.

Total dissolved solids (TDS) is largely unchanged by the process, but hardness is removed. Base exchange produces extremely soft (and probably very aggressive) water, which can only be used when blended. Sodium concentrations after blending must not be allowed to exceed the maximum allowable concentration (MAC) of 150 mg/L, which limits the amount of hardness that can be removed in this way.

Ion exchange softening

Alternatively the calcium and magnesium ions can be replaced by hydrogen, using a resin in a hydrogen form. Such resins are regenerated with strong mineral acids. Adding hydrogen ions is acidification, so the treated water may need to be neutralized. Carbon dioxide may also be produced by treatment, and degassing may consequently be required. These additional complications make this a less popular option than base softening. There are also potential problems with regulatory approval of resins for potable water duty.

Membrane treatment

While it will soften water, membrane technology is unlikely to be the cheapest option, (and is not a chemical treatment). Membrane softening is usually via nanofiltration, which removes only divalent ions. Design heuristics are given in Chapter 7, Clean Water Unit Operation Design: Physical Processes, but this is best undertaken in consultation with specialist suppliers.

IRON AND MANGANESE REMOVAL BY OXIDATION

Reduced manganese and iron salts can be precipitated by oxidation using aeration, chlorine, "greensand," or biological filtration.

Iron removal is generally achieved by either oxidization of soluble ferrous iron (Fe^{2+}) to insoluble ferric iron (Fe^{3+}) in the form of hydrated oxide; and/or removal of suspended iron containing particles via rapid gravity, or pressure filtration. Coagulants may be required to optimize solids removal.

Depending on concentration, ferrous iron removal by oxidation may be achieved by one of the following methods, ranked in order of suitability for low-high iron concentration:

1. Chlorine dosing (for lowest concentrations)
2. Cascade aeration
3. Spray bar aeration
4. Packed tower aeration (for highest concentrations)

The stoichiometry of iron oxidation is

$$4Fe_3^+ + O_2 + 4H_2O \rightarrow 2Fe_2O_3 + 8H^+$$

This means that the absolute minimum oxygen requirement is 0.14 g O_2 for every 1 g Fe_2^+. Design figures will need to be a little higher than this, but it is possible to overaerate, removing too much CO_2, causing scaling, unless acid dosing is provided. The Langelier Saturation Index (LSI) can be used to assess the requirement for this.

Manganese may be present in natural waters in insoluble or soluble inorganic forms or organic complexes. Manganese levels are generally low compared to the total mineral load. As with iron, treatment converts manganese into an insoluble form, subsequently removed by filtration. Unlike iron, there is often a catalytic effect to consider.

Manganese dioxide is the insoluble form of manganese removed in water treatment. Any manganese present must be converted into this form. Manganese dioxide acts as a catalyst for the oxidation of manganese with an oxidizing agent, which is why the use of various forms of manganese dioxide in filtration media to improve manganese removal is relatively commonplace.

Removal of organically bound manganese is more difficult than soluble inorganic species, due to the wide range of complexes possible. Such complexes are, however, commonly removed by upstream coagulation and solid separation stages. Any remaining may be converted by an oxidant into inorganic forms, which are then oxidized to manganese dioxide in the usual way.

Any ammonia in the water to be treated must be oxidized by breakpoint chlorination to allow sufficient chlorine to become available (to a level giving at least 0.3 mg/L measured in the filtrate) to oxidize manganese.

The addition of manganese dioxide to the filter can be achieved in several ways. If water containing high manganese is treated by chlorination prior to sand filtration, a manganese dioxide coating will develop on the sand grains, which is why "old sand" is thought better for manganese removal than "new sand." New sand may take days or weeks to develop catalytic activity. Some think the removal mechanism is adsorption onto the manganese dioxide followed by slow oxidation in situ.

Filtration media can be coated with manganese dioxide by potassium permanganate treatment, and potassium permanganate then used as the dosed oxidant. This process is, however, hard to control reliably, and may send pink water to supply. For this reason, it is not used in the United Kingdom, although it is often used in the United States.

Alternatively granular manganese dioxide (still often referred to by its trademarked name of "Polarite") can be added to the filter media. Uniform distribution is important to ensure adequate contact time between the water and manganese dioxide for manganese removal, but if 18-44 BS mesh (grain size 0.3−0.6 mm) manganese dioxide ore is added at 10%−20% v/v to 14/25 or 16/30 sand, the two materials are mixed completely by fluidization during backwashing. Even though polarite is denser than sand, the grain size will cause it to settle at the same speed as the lighter, smaller sand particles after fluidization.

BOILER AND COOLING WATER TREATMENT
COOLING WATER TREATMENT

Cooling water treatment is concerned with avoiding corrosion, scale, and sludge formation. Its lower temperatures compared with boiler water mean that microbiological growth can be a problem in cooling water systems. Aquatic organisms from the natural environment can, if given the correct conditions, increase exponentially in numbers, making the water unsuitable for potable or industrial use.

Thick biofilms of bacteria, algae, or fungi can grow in unprotected cooling water and other raw water systems. Larger organisms such as barnacles, zebra mussels, and jellyfish can also pose problems. Broad spectrum biocides such as chlorine can control these organisms to some extent, but biofilms can protect microorganisms from even the very high concentrations of chemical disinfectants used in shock dosing. Regular monitoring and general "good housekeeping" will consequently be required to minimize biofilm growth.

Cooling waters can also host the *Legionella* bacteria that can cause human illness if contaminated cooling waters are aerosolized. Corrosion can be caused by microbiological growth as well as the factors discussed later for boiler water systems. Corrosion can lead to failure of critical parts of the system or reduced heat transfer or hydraulic efficiency. Biocides may therefore play a part in biosafety as well as avoiding corrosion in cooling water systems in addition to chemical corrosion inhibitors.

BOILER WATER TREATMENT

Introduction

Dry steam is the gaseous form of water. It is transparent, though released steam has a cloudy appearance in the environment due to entrained water droplets. It is odorless, though additives in industrial steam such as amines may give it an odor.

In process plants steam is generated by controlled boiling of treated water in steam boiler plant, mainly for use as a heat transfer medium, though it is also used to provide motive power for several pump types, and is added as a process constituent.

Steam will always have an "aggressive" nature, as measured by the LSI, and will therefore dissolve other materials readily, which in turn can cause problems requiring chemical treatment. In the case of boilers, corrosion is usually the conversion of steel into soluble or insoluble iron compounds by oxygen pitting corrosion of the tubes and preboiler section, by acid corrosion in the condensate return system or, less commonly, by electrochemical corrosion.

Boiler water treatment is therefore necessary to prevent carryover of water into steam and to prevent reduced boiler performance due to corrosion, sludge, or scale formation, described in the following sections.

Carryover

Carryover of water droplets into steam leads to wet steam, or more seriously, to potentially destructive water hammer.

Carryover can be caused by either mechanically induced "priming," or foaming. The first of these must be controlled mechanically, and the second by maintaining proper water chemistry and using antifoam.

Priming is caused when mechanical conditions promote sudden violent boiling, which carries water along with steam to a relatively short distance out of the boiler, rapidly causing deposits around the main steam header valve.

Foaming can form a stable froth, carried out with the steam to cause blockages through the steam and condensate system over a longer timescale.

Sludge and scale formation

Sludge and scale formation are caused by too high a TDS level and insufficient fluid velocity in the boiler to flush out precipitated solids or corrosion products. Maintaining proper water chemistry through blowdown and dosing, and using scale inhibitor can control the first of these, and to some extent the second.

Sludge

Sludge formation occurs when suspended solids are deposited on equipment surfaces. The suspended solids may be introduced in makeup water, as airborne material in cooling towers, or they may be corrosion byproducts.

Most problematically, solids which adhere to heat transfer surfaces reduce heat transfer and hydraulic efficiency. More generally, sludge causes problems in all parts of the system, from the feed tank through the boiler to the condensate system.

The suspended solids might be minerals, oils, or biological material, though in boiler water systems, high temperatures prevent the growth of most microorganisms, so biological solids deposition is not the problem it is in cooling water systems.

Problems arise both from the quality of the water used within the steam raising system and the way the system is operated. Hence it is not only chemical issues that need to be addressed.

Scale

In addition to its obvious heat transfer purpose, a boiler is a separation process. In creating a stream of pure gaseous water, it also creates a stream of water with all its original contaminants, but with less water. These contaminants (especially calcium salts, known to water engineers as hardness) may exceed their solubility in the remaining water stream, and be deposited as a recalcitrant scale on heat transfer surfaces where it causes the same problems as sludge. The extreme example is supercritical boilers where all solids end up in the steam so ultrapure water is needed as a feed and only volatile additives, e.g., hydrazine are permitted.

Other treatment technologies

There are several electronic, magnetic, or other electrically based units that claim to be able to reduce scale build up, or favorably condition water (and sometimes other liquids such as fuel). There is in my opinion little or no scientifically creditable evidence that they work at all. The manufacturers' explanations for how they think they work are no more than conjecture.

To the extent that purchasers of such equipment turn off other protection systems, they are putting their boiler at risk with an unproven technique that few professional water engineers believe does anything.

My personal opinion is that all such techniques are best described as "magic."

Clean water unit operation design: biological processes

CHAPTER OUTLINE

Introduction 111
Discouraging Life 111
   Disinfection 111
   Sterilization 115
Encouraging Life 115
Biological Filtration 115
Further Reading 116

INTRODUCTION

Nowadays fewer biological processes are used in modern clean water treatment (though ironically in the days of slow sand filtration, there was more biology than chemistry in clean water treatment). This is mainly because clean water usually has fewer of the nutrients that are needed to sustain life. The most important aspect of life, as far as clean water treatment is usually concerned, is how to eradicate it.

There are three levels of cleanliness: cleaning, disinfection, and sterilization. Cleaning water (and equipment) involves removing gross contamination. Disinfection involves removing pathogenic organisms (those which cause disease). Sterilization involves removing or inactivating every single viable or potentially viable organism.

Cleaning is done by the standard water treatment processes, and by maintenance activity, whilst disinfection and sterilization are distinct unit operations. Drinking water is merely disinfected, but water and equipment used in pharmaceutical manufacture must be sterile. Let us start with the more common process.

DISCOURAGING LIFE

DISINFECTION

The addition of either chlorine or ozone is almost always undertaken to disinfect municipal water supplies before they are sent to supply.

An Applied Guide to Water and Effluent Treatment Plant Design. DOI: https://doi.org/10.1016/B978-0-12-811309-7.00009-6

Which is better? British engineers tend to favor chlorine, whereas French engineers favor ozone. Why favor chlorine? The number one reason (other than its far lower price) is that if you dose enough of it, it persists as a chlorine residual that disinfects the main all the way to the tap, unlike ozone. Why favor ozone? Chlorine reacts with organic compounds to produce suspected carcinogens [of which the most common are trihalomethanes (THMs)] as well as things that can taste unpleasant. Ozone is a more powerful oxidizing agent, which destroys organics, leaving no nasty by-products from reaction with organics (though there can be problems with other by-products, e.g., chlorate and bromate).

Chlorine gas is the cheapest and still the most common disinfectant, but the standard 1 tonne drums are a great deal of poisonous gas to be storing under pressure. They are a magnet for saboteurs, a fact that plant designers should bear in mind.

Domestic supplies are sometimes disinfected with UV light, which has the advantage of not requiring the storage of chlorine or its dangerous compounds, and the disadvantage of not producing a disinfectant residual.

Membrane filtration can also be used, with much the same disadvantages and advantages as UV. Note that for manufacturing QA reasons, UF membranes are better at disinfection than RO membranes with far smaller pores.

Chlorination

Chlorination of water involves adding a controlled amount of free or combined chlorine to water at a controlled pH, and retaining the mixture for a controlled minimum time before release to supply.

Care must be taken not to chlorinate highly colored natural waters, as this will lead to excessive concentrations of disinfection by-products such as THMs, suspected human carcinogens.

Chlorine gas, sodium, and calcium hypochlorite are used to add chlorine to water to disinfect it. Alternatively, sodium hypochlorite can be produced on-site from brine via electrolytic chlorination.

Chlorine gas is the cheapest option, but the standard 1 tonne drums of chlorine gas are arguably not intrinsically safe. Lots of small sites use 33, 50, or 70 kg cylinders, and on-site hypochlorination or electrochlorination (OSE) may be an option at large plants, especially those close to the sea's free supply of brine, although issues with brine purity, strength, and bromate concentration would have to be considered. Sodium hypochlorite solution is a practical alternative at small- and medium-sized plants, and solid calcium hypochlorite may be used at the smallest plants. This last is known as "HTH," and is rarely used in the United Kingdom, probably because operational staff hate dealing with the solids handling and dissolution kit.

Chlorine gas is depressurized to vacuum conditions in chlorination equipment to reduce leakage potential and consequences. It is mixed with water using an eductor known as a chlorinator, to produce hypochlorous acid and hydrochloric acid:

$$Cl_2 + H_2O \rightleftharpoons HClO + HCl$$

$$HClO \leftrightarrow H^+ + OCl^-$$

Above pH 8 the hypochlorite ion (OCl$^-$) predominates. Hypochlorous acid is a much stronger disinfectant than hypochlorite ion. pH control is therefore crucial to effective chlorination.

Sodium and calcium hypochlorite solutions are dosed using dosing pumps. These are modified to handle the bubbles that can form in hypochlorite solutions due to decomposition of hypochlorite producing oxygen, which can cause gas locking, especially in low flow conditions.

When it comes to ensuring sufficient contact time, the WRc publication TR60 "Disinfection by chlorination in contact tanks" (see Further Reading) is a useful starting point. Chlorine contact tanks (CCTs) are essentially a plug flow reactor, and TR60 is mostly about how to design them hydraulically to get a tight range of residence times.

While TR60 is now arguably quite outdated the following recommendations still hold good:

- Using inlet baffles, or more simply (and lower in headloss), introducing flow to the tank at right angles to the tank flow direction
- Using a rectangular tank, baffled to produce a long, narrow serpentine flow path
- A channel length to width ratio of 40:1
- Ensuring linear flow rates are high enough to prevent sediment settlement (minimum 1 m/s)
- Ensuring 0.1−0.2 ppm chlorine residual at outlet after 30 minute contact time
- Ensuring pH in the range 6−7 throughout CCT

The chlorine dose rate can be controlled using flow-paced feedback from the chlorine residual, though clearly there will be a long lag in the control loop. Alternatively, chlorine can be deliberately overdosed (known as superchlorination), and the chlorine residual "trimmed" after the CCT by dechlorination to give far tighter control of the residual. A common UK standard is to have a chlorine analyzer close to the chlorine dose with the set point cascaded from another analyzer downstream of the contact tank. Overdosing is these days usually only employed for ammonia, manganese or iron removal or to compensate for inadequate contact time.

This stable residual trace of chlorine is the big advantage of chlorine as a drinking water disinfectant. It disinfects the distribution system all the way to the customer's tap. All UK water companies now use the Ct concept to design and operate chlorine dosing systems, taking account of the residence time distribution (e.g., t0 or t10) and the hypochlorous acid concentration.

Ozonation

Ozone is a stronger oxidizing agent than chlorine and is also more toxic and corrosive. This affects the selection of construction materials for ozonation systems. It also affects how the excess air carrying the ozone with its traces of ozone is handled after it has been passed through the water to be treated. Activated carbon is very effective in removing ozone.

Ozone is made on-site from air or from pure oxygen, most commonly by "corona discharge," in which very high voltages are used to produce ozone from hot electrical sparks. The process also produces oxides of nitrogen, and the air is usually dried before being subjected to corona discharge to prevent the formation of nitric acid in the air handling system.

Dechlorination

Chlorine can also be removed very reliably with activated carbon from either air or water, though this can be expensive at the scale of municipal water treatment. The carbon requires replacement after a couple of years and the process is not suitable for partial removal of chlorine.

Chlorine can be removed more cheaply by reduction with sulfur dioxide or bisulfite. The reactions are:

$$SO_2 + H_2O \rightarrow HSO_3^- + H^+$$
$$HOCl + HSO_3^- \rightarrow SO_4^{-2} + Cl^- + 2H^+$$
$$SO_2 + HOCl + H_2O \rightarrow Cl^- + SO_4^{-2} + 3H^+$$

Gaseous sulfur dioxide is handled and dissolved in the same way as chlorine. Dechlorination is most commonly required after chlorine has been added in excess as an oxidant to remove reduced metal ions or ammonia. Just sufficient bisulfite is added to reduce the excess of chlorine after the reaction has taken place.

There may be a requirement to operate the disinfection process at a different pH from earlier coagulation stages, and then adjust the pH again afterwards.

Historically, dechlorination was commonly achieved by flow-paced feedback controlled dosing of the reducing agent into a static mixer at a dose rate controlled by chlorine residual analyzers or redox probe, although some older works may still use "flash" mixers. The dechlorination reaction is rapid and complete, so mixer hydraulic retention time (HRT) is not a practical concern. The sulfur dioxide dose rate can be controlled using flow-paced feed forward and/or feedback for a sample point just upstream of the dosing point. This is usually controlled by chlorine analyzers nowadays rather than redox probes though a combination of chlorine and sulfite monitors may be used for nominally zero chlorine zero sulfite discharges to a watercourse.

Ultraviolet light

Water clarity is key in ensuring efficient disinfection by UV. The process involves passing the water to be disinfected through a chamber illuminated by UV lamps. It is now applied to drinking water at all scales of supply. UV needs at least as low a solids concentration as chlorine disinfection (i.e., <1 NTU), so it is fine for both surface and borehole water, but it works best with borehole water, due to its innately low suspended solids concentration.

UV is also much used in pharmaceutical water treatment, as it does not involve adding chemicals, which might contaminate or react with pharmaceutical ingredients. Potable water that has been demineralized and UV treated is commonly referred to in the pharmaceutical industry as "purified water."

Membrane filtration

Ultrafiltration (UF) membranes have pores that are too small (5−20 nm) to allow bacteria or even viruses to pass through and, if configured correctly, can be an effective means of disinfection largely by size exclusion. Work in the 1990s by the UK's WRc showed virtually complete virus removal of some wastewaters despite the UF membranes having a pore size much larger than the viruses. Coarser membranes used for microfiltration with pore sizes of 20 nm−1 μm remove only a percentage of these microorganisms.

STERILIZATION

Water can be sterilized by heating and by the same agents used for disinfection, but with multiple stages, higher doses, longer residence times, and posttreatment filtration.

ENCOURAGING LIFE

The only commonly used engineered biological process in clean water treatment is biological filtration. It can be used to remove natural and synthetic organics, iron, manganese, nitrate, and ammonium ions.

Slow sand filtration (see Chapter 7: Clean Water Unit Operation Design: Physical Processes) is the most frequently used process and removes all these substances to varying degrees, along with suspended solids. Removal of organics in granular activated carbon (GAC) filters may also be to some extent biological, especially if ozone activated (BAC).

In practice, only iron, manganese and ammonium ions tend to be removed by dedicated biological filters at commercial scale, and even these are uncommon processes.

BIOLOGICAL FILTRATION

Dedicated biological filters can be used to remove up to 1 mg/L of ammonium, as well as a few milligrams per liter of reduced manganese and iron from drinking water. Surface loadings from 10 to 34 m/h and operating pH around 7 are reported in full-scale plants. Operational experience suggests that individual air injection to each filter should be used, and good quality air blowers. Research indicates that the dissolved oxygen (DO) (or redox potential) needs to be less than saturation to achieve the conditions required for the right biology to develop.

The process is not completely robust and pilot plant testing is strongly recommended if this technology is being considered.

If the ammonium concentration is less than 1.0 mg/L, nitrification will reliably take place on a conventional sand filter, though with concentration higher than 1.5 mg/L a discrete biological nitrification stage of aerated filters is required.

FURTHER READING

McNaughton, J. G., & Gregory, R. (1977). *Water Research Centre Technical Report disinfection by chlorination in contact tanks, TR60.* Marlow: Water Research Centre.
Mouchet, P. (1992). From conventional to biological removal of iron and manganese in France. *JAWWA, 84*(4), 158.

Clean water hydraulics

CHAPTER OUTLINE

Introduction 117
Pump Selection and Specification 118
    Pump Types 118
    Pump Sizing 119
Open Channel Hydraulics 132
    Channels 132
    Chambers and Weirs 133
    Flow Control in Open Channels 134
Hydraulic Profiles 134
Further Reading 135

INTRODUCTION

Clean water hydraulics are the kind most closely related to fluid mechanics, which all chemical engineering students study at university. However, students often find (as I did when I left university) that additional information is needed to turn theoretical knowledge of fluid mechanics into the practical knowledge required to size a pump or a flow control chamber. Judging by the questions I see nearly every week on social media and elsewhere, I believe this is a problem shared by many engineers early in their careers.

The most common task for the process/hydraulic engineer in clean water hydraulics is the selection and specification of pumps. This is relatively familiar, as is to a lesser degree the specification of in-pipe flow control devices. The open channel hydraulics used to design and analyze channels, chambers, and weirs is, however, often a mystery to new graduates.

Let us first consider the most common case, the sizing, selection, and specification of pumps.

An Applied Guide to Water and Effluent Treatment Plant Design. DOI: https://doi.org/10.1016/B978-0-12-811309-7.00010-2
117

PUMP SELECTION AND SPECIFICATION

PUMP TYPES

Pumps can be used to move fluids, which flow from regions of high pressure to regions of low pressure, by increasing the pressure of the fluid. Before purchasing a pump, you must specify the type of pump and the required flowrate at a given pressure.

There are two main pump types: rotodynamic and positive displacement (PD). In a rotodynamic pump, a rotating impeller imparts energy to the fluid. The most common types of rotodynamic pumps are axial-flow, mixed-flow, and centrifugal pumps. The amount of liquid that passes through the pump is inversely proportional to the pressure at the pump outlet. In other words, the outlet flowrate of a rotodynamic pump varies nonlinearly with pressure. Table 10.1 outlines the key features of different pump types.

In a PD pump (Fig. 10.1), a discrete amount of fluid is trapped, forced through the pump, and discharged. A piston diaphragm pump is an example of a PD pump. Its output flow tends to vary little with respect to the pressure at the pump outlet, because the moving displacement mechanism pushes the slug of liquid out at a constant rate.

Most process pumps are, however, rotodynamic pumps. To specify a pump that will provide sufficient flow, you also need to know the required outlet pressure.

Table 10.1 Pump Selection (General)

	Rotodynamic	**Positive Displacement**
Head	Low—up to a few bar	High—hundreds of bar
Solids tolerance	Low without efficiency losses	Very high for most types
Viscosity	Low viscosity fluids only	Low and high viscosity fluids
Sealing arrangements	Rotating shaft seal required	No rotating shaft seal
Volumetric capacity	High	Lower
Turndown	Limited	Excellent
Precision	Low—discharge proportional to backpressure	Excellent—discharge largely independent of backpressure
Pulsation	Smooth output	Pulsating output
Resistance to reverse flow	Very low	Very high
Reaction to closed valve downstream	No immediate damage to pump	Rapid pump damage likely

FIGURE 10.1

Memdos-E positive displacement pump.

Courtesy: Lutz-Jesco.

PUMP SIZING

Pump sizing involves matching the flow and pressure rating of a pump with the flowrate and pressure required for the process. Achieving the required process flowrate requires a pump that can generate a high enough pressure to overcome the hydraulic resistance (system head) of the system of pipes, valves, and so on that the liquid must travel through.

The system head is the amount of pressure required to achieve a given flowrate in the system downstream of the pump. The system head varies proportionately to flowrate and a curve, known as the system curve, can be drawn to show the relationship between flow and hydraulic resistance for a given system.

Pump sizing, then, is often the specification of the required outlet pressure of a rotodynamic pump (whose output flow varies nonlinearly with pressure) with a given system head (which varies nonlinearly with flow). Table 10.2 gives some useful comparative data to aid in the pump sizing and selection process.

Understanding system head

The total discharge head, simply the pressure at the pump outlet at a specified flow, is dependent on the hydraulic resistance of the system which the pump is connected to. The total discharge head is the sum of the static head, the pressure head, and the dynamic head. The system head is equal to the total head on the discharge side of the pump minus the total head on the suction side of the pump.

Table 10.2 Pump Selection

	Relative Price	Environmental/ Safety/Operability Concerns	Robustness	Shear	Maximum Differential Pressure	Capacity Range	Solids Handling Capacity	Efficiency[a]	Seal In-Out	Fluids Handled[b]	Self-Priming?
Rotodynamic											
Radial flow	M	Cavitation	H	H	M/H	L-VH	M	H	M	Low viscosity/aggressiveness[a]	N
Mixed flow	M	Cavitation	H	H	M	M-VH	M	H	M	Low viscosity/aggressiveness[a]	N
Axial flow	M	Cavitation	H	H	M	M-VH	M	H	M	Low viscosity/aggressiveness[a]	N
Archimedean screw	H	Release of dissolved gases	VH	H	L	M-VH	H	L	N/A	Low viscosity/aggressiveness[a]	N
Positive Displacement											
Diaphragm	M	Overpressure on blockage	VH	L	M/H	VL-M	H	L	H	Low/high viscosity/aggressiveness	Y
Piston diaphragm	H	Overpressure on blockage	M	L	VH	VL-M	L/M	M	H	Low/high viscosity/aggressiveness	Y
Ram	H	Overpressure on blockage	H	M	VH	M-H	H	M	H	Low/high viscosity/aggressiveness	Y
Progressing cavity	M	Overpressure on blockage	M	VL	H	M-H	H	H	H	Low/high viscosity/aggressiveness	Y
Peristaltic	H	Overpressure on blockage	L	VL	M	VL-M	H	M/H	M	Low/high viscosity/aggressiveness	N
Gear	L	Overpressure on blockage	L	M	VH	VL-L	L	L	H	Low/high viscosity. Low aggressiveness	Y
Screw	L	Overpressure on blockage	L	M	VH	L-M	L	L	H	Low/high viscosity. Low aggressiveness	Y
Other											
Air lift	L		VH	VL	VL	L-M	H	L	L	Low viscosity, low/high aggressiveness	N
Eductor	M	Blockage of eductor	M	M	VL	L-H	M	H	M	Low viscosity, low/high aggressiveness	Y

[a]Centrifugal pump efficiency reduces as viscosity increases, but PD pump efficiency increases. Centrifugal pump efficiency is more to do with impeller type than anything else; impeller type is determined by process conditions such as any solids handling requirement.

[b]Aggressiveness is related to presence of abrasive particles, undissolved gases, or unfavorable LSI.

The static head is created by the head of water above the pump centerline, as calculated from the elevation of the free surface of the fluid relative to the eye of the pump impeller. The static head exists even under static conditions (hence the name), with the pump switched off, and does not change based on flow.

The pressure head is caused by any pressurized systems attached to the pump, such as a pressurized vessel or heat exchanger. The dynamic, or friction, head varies dynamically with flowrate. It represents the inefficiency of the system: losses of energy caused by friction within pipes, valves, and fittings, as well as at any changes in direction. This inefficiency increases as the average velocity of the fluid increases.

Dynamic head can be further split into two parts. The frictional loss as the liquid moves along lengths of straight pipe is called the straight-run headloss, and the loss caused by liquid passing through pipe fittings such as bends, valves, and so on is called the fittings headloss.

Fully characterizing a hydraulic system is incredibly complex, but to specify a pump, you only need to characterize the system well enough to choose a pump that will perform the job in question. The necessary degree of accuracy depends on the stage of the design process. At the conceptual stage, it may be possible to avoid specifying the pump at all, but experience suggests that rules of thumb should be used to specify certain parameters (such as superficial velocity) to prevent difficulties later. I also recommend designing the process so that it does not have two-phase flow.

Superficial velocity or average velocity is the volumetric flowrate (in m^3/s, for example) divided by the pipe's internal cross-sectional area (e.g., in m^2). A quick way to start the hydraulic calculations is to use the following superficial velocities:

- pumped water-like fluids: <1.5 m/s
- gravity-fed water-like fluids: <1 m/s
- water-like fluids with settleable solids: >1, <1.5 m/s
- air-like gases: 20 m/s

Two-phase flow is difficult to model, and should be designed out if possible, as headlosses can be 1000 times those for single-phase flow. We can design out two-phase flow by methods such as using knock-out drums, and by so arranging pipework that gases are not entrained in liquids and vice versa.

Determining frictional losses through fittings

Dynamic, or friction, head is equal to the sum of the straight-run headloss and the fittings headloss. The fittings headloss is commonly calculated by what is known as the k-value method. Each type of valve, bend, and tee has a characteristic resistance coefficient, or k value, which can readily be found in standard references such as Perry's Handbook, Millers "Internal Flow Systems," and online. Example k values are given in Table 10.3.

Table 10.3 K Values for Pipe Fittings

Fitting Type	K Value
Short radius bends for every 22.5 degrees allow	0.2
Long radius bends for every 22.5 degrees allow	0.1
Open isolation valve	0.4
Open control valve	10.8
Tee (flow from side branch)	1.2
Tee (flow straight through)	0.1
Swing check nonreturn valve	1
Sharp entry	0.5

To use this method, count the number of valves, fittings, bends, and tees on the piping and instrumentation diagram (P&ID). Multiply the number of each type of fitting by the corresponding k value, and add the k values for the various types of fittings to get total k values for the discharge (kDischarge) and suction (kSuction) sides.

Calculating straight-run headloss

At a more-advanced stage of design, it may be necessary to know the pumps' physical size to try out on a plant layout drawing. An easy way to determine the straight-run headloss—the most difficult part of a headloss calculation—is to use a nomogram or a table. Pipe manufacturers (and others) produce tables and nomograms (Fig. 10.2) that can be used to quickly look up headloss due to friction for liquids.

To use the nomogram, a line is drawn with a ruler through any pairs of known quantities to allow unknown quantities for be determined. Using the example in Fig. 10.2, a 25 mm nominal bore pipe with a flow velocity of 1 m/s, straight-run headloss is about 6 m per 100 m of pipe. So the headloss through 10 m of this pipe would be around 0.6 mwG.

However, at an early design stage, it is often necessary to calculate the straight-run headloss multiple times. Rather than referring to a table or nomogram numerous times, it can be quicker to set up a MS Excel spreadsheet and use a formula to calculate the Darcy friction factor and headloss.

Chemical engineering students are usually taught to find the Darcy friction factor using a Moody diagram, which is a summary of many empirical experiments. Curve-fitting equations and software such as MS Excel can be used to approximate the Moody diagram's output, but I prefer to use the Colebrook–White approximation to calculate the Darcy friction factor. Although it is an approximation, it might be closer to the true experimental value than what the average person can read from a Moody diagram.

The Colebrook–White approximation (Eq. (10.1)) can be used to estimate the Darcy friction factor (f_D) from Reynolds numbers greater than 4000.

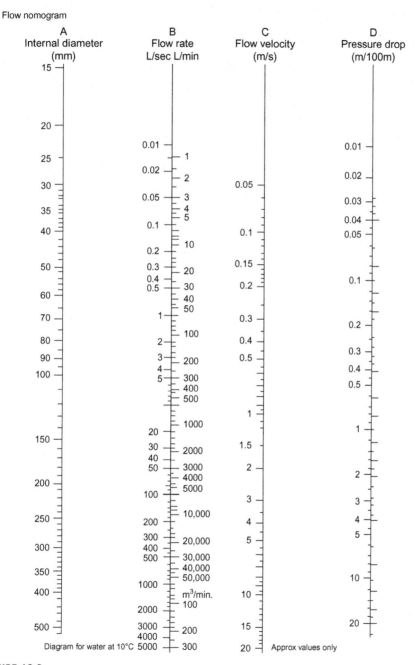

FIGURE 10.2

Example of a pipe nomogram.

Courtesy: Durapipe SuperFLO ABS technical data.

Colebrook−White approximation:

$$\frac{1}{\sqrt{f}} = -2 \, \log\left(\frac{\varepsilon}{3.7D_\mathrm{h}} + \frac{2.51}{Re\sqrt{f}}\right) \qquad (10.1)$$

where

D_h = the hydraulic diameter of the pipe
ε = the surface roughness of the pipe
Re = the Reynolds number, $\rho v D/\mu$

where

v = superficial velocity
D = pipe internal diameter
μ = the fluid dynamic viscosity

The Colebrook−White approximation can be used iteratively to solve for the Darcy friction factor. First, guess a starting value for f_D on the right-hand side of the Colebrook−White equation and solve for the friction factor on the left-hand side, repeating this process until the guessed and calculated values converge. The Goal Seek function in MS Excel does this quickly and easily.

The Darcy−Weisbach equation for pressure loss (Eq. (10.2)) states that for a pipe of uniform diameter, the pressure loss due to viscous effects (Δp) is proportional to length (L).

Darcy−Weisbach equation for pressure loss per unit length:

$$\frac{\Delta p}{L} = f_\mathrm{D} \times \frac{\rho}{2} \times \frac{v^2}{D} \qquad (10.2)$$

where

Δp = pressure loss due to viscous effects
L = length
D = pipe internal diameter
v = average velocity

To calculate the straight-run headloss, determine the f_D and use the Darcy−Weisbach equation to find the pressure loss, Δp, then use that to determine the straight-run headloss.

This iterative approach allows us to calculate straight-run headloss to the degree of accuracy required for virtually any practical application.

I quite recently came across a 2011 paper that suggested there are other equations that provide more accurate results through curve-fitting than the Colebrook−White approximation. Although uncertainty about the roughness makes discussions about the best equations rather academic, if you are producing your own spreadsheet for this purpose, you may wish to consider the Zigrang and Sylvester or Haaland equations which this paper discusses (see Further Reading: Genić et al (2011)). These equations apply for Reynolds numbers greater than 4000.

Determining pump power

After the system head has been calculated, it can be used to calculate an approximate pump power rating for a centrifugal pump (Eq. (10.3)).

Approximate centrifugal pump power rating:

$$P = \frac{Q\rho g H}{(3.6 \times 10^6)\eta} \tag{10.3}$$

where

P = pump power (kW)
Q = flowrate (m^3/h)
H = total pump head (m of fluid)
η = pump efficiency (if you do not know the efficiency, use $\eta = 0.7$)

The pump manufacturer will provide the precise power ratings and motor size for the pump, but electrical engineers need an approximate value of this (and pump location) early in the design process to allow them to size the power cables. It is important to err on the side of caution in this rating calculation (as electrical engineers will find it easier to accommodate a request further down the line for a lower power rating than a higher one).

The preliminary drawings are then modified to match likely hydraulic conditions across the design envelope. This may require many approximate hydraulic calculations before the design has settled into a plausible form.

After completing the hydraulic calculations, the pump and possibly the pipe sizes might need to be changed, as might minimum and maximum operating pressures at certain points in the system. As the system design becomes more refined, there might even be a requirement to change from one pump type to another.

Suction head and net positive suction head

Even at an early stage, I also recommend determining a prospective pump's required net positive suction head (NPSHr) and calculating the net positive suction head available (NPSHa), as they can affect much more than pump specification. The NPSHa of the system should always exceed the NPSHr of the pump to avoid cavitation in the pump.

I recommend creating a MS Excel spreadsheet that uses the Antoine equation (Eq. (10.4)) to estimate the vapor pressure of the liquid at the pump inlet and then calculates the NPSHa at that vapor pressure.

Antoine equation:

$$\log(Pv) = A - \frac{B}{C + T} \tag{10.4}$$

where

Pv = vapor pressure of the liquid at the pump inlet (mmHg)
T = temperature (K)
A, B, C = coefficients (obtainable from the NIST database among other places, see Further Reading)

Table 10.4 Example Antoine Equation Calculation in MS Excel

Material	Density (kg/m³)	ANT A	ANT B	ANT C	Temp. (°C)	VP (mmHg)	VP (Pa)
Water	1000	18.3036	3816.44	−46.13	40	54.7542132	7298.736615

An example for water at 40°C is shown in Table 10.4.

The available NPSH is given at Eq. (10.5).

Available NPSH:

$$\text{NPSHa} = \frac{Po}{\rho g} - \frac{Pv}{\rho g} + ho - hf \tag{10.5}$$

where

Po = absolute pressure at suction reservoir
Pv = vapor pressure of the liquid at pump inlet
ho = reservoir liquid level relative to pump centerline
hf = headloss due to friction on suction side of pump

Pump manufacturers will provide curves, which allow the NPSHr of any pump to be determined, of which more later.

Note that NPSH is calculated differently for centrifugal and PD pumps, and that it varies with pump speed for PD pumps rather than with pressure as for centrifugal pumps. The earlier equations should only be used with centrifugal pumps.

Hydraulic networks

The previous sections describe how to calculate the headloss through a single line, but what about the common situation where the process has branched lines, manifolds, and so on. When each branch handles a flow proportional to its headloss, and its headloss is proportional to the flow passing through it, producing an accurate model can become complex very quickly.

A useful approach to this is to first simplify and then improve the design as much as possible with a few rules of thumb:

- Avoid manifold arrangements that provide a straight-through path from the feed line to a branch. Entry perpendicular to branch direction is preferred
- Size manifolds such that the superficial velocity never exceeds 1.5 m/s at the highest anticipated flowrate
- Specify progressively smaller manifold diameters to accommodate lower flows to downstream branches
- Include a small hydraulic restriction in the branch so the branch headloss is 10−100 times the headloss across the manifold
- Design-in passive flow equalization throughout the piping system wherever possible by making branches hydraulically equivalent
- Perform headloss calculations for each section of the simplified plant design at expected flows to find the flow path with the highest headloss

Use the highest-headloss path to determine the required pump duty—calculate the pump duty at both the average flow with working flow equalization and full flow through a single branch. Usually these do not differ much, and the more rigorous answer lies between them. Only if the results of this approach seem problematic will a more rigorous (and time-consuming) analysis be required.

If such a rigorous analysis is needed, I create a MS Excel spreadsheet based on the Hardy Cross method—a method for determining the flow in a pipe network when the flows within the network are unknown but the inputs and outputs are known—and solve for individual pipe flows. MS Excel's Solver function can be used to find the change in flow that gives zero loop headloss. In the unlikely event that this is needed, Huddleston (see Further Reading) gives an explanation of how to carry out the method. There are many computer programs available to do these calculations.

Pump curves

A pump curve is a plot of outlet pressure as a function of flow and is characteristic of a certain pump. The most frequent use of pump curves is in the selection of centrifugal pumps, as the flowrate of these pumps varies dramatically with system pressure. Pump curves are used far less frequently for PD pumps. A basic pump curve (Fig. 10.3) plots the relationship between head and flow for a pump.

On a typical pump curve, flowrate (Q) is on the horizontal axis and head (H) is on the vertical axis. The pump curve shows the measured relationship between these variables, so it is sometimes called a Q/H curve. The intersection of this curve with the vertical axis corresponds to the closed valve head of the pump.

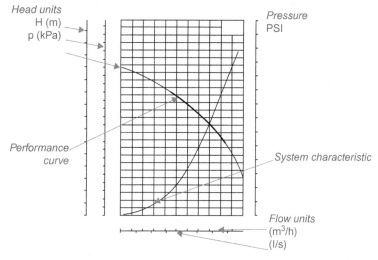

FIGURE 10.3

Basic pump curve.

Courtesy: Grundfos.

These curves are generated by the pump manufacturer under shop test conditions and ideally represent average values for a representative sample of pumps.

A plot of the system head over a range of flowrates, from zero to some value above the maximum required flow, is called the system curve. To generate a system curve, complete the system head calculations for a range of expected process flowrates. System head can be plotted on the same axes as the pump curve (shows as "system characteristic" in Fig. 10.3). The point at which the system curve and the pump curve intersect is the operating point, or duty point, of the pump.

A system curve applies to a range of flows with a given system configuration. Throttling a valve in the system will produce a different system curve. If flow through the system will be controlled by opening and closing valves, you need to generate a set of curves that represent expected operating conditions, with a corresponding set of duty points.

It is common to have efficiency, power, and NPSH plotted on the same graph as in Fig. 10.4.

Each of these variables requires its own vertical axis. To obtain the pump efficiency at the duty point, draw a line vertically from the duty point to the

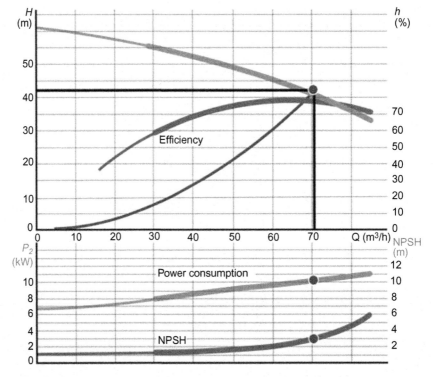

FIGURE 10.4

Intermediate pump curve.

Courtesy: Grundfos.

efficiency curve, and then draw a horizontal line from there to the vertical axis that corresponds to efficiency. Similarly, to obtain the motor power requirement, draw a line down from the duty point to the motor duty curve.

More sophisticated curves may include nested curves representing the flow/ head relationship at different supply frequencies or rotational speeds, with different impellers, or for different fluid densities. Curves for larger impellers or faster rotation lie above curves for smaller impellers or slower rotation, and curves for low-density fluids lie above curves for high-density fluids. A more-advanced pump curve might also incorporate impeller diameters and NPSH. Fig. 10.5 has pump curves for four different impellers, ranging from 222 to 260 mm. Corresponding power curves for each impeller are shown on the bottom of the figure.

These curves can become confusing, but the important point to keep in mind is that, just as in the simpler examples, flowrate is always on a common horizontal axis, and the corresponding value on any curve is vertically above or below the duty point.

These more-advanced curves usually incorporate efficiency curves, and these curves define a region of highest efficiency. At the center of this region is the best efficiency point (BEP).

Choose a pump that has an acceptable efficiency across the range of expected operating conditions. Note that we are not necessarily concerned with the entire design envelope—it is not crucial to have high efficiency across all conceivable conditions, just the normal operating range.

The optimal pump for your application will have a BEP close to the duty point. If the duty point is far to the right of a pump curve, well away from the BEP, that is not the right pump for the job.

Even with the most cooperative pump supplier, sometimes the curves needed to make a pump selection may not be available. This is commonly the case if an inverter to control pump output based on speed is required.

However, it is often possible to generate acceptable pump curves using the curves available and approximate pump affinity relationships (Eq. (10.6)).

Pump affinity relationships:

$$\frac{\text{Flowrate}_2}{\text{Flowrate}_1} = \frac{\text{Impeller diameter}_2}{\text{Impeller diameter}_1} = \frac{\text{Pump speed}_2}{\text{Pump speed}_1}$$

$$\frac{\text{Dynamic head}_2}{\text{Dynamic head}_1} = \left(\frac{\text{Impeller diameter}_2}{\text{Impeller diameter}_1}\right)^2 = \left(\frac{\text{Pump speed}_2}{\text{Pump speed}_1}\right)^2$$

$$\frac{\text{Power rating}_2}{\text{Power rating}_1} = \left(\frac{\text{Impeller diameter}_2}{\text{Impeller diameter}_1}\right)^3 = \left(\frac{\text{Pump speed}_2}{\text{Pump speed}_1}\right)^3 \qquad (10.6)$$

$$\frac{\text{NPSH}_2}{\text{NPSH}_1} = \left(\frac{\text{Impeller diameter}_2}{\text{Impeller diameter}_1}\right)^x = \left(\frac{\text{Pump speed}_2}{\text{Pump speed}_1}\right)^y$$

where

subscript$_1$ = an initial condition on a known pump curve and
subscript$_2$ = some new condition

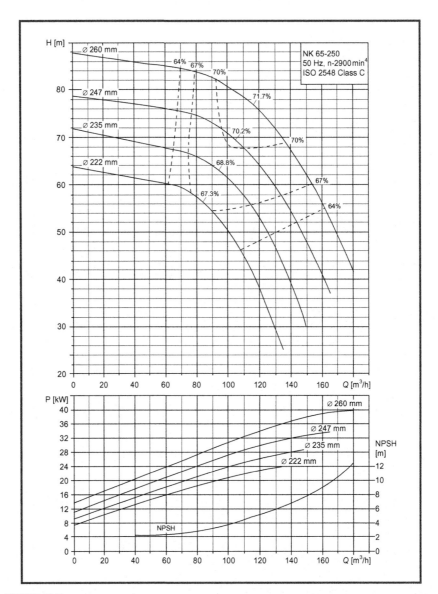

FIGURE 10.5

Complex pump curve.

Courtesy: Grundfos.

The NPSH relationship is more of an approximation than the others. The value of x lies in the range of -2.5 to $+1.5$, and y in the range of $+1.5$ to $+2.5$.

Water hammer/surge analysis

Water is both heavy and incompressible, and water engineering infrastructure can be very large. Water engineering can therefore potentially involve large forces produced by the rapid acceleration and deceleration of thousands of tons of water. Transient pressures of hundreds of bar are readily produced by inexpert hydraulic design of large water pipework. A rigorous analysis of transient pressure spikes in pipelines can be incredibly challenging and time-consuming. There is, however, a shortcut method.

Much of the effort of a rigorous analysis goes into finding an exact answer to the maximum size and duration of pressure spikes, but we do not actually need to know this. What we need to know as engineers is if there are likely to be conditions under which it is likely that our design will subject pipes to pressure greater than their rating, even transiently.

So we can carry out two calculations, representing the boundaries within which the rigorous answer will lie.

First we need to calculate P0, the steady state total pressure (dynamic and static head) for the pipeline by the usual method, expressing our answer in psi.

Then we calculate transient pressure assuming incompressible conditions according to Eq. (10.7).

Transient pressure equation in incompressible conditions:

$$P1 = 0.07 \times V \times L/t + P0 \tag{10.7}$$

where

V = fluid velocity in ft/s
L = pipe length in ft
t = valve closing time in s

Then we calculate transient pressure assuming compressible conditions, P2. This is a two-step process as set out in Eq. (10.8).

Transient pressure equation in compressible conditions:

1. Calculate wave speed (celerity)

$$\text{Celerity} = \sqrt{((1/(1/Ef + D/(W^*Ep))/\rho))} \tag{10.8}$$

where

Ef = modulus of elasticity of fluid
D = pipe diameter
w = pipe wall thickness
Ep = modulus of elasticity of pipe
ρ = density kg/m^3

2. Apply the result as follows:

$$P2 = \text{Celerity} \times \rho \times (\Delta v/10,000)$$

where

Δv = steady state velocity

If P1 and P2 are both lower than pipe pressure rating, no surge protection measures are required. If such measures are required, the system can be reanalyzed by this means to check if they are likely to be sufficient, although it might be better to carry out a more rigorous analysis at the design for construction stage.

OPEN CHANNEL HYDRAULICS

Although the use of open channels is less common in clean water applications than in sewage treatment, we do still use them. In my own professional experience, I have used open channel hydraulics in the design of large water features as well as the water treatment plants associated with them. This can become complicated, but as usual, it is best to keep it simple. For further information on hydraulics, see Chapter 15, Dirty Water Hydraulics.

CHANNELS

The Manning equation (Eq. (10.9)) may be used to calculate straight-run headloss.

Manning equation:

$$Q = VA, \quad V = \frac{k}{n}\left(\frac{A}{P}\right)^{2/3} S^{1/2} \tag{10.9}$$

where

$k = 1.49$ for "British" units (ft/s), $k = 1.0$ for SI units (m/s)
A = cross section area of flow in the channel
P = wetted perimeter (length of channel cross section in contact with water)
Q = discharge (volumetric flowrate)
S = unit fall in channel bed per of unit channel length
V = average velocity

In addition to straight-run headlosses, shock losses from bends, contractions, and expansions may be calculated using the rules of thumb in Eq. (10.10).

Rules of thumb for shock losses:

> If Entry $B1 > B2$ then approximate headloss $(m) = 0.015(V1^2 - V2^2)$
> If Entry $B1 < B2$ then approximate headloss $(m) = 0.026(V1^2 - V2^2)$ (10.10)
> Bend approximate headloss $(m) = 0.015\ V^2$

where

> B = channel width (m)
> $B1$ = upstream of $B2$
> $V1$ = velocity at $B1$ (m/s)
> $V2$ = velocity at $B2$ (m/s)

CHAMBERS AND WEIRS

We commonly divide flow in a "flow splitting chamber," using the constant flow per linear meter of weir to reliably divide flow into several equal streams over the full range on incoming flows.

If we need to estimate depth of flow over a weir more accurately, equations are available for several different weir cross-sections. Note that these standard calculations may not apply to weirs over 10 m long.

Weirs are generally divided into two classes, being "thin/knife-edge" and "broad." This distinction may be defined by means of whether the nappe is free or adheres as it leaves the weir edge. This in turn is a function of the relationship between the head of water passing over the weir, and the breadth of the weir. We can define broad weirs as being those where weir breadth is greater than three times the head upstream of the weir.

Thin weirs

A knife-edge or thin weir will have a flowrate at a given depth over the weir according to the formula in Eq. (10.11), if approach velocity is less than 0.6 m/s.

Thin weir equation:

$$Q = 1.84 \times (L - 0.2\ H) \times H^{1.5} \qquad (10.11)$$

where

> L = weir length (m)
> H = head upstream of the weir (m)
> Q = flowrate (m³/s)

For deeper channels, and higher approach velocities, Bazin's formula may be used.

Broad weirs

Determination of flow over broad crested weirs is more difficult than for knife-edge weirs, and there are several complicating factors, such as the degree of detachment of the nappe. A clinging nappe gives a higher discharge for a given head than a free nappe.

The flow at a given weir breadth and head may be calculated from Eq. (10.12).

Broad weir calculation:

$$Q = C \times b \times H^{1.5} \tag{10.12}$$

where

b = weir breadth (m)

C varies from 1.4 to 2.1 according to weir shape and discharge condition. A value of 1.6 may be taken for estimation purposes.

FLOW CONTROL IN OPEN CHANNELS

Penstocks or sluice gates are the equivalent of valves used in open channels (though both of these terms are sometimes used to mean other hydraulic control devices and structures). There are types in which flow goes over the gate (weir penstocks), and types where flow goes under the gate (submerged sluice gates).

The first category can be treated as a knife-edge weir with variable depth of submergence. Submerged sluice gates are more difficult to analyze. The Bernoulli equation cannot be used if there is a "hydraulic jump" after the sluice gate. This is discussed in more detail in Chapter 15: Dirty Water Hydraulics.

HYDRAULIC PROFILES

A hydraulic profile (Fig. 10.6) is a simplified longitudinal section through the water retaining structures, usually showing important hydraulic details at each stage. It is always a scale drawing in the vertical dimension. If it is to be to scale horizontally, there are often two different scales, with vertical and horizontal scales differing.

It should show the following levels at a minimum of dry weather and maximum flows:

- any critical levels in existing plant
- TWL, soffits, inverts, gradients of tanks and channels
- levels of all bellmouths, weirs, and chambers
- width of channels
- any special conditions

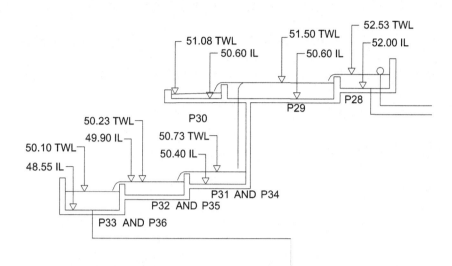

FIGURE 10.6

Example of a hydraulic profile.

FURTHER READING

Genić, S., Arandjelović, I., et al. (2011). A review of explicit approximations of Colebrook's equation. *FME Transactions*, *39*, 67–71.

Huddleston, D., et al. (2004). A spreadsheet replacement for Hardy-Cross piping system analysis in undergraduate hydraulics. *Critical transitions in water and environmental resources management* (2004, pp. 1–8).

Miller, D. S. (2014). *Internal flow systems* (3rd ed.). Shannon: Mentor Graphics (Ireland) Ltd.

US, Department of Commerce (2017) National Institute of Standards and Technology. In *NIST chemistry webbook*. <http://webbook.nist.gov> Accessed 09.10.17.

SECTION

3

Municipal dirty water treatment engineering

INTRODUCTION

Water contaminated with domestic wastes and in some cases combined with surface water runoff is generally known as "wastewater" or "sewage." Water contaminated because of its use in industrial applications is usually called "effluent." Both kinds of "dirty water" are a constantly changing mixture of water with contaminants, which might be a mixture of solids, liquids, or gases.

To the designers and operators of dirty water treatment plants, the precise chemical composition of the water is of no interest. If they are discharging to the environment, the treated water quality standards that environmental regulators set them are related to the effect of the water's contaminants on the environment. If the water is to be reused, the specification to be reached will be to do with protecting the process or people who are to be exposed to the treated product.

137

Municipal wastewater is the water used by a community for bathing, washing, in toilets, and so on, with added environmental waters. These waters are collected and usually treated before they are discharged to the environment. The degree of treatment varies from region to region, and often within regions depending on the standard of the receiving body. In many regions of the world, there is a continuing process of tightening environmental standards, resulting in a similar tightening of discharge standards from wastewater treatment facilities.

In addition to the increase in environmental concerns that drives the tightening of effluent standards, new chemical compounds are continuously being synthesized. Some of these will find their way into the wastewater treatment system and their presence will therefore have to be accommodated within the design of the plant.

The treatment of sewage and effluent can be understood as a process of the removal of smaller and smaller particles. Hence we proceed via removal of gross solids and easily removed grits and sludges in primary treatment through to removal of solids and dissolved organic matters in secondary treatment, then to the removal of ultra-fine particles and increasingly, specific chemical species in tertiary treatment.

Dirty water characterization and treatment objectives

11

CHAPTER OUTLINE

Wastewater Characteristics 139
   Solids...140
   Biochemical Oxygen Demand 140
   Ammonia..140
Flowrate and Mass Loading 140
   Selection of Design Flowrates...141
   Selection of Design Mass Loadings 143
Treatment Objectives...144
   Discharge to Environment 145
   Industrial Reuse..146
   Reuse as Drinking Water 146
Further Reading 146

WASTEWATER CHARACTERISTICS

Although wastewater treatment plants are designed to meet future conditions, it is rare that the engineer engaged to design a wastewater treatment plant even has comprehensive data about past flows and compositions of the wastewater coming into the plant. It is rarer still for the engineer to have reliable future projections.

Consequently, various empirical methods for estimating flows and composition of future wastewaters have arisen over the years, allowing designs to proceed based on whatever information is available.

The most important characteristics of wastewater coming into the treatment plant from the point of a designer are its levels of solids, biological treatability, temperature, and nutrient levels. These form the basis of the main design considerations for most plants, and are therefore tend to be the first questions on the designer's mind. From the point of view of the discharge location, microbiological characteristics may also be of interest. In addition to the concentrations of suspended solids, BOD, and ammonia that the designer is working to reduce, there is frequently a desired microbiological quality, often measured by the presence of such indicator organisms as *Escherichia coli*.

Eutrophication of watercourses may be caused by nitrate and/or phosphate added by agriculture and effluent discharge may lead to overgrowth of undesirable algae, and bacteria. This can reduce oxygen levels and hence kill oxygen-requiring organisms such as fish. It is therefore becoming more common for levels to be set for nitrate and phosphate in discharge consents. Special treatments may be necessary to ensure discharged effluent meets these consents.

SOLIDS

To the designer, the density and particle size of the solids are their most notable features, these being the characteristics that affect their behavior in settlement processes. A secondary characteristic is the behavior of the solids when heated to about 550°C. Organic components are driven off or oxidized, and are therefore referred to as Volatile Suspended Solids (VSS or VS). This is sometimes used as a measure of the treatability of a wastewater.

BIOCHEMICAL OXYGEN DEMAND

The BOD is commonly used as a measure of the amount of biologically degradable material in the wastewater, although it is really a measure of the amount of oxygen used by bacteria suspended in the wastewater over a given period. The most common measure used is the 5-day BOD (BOD_5), with incubation for 5 days at 20°C.

AMMONIA

There are several forms of nitrogen in wastewater. Ammonia is the commonest measure of interest, being potentially highly toxic in the environment, and consequently being in need of regulation in discharges. Ammonium (NH4) levels in a wastewater are easily determined by chemical means, by distilling off ammonia after pH increase, and determining ammonia content by standard chemical methods.

FLOWRATE AND MASS LOADING

Unlike the design of other process plant, where feedstock is specified as part of design, sewage treatment plants must treat whatever arises in the sewerage system.

We may be in the position where we have access to extensive historical records of flow and loading at the works site but, more often, we must estimate the flows and loadings. Even when we have a body of data, we still must forecast future expansion at the site. We tend to proceed from estimates of the average flows and loads via peaking factors to several design cases. The works must be able to deal with the highest flow condition with very dilute effluent, low flow conditions where influent is very concentrated, and a maximum load condition, probably intermediate between the two previous conditions in terms of flow.

The handling of these various conditions must be considered in conjunction with how we handle peak flowrates upstream of the plant (in storm conditions for example).

SELECTION OF DESIGN FLOWRATES

In addition to the expected flows of wastewater arising from households connected to the wastewater system, there are flows arising from industrial facilities, infiltration, and storm flows to consider in setting plant design flowrates.

These design flows are considered in both process and hydraulic design. It should be noted that part of the design might involve treating the highest flowrates less than the lowest flowrates, because at high flowrates, dilution has reduced the concentrations of contaminants.

The common techniques for estimating industrial and domestic flowrates are:

- *Based on actual flow data*: This is the best method, but good data of this sort is rarely available to the designer. Even when such data are available, it is usually required to estimate for a future expansion in population or industry
- *Based on water supply figures*: This is one of the more accurate methods. If data are available for water supplied to the catchment area for the wastewater treatment facility, it may be used to estimate the domestic and possibly the industrial components of the flow
- *Based on empirical data* for either the expected population in the area, or the expected number of facilities, or some combination of the two

These techniques allow the estimation of one or more components of the flow, averaged over a period, usually 1 day.

Infiltration is usually estimated as a percentage of these flows, in the usual absence of survey measurements. Detailed information may be obtained by graphical means by measurement of nighttime flowrates.

Storm waters are usually estimated based on areas of hard and soft standing connected to the drainage system. A storm return period is selected for the design, giving a depth of rain in a given period to be multiplied by the area of drainage.

Variability in all the above factors may be very great. The size and density of the community served, proportion of flows arising sent to treatment, local climate, degree of societal affluence, whether supplies are metered or not, supply quality and dependability, and water conservation measures may all impact on the degree of variability around the mean figures calculated by the techniques previously given. A peaking factor will be applied to the dry weather flow (DWF) previously determined. Various multiples of the DWF will be destined for varying degrees of treatment.

For example, up to $6 \times DWF$ may be screened, but only $3 \times DWF$ may be given secondary treatment, and only a proportion of this may pass to tertiary treatment.

There are several common formulae used to calculate flows to each stage of treatment.

Domestic Component

Domestic flows are often estimated based on a flow per head of served population per day, as described previously. The quantity of water used per head per day varies between countries, mainly as a function of degree of industrialization. An estimate of UK usage in 2011 is 185 L/h/d including unmeasured flows.

Industrial Component

The following factors should be considered when estimating industrial and commercial flows:

- abstraction licenses from private boreholes
- metered flows of water and sewage to sewer
- trade effluent consent conditions
- onsite sewage treatment
- any rainwater flows

In the absence of reliable data, the rules of thumb in Table 11.1 can be used.

Infiltration/Exfiltration

Sewerage pipes leak, and they tend to become leakier over time. Whether this results in net influx or efflux of water depends on ground conditions local to the

Table 11.1 Rules of Thumb for Estimating Effluent Flows

Industry Type	Average (L/s/ha)	Peak (L/s/ha)
Heavy industrial water user	1.35–2.7	4.0–8.0
Intensive light industry	0.5–0.8	1.5–2.4
Open light industry	0.4–0.7	1.2–2.1
Warehousing	0.3–0.5	0.9–1.5

Source	Design Flow (L/h/d)
Boarding schools	150–200
Camp sites	50–100
Day schools	50–60
Factories	40–80
Hotels	150
Offices	40–50
Restaurants (per cover)	30–40

leak. In the United Kingdom, we tend to allow 20%−100% (average 50%) of the domestic component as an allowance for infiltration.

Infiltration can be estimated based upon minimum nighttime flows, measured not less than 3 (ideally 7) days after the last rainfall. Due allowance must be made in the calculation of any remote pumping and nighttime industrial discharges.

Peaking Factors

The size of a community is a major factor to consider when deciding on the peaking factor to apply to the sum of the earlier figures. For equivalent populations up to 5000, peaking factors of around 4 are used, dropping to around 1.75 at 1.5M PE level.

Where industrial loadings are 25% or more of the total flow, the patterns of industrial use should be analyzed separately, and overlaid on the domestic pattern.

Upstream Flow Equalization

The sum of the earlier elements is an estimate of the total likely peak arisings at the works. It would be extremely unusual to completely treat these. Instead, hydraulic controls are applied such that flows more than the treatment plant design flow are sent to storm systems on larger works.

In the United Kingdom, the maximum forward flow to treatment plant is usually defined by 'Formula A'.

$$\text{Formula A} = PG + E + I + 1360P + 2E \; l/d$$

where

P = population
G = average daily consumption (L/h/d)
E = industrial and commercial discharges to sewer (L/d)
I = infiltration (L/d)

For larger works, with storm tanks, flow passed forward to full treatment will be limited to 'full treatment flow' (FTF):

$$\text{FTF} = 3PG + 3E + I$$

Formula A-FTF passes to storm tanks.

SELECTION OF DESIGN MASS LOADINGS

The simplest and most common technique in determining average mass loadings for biochemical oxygen demand (BOD) and Suspended Solids Levels is the population equivalent (PE).

Table 11.2 Per Capita Contributions

Parameter	Daily Load
BOD	50–70 g/d
SS	60–85 g/d
Ammonia	3–8 g/d

Table 11.3 UK Sewage Strength Classification

Sewage Strength	BOD
Weak	<200
Medium	350
Strong	500
Very strong	>650

Table 11.4 International Variations in Average Sewage Strength

Country	Average BOD
Qatar	As low as 20
Peru	195
The United States	215
India	280
Israel	385
The United Kingdom	325
Kenya	>450

60 g/h/d is used as an estimate of both BOD and Suspended Solids per capita figures in the United Kingdom (Table 11.2). Industrial loads are commonly converted to their PE for plant design.

In the United Kingdom, we characterize sewage as relatively "strong" or "weak" based on BOD level as given in Table 11.3.

Different countries, however, have different average sewage strengths, as shown in Table 11.4.

The determination of design mass loadings is, however, more complex. There are seasonal, diurnal, and industrial load variations. As with flow, smaller catchments tend to have greater variability. Generally, peaking factors tend to range from 1.5 to 4.

TREATMENT OBJECTIVES

Wastewater treatment produces a cleaned water stream. The cleaned stream is usually discharged to some body of water, to further reduce its environmental

Table 11.5 Sewage Quality at Different Stages of Treatment

Quality Parameter	Raw Sewage	Primary Effluent Settled Sewage	Secondary Effluent (20:30)	Sand Filtration	Lagooning
Suspended solids (SS) (mg/L)	300–500	120–250	<30	10	5
Chloride	100	100	100	100	100
BOD (mg/L)	250–400	150–300	<20	10	5
COD	640–800	360–600	100	70	50
BOD:COD	0.5	0.5	0.2	0.15	0.1
Ammonia (NH$_3$)	30–40	30–40	5–20	5	1
Nitrate (NO$_3$)	<5	<5	5–15	20	25
E. coli (MPN[a] per 100 mL)	10^7	10^6	10^6	10^5	10^3

[a]MPN, most probable number. "It is customary to report the results of the multiple fermentation tube test for coliforms as a most probable number (MPN) index. This is an index of the number of coliform bacteria that, more probably than any other number, would give the results shown by the test. It is not a count of the actual number of indicator bacteria present in the sample." (WHO).

impact by dilution and natural biological treatment. The effect of this discharge on the body of water is usually the criterion by which the operating standards for the treatment plant have been selected.

Table 11.5 summarizes how successive stages of sewage treatment reduce levels of suspended solids, BOD, and to some extent pathogens.

DISCHARGE TO ENVIRONMENT

The objective of most wastewater treatment is to maintain or improve the quality of the receiving body of water. There has been a trend over time to tighten the requirements from the initial simple removal of gross debris, through biological treatment for BOD and Suspended Solids removal, more recently to removal of dissolved nutrients. As well as this qualitative change of new processes being introduced, the existing processes are increasingly required to give better performance, with acceptable levels for BOD, SS, and ammonia having been reduced continuously over time.

In drier parts of the world, wastewater may be used for crop irrigation, indirectly for production of drinking water by aquifer recharge, or even sometimes processed directly into drinking water.

INDUSTRIAL REUSE

The unique properties of water make it ideal for use in industry, two of its most common applications being:

- As a transporter of energy, for example: in steam raising plant, cooling systems, and heating circuits
- As an integral part of a process, for example: in brewing and soft drinks manufacture, chemicals production, and the food industry

There is no other relatively cheap, readily available, nontoxic material capable of carrying out the tasks we require and expect of water. As these tasks become more complex, and the plant employed more sophisticated, it becomes apparent that the use of water is not as simple as merely piping up to the nearest supply. Impurities in water can cause problems within the plant resulting in reduced efficiency, increased maintenance costs, and lost production time.

Although the need for water remains, the quality of the water must be increasingly controlled.

REUSE AS DRINKING WATER

Common for some time now in China and elsewhere, reuse of treated effluent as drinking water is becoming more common. This may be referred to as reclaimed water or sometimes NEWater, but in fact it is highly treated sewage. In California, the Advanced Water Purification Facility (AWPF), the largest water reclamation plant in the United States, yields 70 million gallons per day of drinkable water from sewage. That is about 10% of the district's daily water demand for its 2.3 million residents. Although AWPF's purification process is complex, it produces clean, pure water that meets or exceeds all drinking water standards.

FURTHER READING

Bartram, J., & Balance, R. (1996). *Water quality monitoring—a practical guide to the design and implementation of freshwater quality studies and monitoring programmes.* London: Chapman & Hall, United Nations Environment Programme.

Dirty water unit operation design: physical processes

12

CHAPTER OUTLINE

Introduction 147
Flow Measurement 148
Pumping 149
   Airlifts 149
Racks and Screens 151
   Bar Screens 153
   Screens 153
   Comminution 153
Grit Removal 154
   Grit Chambers 154
   Vortex Separators 154
   Suspended Solids Removal 154
   Screens and Filters for Residual Suspended Solids Removal 155
Flow Equalization 155
Primary Sedimentation 156
Final Settlement Tank Design 158
Flotation 160
Filtration 160
Gas Transfer 161
   Stripping 162
Adsorption 162
Other Physical Processes 163
   Physical Disinfection 163

INTRODUCTION

As previously noted, the treatment of wastewater is to some extent the process of removal of ever small-sized particles. Physical processes tend to be the most important in the first stages of this process, although there are physical aspects to every stage of treatment.

FIGURE 12.1

Flume.[1]

Courtesy: LMNO Engineering, Research, and Software, Ltd.

FLOW MEASUREMENT

Flow measurement devices for wastewater treatment plants may be conveniently divided into open channel and pipeline devices. The inlet and often the outlet of a wastewater treatment works will tend to carry wastewater in open channels, and pipelines will tend to be used within treatment processes after the inlet works.

The most common flow-measuring device on wastewater treatment plants is the flume (Fig. 12.1), combined often with an electronic level measurement instrument. The flume converts a variable flow of effluent into a variable height, which may then be measured directly from a scale within the channel, or indirectly by means of an electronic level instrument.

A weir-type device works in much the same way. There are other types of open channel devices, including magnetic, velocity head, and dilution types.

There are any number of flow measurement devices for closed systems, based on differential head over an obstruction, moving fluid physical effects, or positive displacement. These have a range of applications based on their type. For example, units based on obstructions to flow or small orifices are unsuitable for early stages of treatment, where there are high levels of gross solids still within the wastewater.

---

[1]https://www.LMNOeng.com/Flumes/flumes.php, accessed 22 November 2017

PUMPING

Pumps used for dirty water are rated by solids handling capability, as well as flow and head requirements. The larger and harder the solids a rotodynamic pump can handle, the lower its efficiency. Rotodynamic pumps are still, however, usually favored where possible, and are often of the submersible type rarely used in other areas. The tables in Chapter 10, Clean Water Hydraulics, will help with pump-type selection.

AIRLIFTS

An unusual pump type common in this area is the airlift (Fig. 12.2), used to recirculate activated sludge on small plants. While it is not the most efficient, it has no moving parts, and allows a blower to be used as the sole item of rotating equipment to produce a small, economically low-maintenance design.

Although the airlift pump was invented in 1797, knowledge of how to design it is not widely held, so I have included guidance here, based upon long-standing heuristics.

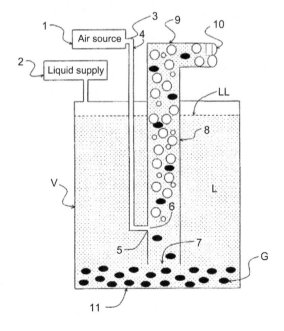

FIGURE 12.2

Airlift pump.

In the following discussion:

Q = required liquid flow
$H1$ = vertical distance from water surface (LL in Fig. 12.1) to air pipe inlet (point 6 in Fig. 12.2)
$H2$ = vertical distance from water surface (LL) to centerline of discharge pipe (point 10 in Fig. 12.2)
H = total head required to produce required liquid flow of Q

Design constraints

Line size

- The minimum pipe size is 4″/100 mm Nominal Bore (NB) if large solids are to be lifted
- If used with RAS, or similar, the line size can be selected from the Fig. 12.3
- The discharge line velocity should not exceed 1.1 m/s

Required percentage submergence

- The percentage submergence $(H1 \times 100)/(H1 + H12 + H)$ should be greater than 65% (Fig. 12.4)
- The fluctuation in percentage submergence in service should not be less than 1%

Other considerations

- The distance from the tank invert (point 11 in Fig. 12.2) to the bottom of pump should be 150 mm
- The distance from the airline entry (point 5 in Fig. 12.2) to the bottom of pump should be 150 mm
- The pipe at point 9 of Fig. 12.2 is commonly a tee, rather than a bend, with an open vertical leg of height from the discharge branch centerline of $1.5 \times H2$

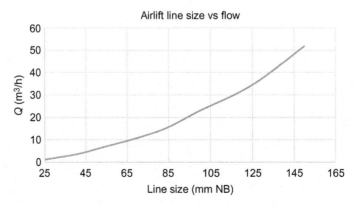

FIGURE 12.3

Airlift line size graph.

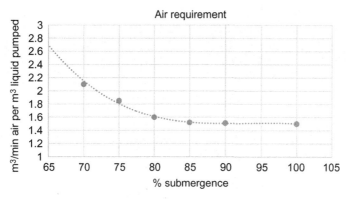

FIGURE 12.4

Air lift percentage submergence.

- A pressure-reducing valve or air bleed valve should be provided on the incoming airline (point 1 on Fig. 12.2) to allow flow to be set at commissioning

   Now, we can size the airlift pump as follows:

1. Choose a suitable line size from Fig. 12.3
2. Choose a degree of submergence from Fig. 12.4
3. Find the ratio of air required to liquid flow at this degree of submergence from the other graph above
4. Knowing Q, use the ratio to work out theoretical air supply requirement
5. Add 25% to the theoretical requirement
6. Allow the blower head requirement at point of supply in mWG as depth of submergence
7. Allow an additional 0.5 mWG for line losses in a line less than 6 m long; calculate and add the line and fitting losses for longer lines

RACKS AND SCREENS

Often the first process in treatment is to remove gross particles by passing the wastewater through holes either in a sheet, or between wires. The holes tend to be of uniform size, usually a few millimeters.

   The requirement for screening of wastewater prior to treatment is becoming progressively finer over time, from perhaps 15 mm holes 20 years ago, down to 6 mm or even less nowadays. Although this is a good thing from a point of view of protection of downstream equipment, in general, the finer the screen, the greater the headloss. It is also notable that the finer the screen, the higher the percentage of

FIGURE 12.5

Bar screen.

putrescible matter recovered, and therefore the more offensive the screenings. Screenings from modern fine screens are often washed prior to storage.

Modern screens tend to be mechanically cleaned, often either by raking, or by means of water jets. Hand-cleaned screens tend to be limited to smaller plants and the coarse screens used in storm systems.

In addition to the screens used in pretreatment, finer screens are gaining popularity for tertiary treatment applications.

The oldest type of screen is the bar screen (Fig. 12.5), a coarse screen formed by the gaps between parallel bars.

There are several variants on the inclined screen, standing at an angle within a channel used for pretreatment. These may be fixed, usually mechanically raked, or rotary, with cleaning usually by water jets. Less commonly, drum screens may be used for pretreatment.

Variants on the inclined screen (Fig. 12.6) and drum screen may be used for primary treatment, removing finer particles.

The different grades of screens, from coarse bar screens to fine inclined or drum screens, remove successively finer particles. Hence the bar screen removes plastic, wood, paper, and rags, with some fecal contamination, but fine screens remove all the above plus fine solids, fats, oils, and greases.

Screenings tend nowadays to be washed of fecal material, compacted, and bagged by automated processes, to avoid offensive odors, and manual handling.

FIGURE 12.6

Inclined screen.

BAR SCREENS

Hand-raked screens rarely find much application on modern works of any size. They are, however, sometimes to be found on smaller works or, sometimes, on storm overflows.

Much more common nowadays are mechanically raked units. The most common of these is the chain-raked type, of which there are many variants.

There are also reciprocating-rake, catenary, and cable-raked types. Each of these types and their variants suffer to a greater or lesser degree from problems of jamming, corrosion, and the problems of having underwater bearings chains and sprockets.

SCREENS

Screens comprising either perforated plates or wedge wire panels range from the simple rundown screen to inclined or drum types. These can give far finer screening than bar racks, but can have additional problems of grease build up. Finer and finer screening prior to treatment is becoming the norm.

The finer screens used more recently trap grit as well as the other components and, in the case of rotary disc and drum types, they may be used as an alternative to primary settlement. Experience, however, suggests that such substitution does not always work as well in practice as hoped.

COMMINUTION

Comminutors can be installed instead of screens, although this is a less common approach in recent years. Rather than removing the solids, they are shredded to the point where, theoretically, they will pass through pumps and so on. In practice, however, the shredded solids recombine to form masses that will not pass through pumps and finer pipelines. The maintenance problems this causes mean that comminutors are rarely specified in place of screens.

GRIT REMOVAL

Hard, dense (mostly mineral) particles that can cause excessive wear of downstream process units and settle out in pipelines are removed either by a form of gravity settlement or by a combination of gravity and centrifugal effects. The first principle is used in grit chambers, and the second in vortex separators.

Velocity-associated "centrifugal" forces may be used to enhance gravity separation. As well as the centrifuges used in sludge thickening, this is an important process in the "Pista" type grit separator.

Some form of washing is recommended for collected grit. Unwashed grit may be 50% organic matter, and rapidly becomes obnoxious if stored in unwashed form.

GRIT CHAMBERS

These units are basically designed to provide a large enough area to slow down particles to the point where they settle out under gravity. They may be of a simple rectangular, or more commonly nowadays a raked circular design. Grit removal and washing is commonly by means of a reciprocating-rake mechanism.

A common variant is the aerated grit chamber, which is like a rectangular grit chamber, except it has a line of air diffusers down one side. The rolling motion induced by one-sided aeration, combined with the forward motion of the wastewater, produces a spiral movement of the channel contents. This results in both grit removal and washing in a single operation.

VORTEX SEPARATORS

In these circular units, a vortex is produced either by tangential introduction of wastewater, and/or by means of a turbine blade. In either case, grit is thrown to the walls of the unit. The grit collects at the base of the unit, and is either removed by airlift pump, which has the advantage of further washing the grit, and is removed by grit pump or grit conveyor.

SUSPENDED SOLIDS REMOVAL

Engineering the removal of solids from treated wastewater is complicated by the nature of these solids. These solids are largely composed of the microorganisms that treated the effluent rather than original effluent solids, which have escaped the preceding settlement stage.

They are consequently fine, soft, even sticky particles. This makes them hard to filter out, and makes subsequent cleaning of the filter of the collected solids a challenge.

Very fine mesh screens, sand filters, and membranes have all been used successfully for tertiary solids removal. In each case, they tend to be bought as a complete proprietary system incorporating the associated cleaning mechanism.

SCREENS AND FILTERS FOR RESIDUAL SUSPENDED SOLIDS REMOVAL

Granular medium filtration

Sand or multimedia filters may be used to remove residual suspended solids and solid biochemical oxygen demand (BOD), as described earlier.

Microscreening

Less common than granular media filtration, design of microscreens will be a matter for the supplier of the units.

FLOW EQUALIZATION

The considerable variability in both flow and load already discussed must be accommodated in the design of the plant. It is, however, not likely to be economically viable to arrange for a plant to treat the maximum possible flow arising at the plant under all conditions. While it would be more likely to be practical to design the plant to cope with the maximum load possible, the variability in concentration between the maximum load and minimum load situations would be likely to become a problem. In flow equalization, we therefore have two aims: accommodating maximum flow and, ideally, controlling the range of load concentrations we must deal with.

It is possible to have an inline flow equalization system, where we blend incoming effluent with a large volume of stored effluent. This gives a high degree of load balancing, but may have problems when dealing with high storm flows. Any discharges from the mixed tank will have relatively high strength, even when the incoming wastewater to the tank is highly dilute storm water. This system also usually involves pumping the whole process flow to treatment. This system is therefore more common in systems with high load variability, and small storm flows, such as onsite treatment plants for industrial facilities.

It is most common for municipal effluents to have an offline system of storm tanks, where flows above a certain limit are sent to tanks where they are held until the flow through the works has dropped back to the point where the tanks can be pumped back through the system. Only these returned flows need to be pumped, and any overflow arising from the tanks will have received some settlement in the storm tanks, and will therefore be more dilute than if arising from inline tanks. The water and wastewater treatment plant must be designed to accommodate quite a high degree of variation in mass loadings, but this still usually represents the best solution on economic grounds.

Storm tanks and other flow equalization tanks usually have a sloping bottom, to facilitate removal of any settled sludge, and may have one or more inclined sloped sides to allow vehicle access.

The main difference between offline and inline systems is that offline systems are often designed to operate as settling tanks, so that flows greater than their capacity are discharged to the watercourse having had substantial solids removal. Inline tanks are usually designed to balance incoming load, and hence approximate completely mixed systems. Complete mix tanks whether online or offline will tend to be either circular or square, and tanks providing a degree of settlement will tend to be an elongated shape.

Mixing may be by means of propeller, surface aeration, or venturi jet mixer. These last two types can also answer the requirement for aeration to prevent septicity of the stored effluent.

Whether designed in or not, all types will see some solids settlement, and several systems are available to remove any accumulation of sludge in the tank. These may be tipping buckets, water, or water/air jets, or simply vehicle access to allow a tractor with a blade to enter the tank.

PRIMARY SEDIMENTATION

Having removed gross solids by screening and coarse settlement, it is then desirable to remove as much of the denser material as is practical by simple settlement.

There are two design considerations. Firstly, the upflow velocity needs to be less than the natural settlement velocity of the particles we are trying to remove. Secondly, the hydraulic retention times (HRT) should be sufficient to allow natural flocculation to increase the size of a substantial proportion of particles in the wastewater to the point where they settle out.

Primary tanks may be circular or rectangular. The circular design (Fig. 12.7) is more common in the United Kingdom. Circular tanks are supplied with a bridge scraper, scum collection facilities, and facilities to distribute incoming and collect outgoing flows.

The commonest arrangement is for incoming flow to enter a central "stilling baffle" from where it is distributed radially into the tank, to be collected by a peripheral weir. The bridge scraper and the sloping tank floor drive sludge into a central hopper, from where it is collected either by pumping or more commonly by hydrostatic means. Smaller units use a half-bridge scraper (as shown in Fig. 12.7), and larger units use a full bridge.

Rectangular designs have either a traveling bridge scraper (analogous to that used in circular tanks except it moves back and forth, rather than round and

FIGURE 12.7

Primary tank.

Table 12.1 Design Criteria for Primary Settlement Tanks

Design flow (max. flow if no storm separation)	$3 \times (PG + E) + I +$ works recycles
HRT	1–2 h
Surface loading/h	1.2–1.5 m/h
Surface loading/d	30–36 m/d
Minimum sidewall depth	2 m
Floor slope	12.5 degrees (ranges from 7.5 to 30 degrees)
"V" notch or castellated weir overflow rate	100–150 m^3/m/d

round), or a chain and flight scraper mechanism to collect sludges. Scum collection systems like those used on the circular design are provided. Inlet distribution arrangements are more varied and crucial to proper tank operation than in circular tanks. They are usually some combination of weirs and orifices, designed to minimize any shortcut streaming through the tank.

Primary settlement tanks designed at a rise rate of no more than 1.5 m/h and HRT of at least 1 h would be expected to remove 50% of SS, 30% of BOD, and yield a sludge of 1%–5% dry solids.

Design criteria for primary settlement tanks are summarized in Table 12.1.

Expert opinion differs on whether or not surplus activated sludge (SAS) should be co-settled with primary sludge. Some think SAS should ideally go direct to consolidation and storage, but there are sites where co-settlement has been practiced historically without causing problems. The practice may be continued at such sites.

In 1975, the United Kingdom's Water Research Council produced a Technical Report (TR11) about settling of activated sludge, which many engineers of my generation still use to design settling tanks. I am not aware of anything that has superseded it, although it can be hard to obtain a copy of TR11. It discusses the basic approach based on HRT and rise rate, a more sophisticated approach based upon a common and easy laboratory test known as sludge volume index (SVI), and recommends an approach based on a novel laboratory test (SSV) and mass flux theory. While its discussion of theory is excellent, the most sophisticated approach suggested in TR11 may be very difficult to put into practice, as a budget will almost never be available for the requisite laboratory tests. In practice, I am rarely even able to determine SVI, never mind to carry out novel tests.

Designing on the basis of SVI (volume of sludge after 1 h settlement) is, however, possible despite the drawbacks explained in TR11.

The minimum rate of sludge return required can be calculated by Eq. (12.1).

Minimum rate of sludge return:

$$\frac{Q_u}{(Q_o + Q_u)} = \text{SVI} \times C_f \times 10^{-6} \qquad (12.1)$$

where

Q_u = volumetric flow of RAS
Q_o = volumetric flow of settled sewage
C_f = concentration of suspended solids in the mixed liquor

FINAL SETTLEMENT TANK DESIGN

The following approach should be used with caution, and examined in the light of commonsense.

Tank shape, flow regime, and SS of the mixed liquor (MLSS) concentration are determined during the design of the aeration tank. Once the type of activated sludge has been selected, design values may be selected from Table 12.2.

The design recycle ratio should now be selected from Table 12.3 and used to determine the capacity of the returned activated sludge pumps required to maintain a stable concentration of mixed liquor in the aeration tank.

The recycle ratio from the table is used as follows:

RAS pump capacity = recycle ratio R × maximum design flow

The required area of the final settlement tank should be determined based upon compatibility with aeration tank, and required final effluent quality.

The lesser value of U_{max} (surface loading) should be selected from either of Tables 12.4 or 12.5.

Table 12.2 Values of SSVI (mg/L) to be Used in Design

Plant Configuration	Type of Wastewater	
	Mostly Domestic Sewage	Significant Proportion of Industrial Effluent
1. Plug flow:		
$L{:}W < 12$	120	140
$L{:}W > 12 < 20$	110	130
$L{:}W > 20 < 30$	100	120
2. Completely mixed	130	150
3. Partial treatment	180	200

Table 12.3 Values of Recycle Ratio R to be Used in Design

SSVI (L/mg)	RAS Concentration (mg/L)	MLSS mg/L					
		2000	2500	3000	3500	4000	4500
200	5000	0.7	1.00	1.50	2.3	4.0	9.0
170	6000	0.5	0.7	1.00	1.4	2.0	3.0
140	7000	0.4	0.6	0.75	1.0	1.3	1.8
130	8000	0.35	0.5	0.60	0.8	1.0	1.3
110	9000	0.3	0.4	0.50	0.6	0.8	1.0
100	10,000	0.25	0.35	0.45	0.5	0.7	0.8

Table 12.4 Values of U_{max} (m/h) Compatible With Aeration Tank (Derived From WRc Equations With Factor of Safety = 1.25)

MLSS (mg/L)	SSVI (mg/L)							
	100	110	120	130	140	150	180	200
2000	2.0	1.8	1.6	1.5	1.3	1.1	0.7	0.4
2500	1.6	1.5	1.3	1.2	1.0	0.9	0.5	0.35
3000	1.3	1.2	1.1	0.9	0.8	0.7	0.4	0.3
3500	1.1	1.0	0.9	0.8	0.7	0.6	0.3	0.2
4000	0.9	0.8	0.7	0.6	0.5	0.5	0.3	0.15
4500	0.7	0.7	0.6	0.5	0.4	0.4	0.2	0.1

Table 12.5 Values of U_{max} (m/h) for Final Effluent SS Required (Derived From German Design Guidelines)

Final Effluent SS 95%ile (mg/L)	U_{max} (m/h)
SS < 20 mg/L	$\frac{300{,}000}{MLSS \times SSVI}$ or 1 m/h max
SS < 30 mg/L	$\frac{450{,}000}{MLSS \times SSVI}$ or 1 m/h max
(MLSS in mg/L, SSVI in mL/g)	

In all cases where final effluent SS standards better than 15 mg/L is specified, or SS < 20 mg/L with large fluctuation in flow, tertiary treatment should be specified.

Should the minimum retention at maximum flow be less than 2 h, the tank diameter should be increased to give a retention time of 2 h.

FLOTATION

Flotation is the process in which particles are floated to the surface of water by means of attachment to the particles of fine gas bubbles. The main deliberate application of this process in wastewater treatment is the Dissolved Air Flotation sludge thickener, although nondeliberate use of the process may lead to problems in aerated biological systems, and final settling tanks.

In dissolved air flotation, the differential solubility of air with respect to pressure is used to create microbubbles of air within the body of the sludge to be treated, creating a floating high-solids layer that may be skimmed off after allowing it to self-thicken further by drainage. It is common to add polymeric substances to enhance sludge cohesiveness, and therefore dryness of the floating layer produced.

A recycle stream of clarified underflow is taken from the body of the thickener, saturated with air at up to 9-bar pressure, and released through small nozzles within the body of the unit in such a way as to create bubbles of a few microns in size. These bubbles have high surface charges, and adhere readily to sludge solids particles.

Coarser bubbles have been used to float sludge solids less effectively, in a process called Induced Air Flotation. This process is, however, of most use in enhancing scum removal in grit traps.

A vacuum may be used to generate bubbles from a wastewater or sludge previously saturated with gas. This may be especially useful in applications like the deep shaft process, where the arisings are already naturally saturated with gas by the high operating pressures.

Sludge thickening dissolved air flotation has far higher recycle ratios than when the process is used in drinking water treatment. The recycle stream should have at least 100% of the incoming flowrate.

FILTRATION

The introduction of filtration to wastewater treatment is relatively new. The most common application is at present the enhanced removal of suspended solids (including particulate BOD) from effluent prior to discharge.

Sand filters are most commonly used for these operations. The types of sand filters used are largely based on those that are commonplace in potable water treatment. There may be additional media used in conjunction with the sand, such as anthracite or garnet. These media differ in density from sand, and may be graded in size such as to arrange in a stratified way upon backwashing. This arrangement of different particle sizes allows extended filter run times by preventing premature clogging of the sand bed.

The three types of sand filter on which wastewater filters are based are known as pressure filters, rapid gravity filters, and slow sand filters. The first two types operate in an intermittent basis, and the last type on a continuous basis.

Rapid gravity filters operate in downflow mode for a period, until either a maximum time has elapsed or the headloss through the filter or outlet turbidity becomes excessive because of solids deposited within the body of the filter medium. The filter is then taken offline, and backwashing is carried out to recover the filter to a clean state, ready to be brought back into service.

Sand filters should be expected to reduce SS by 80%, BOD by 60%, and coliform bacteria by 30%.

Pressure filters operate in the same way, but instead of being open to the atmosphere, the filter is contained within a pressure vessel and pump pressure rather than static head drives operation.

Slow sand filters build up a surface filtration layer that is removed periodically, the sand washed clean of dirt and redeposited. These are used in an unmodified form to treat effluent from oxidation ponds. The most common wastewater treatment version of this process, the traveling bridge filter, fully backwashes individual areas of the bed while the remainder of the bed remains flooded and in use. A more recent refinement operates in upflow, with sand passing down the column toward the dirty end of the process, and wastewater passing upward. Sand that has reached the bottom is carried upward through the center of the bed to washing at the top of the filter, prior to redeposition on the top of the moving bed.

GAS TRANSFER

The most common gases deliberately transferred in wastewater treatment are oxygen, carbon dioxide, methane, and chlorine. The first two of these are of importance in aerobic treatments, the second and third in anaerobic treatment, and the last in disinfection.

Aeration of sewage may be achieved by bubbling air through water, or by throwing water through air by means of surface aerators (Fig. 12.8).

When bubbling air through water, the finer the bubbles, the more efficient aeration is, as summarized in Table 12.6.

Where consents to discharge specify high dissolved oxygen levels, aeration of the outgoing effluent may be required. Where there is sufficient head available,

FIGURE 12.8

Surface aerators.

Table 12.6 Efficiency of Aeration Systems

Aeration System	Oxygenation Efficiency (kg/kWh)
Fine bubble diffused air	1.5–3.6
Coarse bubble diffused air	0.9–1.2

cascade aeration, where the effluent is simply passed over a succession of weirs, will carry out the requirement with no local energy input. Otherwise, the techniques used in aeration basins must be employed.

STRIPPING

The most important application for deliberate stripping of gases from wastewater is the reduction in levels of ammonia. This is especially useful in reducing ammonia levels in returned liquors from sludge treatment, which may be very high in ammonia. It will be necessary to re-dissolve the stripped ammonia in water or acid, rather than simply venting to atmosphere, to avoid odor problems.

The unintended stripping of odorous compounds and VOCs (especially those having greenhouse gas potential) by aeration processes is becoming something of a concern.

ADSORPTION

The most common adsorption process in wastewater treatment is granular activated carbon treatment.

Activated carbon is produced from charring coconut or other nutshells, coal, or wood in a controlled manner. This process results in a very open-textured carbon material, full of macroscopic and microscopic channels throughout its bulk. The action of the carbon in removing dissolved substances from solution and attaching them to its surface happens mostly within the microscopic channels, which make up the bulk of its surface area.

The material is used either as a fine powder ("powdered activated carbon" or PAC) or alternatively in a granular form, ("granular activated carbon" or GAC). PAC tends to be added directly into the water to be treated, and GAC held within a vessel, and a flow of wastewater passed through the vessel. The retention of GAC means that it may be recovered, and regenerated for further use, whereas PAC used is lost, usually in sludge production. Onsite regeneration is, however, rare. Usually at a site level, the process is one of removal of "spent" GAC, and its replacement with virgin material.

The process is one of nonselective adsorption of all substances that are adsorbed by GAC, until the available binding sites are exhausted, and contaminants break through the end of the bed. The time to breakthrough is the most important consideration, and is theoretically determined by means of adsorption isotherm data available from GAC suppliers. In practice, empty bed contact time is the main design parameter, as theoretical approaches are overwhelmed by the number of wastewater contaminants.

OTHER PHYSICAL PROCESSES

In addition to the fine screens previously discussed, flotation might be used as an initial solids removal technique. This is a very unusual application of the technique, more commonly used to treat smaller, stronger flows, but it can be used in the primary stage. The units would be far smaller than the equivalent gravity settlement tanks, but energy costs would be far higher.

Preaeration may be carried out for septicity control, or as part of other activities such as grit removal.

Mechanical coagulation/flocculation may be used but it is far less common than chemical dosing (see Chapter 13, Dirty water unit operation design: chemical processes) to enhance primary settlement.

PHYSICAL DISINFECTION

Physical treatments tend to be energy intensive, but have the virtue of leaving no residual toxicity to contaminate the discharge point. Heat has been used at small scale, and is an intrinsic feature of sludge digestion processes. UV light is becoming the most popular technique for final disinfection, and is becoming increasingly specified as a discharge consented parameter in Europe. Other electromagnetic radiation, including gamma radiation, has been used, but this last is unpopular, despite its effectiveness, due to the difficulty in handling radioactive materials.

Dirty water unit operation design: chemical processes

13

CHAPTER OUTLINE

Introduction 165
Coagulation 165
Chemical Precipitation 166
Disinfection 166
  Chemical Disinfection...167
  Dechlorination...167
Nutrient Removal 167
  Chemical Removal of Nitrogen...167
  Chemical Removal of Phosphorus 168
Removal of Dissolved Inorganics...168
Other Chemical Processes 169

INTRODUCTION

Chemical processes have been less important historically in municipal dirty water treatment than in municipal clean water treatment. Cost is all important in this area, and chemicals cost money. However, as regulators drive treatment requirements ever higher, chemical addition is becoming required to achieve them.

COAGULATION

In standard engineered coagulation processes, coagulant chemicals are added and pH adjusted to minimize the solubility of the coagulant reaction product. In most cases, pollutants are removed either by adsorption onto the surface of the coagulant, or entrapment within a matrix of precipitated coagulant. In some cases, the process is one of direct reaction of the coagulant with the pollutant to produce an insoluble salt. It should be noted that adding chemicals always results (in the first instance) in a net increase in dissolved solids, which may be a consideration when designing systems for water reuse.

An Applied Guide to Water and Effluent Treatment Plant Design. DOI: https://doi.org/10.1016/B978-0-12-811309-7.00013-8

The chemicals used in wastewater treatment are largely the same as those used in potable water treatment: salts of aluminum and iron—albeit less fine grades. These both form insoluble gelatinous hydroxide precipitates with (at optimum pH) high surface charge density, attracting both solid and dissolved pollutants. The pH of the wastewater is crucial in determining the effectiveness of precipitation with these substances, and pH correction may be required, unless there is sufficient alkalinity present to provide natural neutralization of the acid salts used for dosing.

Lime is also added alone as an aid to settlement, but the process is one of weighting the sludge with high-density particles, rather than coagulation/precipitation. Lime is reactive with acids, and will not start to work until all acids have been neutralized. An additional benefit is conferred when lime is used in conjunction with sulfate salts: the production of insoluble calcium sulfate, rather than soluble calcium chloride, further weights the flocs produced.

One particular reduced iron salt, ferrous sulfate (copperas) must always be used in conjunction with lime. It requires oxygen for the reaction leading to precipitation as ferric salts and thus results in a large drop in dissolved oxygen. It may not be used unless the required oxygen is available. It may, however, be locally available as a cheap by-product of steel processing and this factor may outweigh the problems associated with its use.

CHEMICAL PRECIPITATION

The addition of chemical precipitants to wastewater entering primary settlement enhances sludge yield. It can also control septicity and odor, and remove phosphorus.

Lime, iron, and aluminum salts are the usual chemicals used. Each has advantages and disadvantages. Iron and aluminum salts tend to acidify, and lime to alkalize the effluent. Iron and aluminum require sufficient alkalinity in the water to work effectively.

Iron has the advantage of having insoluble phosphate and sulfide salts, and can therefore remove phosphate and odorous sulfides. Lime can remove phosphate as insoluble calcium phosphate.

DISINFECTION

Chemical predisinfection of wastewater is not very common. Where it is carried out for control of odor or filamentous organisms, chlorine or hypochlorite is most commonly used. Other halogens, or combinations of halogens, are also used to a lesser degree.

CHEMICAL DISINFECTION

Chemical disinfection is not as popular as it once was due to the likelihood of leaving residual disinfectant chemicals or their products in the wastewater.

Halogens and their compounds have been used, as well as other oxidizing agents such as ozone and hydrogen peroxide, detergents, phenolic compounds, alkalis and acids, heavy metals, and alcohols—in summary almost all known disinfectants.

Chlorine and its compounds are the most commonly used of these, due to their cheapness, ready availability, and ease of control of residual. Chlorine gas is the cheapest form of the chlorine-containing disinfectants and was therefore favored historically. Safety considerations have led to a trend away from chlorine toward its compounds.

Having dosed and effectively mixed in the chlorine, the wastewater must be impounded for a known period to allow the required exposure to be reached. Care must be taken in designing these holding facilities to facilitate consistency in residence times for wastewater passing through.

Ozone may also be used for this duty, but it is energy intensive to make and highly toxic.

DECHLORINATION

Chlorination is perhaps the best of the chemical disinfection techniques, as the addition of sulfur dioxide may be easily controlled to give zero chlorine and sulfite residuals.

The control of this addition by means of combined chlorine residual monitoring leads to very good control of release of chlorine and its compounds, if monitoring and mixing processes are well designed and reliable.

An alternative technique for removal of chlorine in free and combined forms is GAC contact. This is more expensive than SO_2 dosing, but is 100% reliable if the bed is not exhausted.

NUTRIENT REMOVAL
CHEMICAL REMOVAL OF NITROGEN

Removal of ammonia by air stripping is less popular now that it is no longer possible to simply vent the ammonia to the atmosphere. There are systems available where the ammonia removed is reabsorbed into acid, but these will be unlikely to be economic unless used on very high strength returns, such as centrate from centrifugation of sludges.

Breakpoint chlorination is another largely obsolete process, most countries preferring to move away from chlorine use wherever possible. The process reacts chlorine with ammonia. In the first instance, this yields chloramines but at a certain level, known as the breakpoint, the reaction proceeds to an endpoint of nitrogen, water, and hydrochloric acid. It therefore yields a decreased pH in addition to nitrogen removal. Chloramines are potent disinfectants, and the process is usually therefore followed by dechlorination where it is used to ensure a (both bound and free) chlorine free effluent.

CHEMICAL REMOVAL OF PHOSPHORUS

In addition to the metal salts and lime described previously for this application, organic polymers may be used to enhance settlement of produced insoluble phosphorus salts. These are used when phosphorus removal is combined with secondary settlement of poorly settling sludges such as those yielded by trickling filters or extended aeration activated sludge plant.

The unwanted precipitation of magnesium ammonium phosphate (struvite) has historically been a nuisance in the handling of sludge treatment liquors. There is however now a commercial process that deliberately precipitates struvite from these liquors in a fluidized bed crystallizer by increasing pH and magnesium levels. As struvite has commercial value as a fertilizer, this might prove economic, although it is still at an early stage of development.

REMOVAL OF DISSOLVED INORGANICS

Some dissolved inorganics such as ammonium compounds can be removed by conventional primary or secondary treatment, and some can be removed by the processes described earlier. There are, however, others whose removal is more problematic. The removal of such dissolved inorganic compounds is generally speaking very expensive, and requires processes with components that can be irreversibly damaged by substances found in untreated effluents.

The cheapest way to treat such substance is to render them insoluble, so that they may be removed by settlement (or, more expensively, by filtration if unavoidable).

Some inorganic compounds can be reduced or oxidized to insoluble forms removable by precipitation. Chromium 6 can, for example, be reduced to insoluble chromium 4. Many metals can be removed by changing the pH, and/or adding hydrogen sulfide. Conversely, sulfides can be removed by adding iron salts.

There are several processes that can precipitate sulfates but, as discharge consents tighten, the cost of meeting them becomes ever more expensive, and the reliability of the processes lower. Sulfate precipitation as the calcium salt is a well-understood and reliable process, but it cannot attain levels low enough for many regulatory authorities.

There are many research-led approaches to attaining lower levels of sulfates. My own experience is that several of these novel approaches have failed to perform as promised in practice. Consequently, I would urge great caution in this area. Variations in temperature and interference from other ions can make something that works well in the laboratory useless in practical applications.

Some other compounds—such as cyanides—can be chemically oxidized, but like all the following processes (despite pitches you may hear to the contrary from interested parties), treatment by chemical oxidation is more expensive and less reliable than conventional secondary treatment. These more expensive processes should only be used if cheaper ones are not available. Many of them—such as pure oxygen fed activated sludge—are purely "marketing breakthroughs."

For the substances like chlorides that we cannot precipitate, oxidize, or reduce to readily separable forms, we are left only with ultrapure water technologies such as ion exchange, distillation, and reverse osmosis membranes. These processes are relatively costly and lacking in robustness. The water produced will be purer than drinking water, so there needs to be a very good reason to use such technologies.

Often, the rationale for the use of these has had a political basis. "Greenwash" is a term used for "Green" sounding initiatives which do not stand up to deeper analysis. Ideas such as "Total Recycle" and "Zero Discharge," are examples of greenwash because they always exceed the point at which rational measures of resource economy are met. The optimum level of resource recovery is never 100%, as each small increment closer to 100% costs proportionally more, whether we are working in dollars, measures of environmental benefit, or a combination of the two.

OTHER CHEMICAL PROCESSES

Many of the chemicals used in the preceding sections have alternative uses, for example, in controlling sludge bulking by filamentous organisms. The control of pH by addition of lime or other alkaline substances may be carried out to prevent sludges becoming obnoxious, or to give optimal conditions for biological growth. Iron compounds may be used to bind H_2S preventing odors and corrosion.

There are several additional processes coming into use in tertiary treatment applications, such as the use of ozone to remove reactive dyes, and other refractory organics. Advanced oxidation may be by means of chlorine and its compounds, ozone, or alternatively by means of UV-activated peroxide. The first is of declining popularity, and the latter are very expensive to operate.

There are also chemicals to inhibit H_2S formation, micronutrient preparations, and so on, which may be of greater or lesser value.

Dirty water unit operation design: biological processes

14

CHAPTER OUTLINE

An Overview of Biological Processes 172
   Aerobic Attached Growth Processes 172
   Aerobic Suspended Growth Processes 175
   Anaerobic Attached Growth Processes 176
   Anaerobic Suspended Growth Processes 176
   Combined Aerobic Treatment Processes...176
Secondary Treatment Plant Design 177
   Suspended Growth Processes 178
   Activated Sludge Physical Facility Design and Selection 178
   Activated Sludge Process Design 179
Trickling Filters 185
   Hydraulic Loading Rate...186
   Maximum Organic Loading Rate 187
Rotating Biological Contactors 188
Natural Systems 188
   Pond Treatment...188
   Constructed Wetlands 189
   Aerated Lagoons...190
   Stabilization Ponds..191
Tertiary Treatment Plant Design 191
   The Need for Tertiary Treatment..191
   Treatment Technologies..191
   Tertiary Treatment Plant Design 191
   Biological Nutrient Removal 192
Novel Processes 195
   Bio-Augmentation..195
   Deep Shaft Process 196
   Pure or Enhanced Oxygen Processes 196
   Biological Aerated Flooded Filters 196
   Sequencing Batch Reactors..197
   Membrane Bioreactors 197

Upflow Anaerobic Sludge Blanket Process 200
Anaerobic Filter Process...201
Expanded Bed Anaerobic Reactor 201
Two and Three Phase Anaerobic Digestion 201
Further Reading 202

AN OVERVIEW OF BIOLOGICAL PROCESSES

Biological secondary and tertiary treatment processes are used to remove most soluble and colloidal pollutants in municipal wastewater treatment plants. Experience at industrial effluent treatment plants shows that a very high proportion of manmade organic compounds can also be treated biologically.

The biological treatment process captures colloidal matter, and converts soluble (and some insoluble) organic compounds into suspended solids, carbon dioxide, and water. The biomass produced forms a high proportion of the outgoing suspended solids, and these suspended solids are said to be more "stabilized" or "mineralized" that the original feed materials. This implies that the solids are less susceptible to putrefaction than the parent material, and have a lower concentration of organic compounds. These suspended solids are ideally produced in a form readily separable from the surrounding liquid by settlement (or sometimes other processes). This solids separation stage is responsible for the actual production of effluent freed from its original organic load. Without this separation, reduction in load would be perhaps only 50% at best.

There are two main variants on biological treatment facilities:

- the presence or absence of air and
- whether the organisms used are attached to some substrate, or free-floating in the medium.

This leads to processes being called aerobic or anaerobic, suspended- or attached-growth processes (a further distinction may also be made, between batch and continuous processes). Table 14.1 sets out the advantages and disadvantages of anaerobic processes.

It is worth mentioning that, before the introduction of membranes into wastewater treatment, all available processes were at most refinements of systems that have been used for the better part of a century. Despite all claims to the contrary, there is very little novelty in effluent treatment.

AEROBIC ATTACHED GROWTH PROCESSES

Historically, the most important type of attached growth treatment process is the low-rate trickling filter (Fig. 14.1). This process is now well over a century old. A shallow bed of some porous media is built over a highly permeable underdrain

Table 14.1 Advantages and Disadvantages of Anaerobic Treatment

Advantages	Disadvantages
• Significant COD reduction for high strengths	• Generally higher capital cost
• Can deal with some compounds that aerobic cannot	• Effluent requires heating (in temperate climates) to around 37°C
• No aeration required—low energy costs	• Retention time of at least 1 day for traditional process
• Low sludge volumes and compact plant	• Intermediary products (H_2S) can be corrosive/toxic
• Biogas can be used as fuel	• Final effluent may need polishing to remove residual COD and hence meet consent
• Contained process—low odors/no aerosols	• Visual impact
• Not very susceptible to loading and environmental changes	
• Greenhouse gas reduction	
• Fertilizer product	
• Tax breaks and other incentives	

FIGURE 14.1

Low-rate trickling filter.

system. This underdrain system functions both as a collection system for treated effluent, and an entry point for the passive distribution of air. Effluent is spread over the media by some means, and a community of organisms grows up on the surface of the media to take advantage of the food source.

The growth will comprise several quite discrete layers of a mixture of bacteria, protozoa, fungi, and so on, with the outermost layer seeing the highest levels of nutrients and oxygen, and the layers closest to the media having the lowest. The filter will also have decreasing levels of nutrients from top to bottom of the structure. Decreasing levels of nutrients (and to some extent oxygen) at the support media surface eventually reach the point where the bacteria enter an endogenous phase, and lose their attachment, resulting in sloughing of the layer, along with all supported layers. Higher organisms, such as snails, worms, and insect larvae will also graze on these layers, ensuring a certain turnover and destruction of biomass. These processes ensure that much of the biomass is maintained in log phase growth. This may be undesirable on certain specialized types of filters, and desirable but absent on high-rate filters. As the system is unmixed, stratification of species occurs, both within the layers on the media particles, and from top to bottom of the bed.

The effluent from the filter is passed through a settling tank, known as a humus tank, where biological solids are removed prior to discharge. Some of the clarified effluent may be recycled to dilute incoming wastewater, or simply to maintain the bed in a wetted condition.

There are several modern variants of the traditional rock-filled trickling filter design, using plastic media, powered distributors, and sometimes forced aeration for high-rate treatment.

High rate filters (also called roughing filters) are usually very tall tower type systems, filled nowadays most often with structured or random plastic packing. Their high effluent flowrates, often through powered distributors, mean that biomass sloughing is almost continuous. Typical performance data for high-rate filters are given in Fig. 14.2.

Rotating biological contactors (see later) are a variant on the system, with a rotating plastic media pack and sometimes forced aeration.

A more recent variant on the system is the biological aerated flooded filter, where the system is filled with water rather than air, and aerated like a suspended

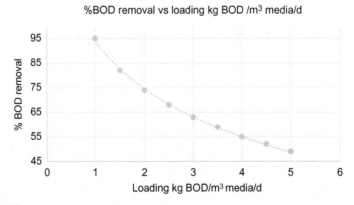

FIGURE 14.2

Typical performance chart for high-rate biofiltration.

growth process. This system also offers some solids removal by filtration, and hence disposes of the humus tank requirement.

AEROBIC SUSPENDED GROWTH PROCESSES

Aerobic suspended growth processes are most commonly used in sewage treatment, and include the activated sludge (AS) system and its many variants. Like the previous process, it was originally developed in England, in the early 1900s.

The central feature of the aerobic suspended growth process is the collection and recirculation of the biomass (or "AS") collected by settlement from the outgoing treated effluent for mixing with the incoming wastewater to be treated. This allows the age of the organisms within the sludge to be manipulated independently of the system hydraulic residence time. The standard form of the system approximates complete mixing in the aeration tank. Separate compartments must be supplied if it is desired to have more than one regime of substrate or oxygen concentration, for example for denitrification.

There are very many types of AS process, but all are just variations on a theme. Sequencing batch reactors (SBRs) are one of the more recent variants, operating on a batch basis rather than the conventional continuous one. Membranes may replace gravity settlement and return of AS in membrane bioreactors (MBRs), but other than higher achievable biomass concentration in the aeration basin, nothing new is happening. Pure oxygen is easier to transfer to water than air, and producers of oxygen are understandably keen to sell people pure oxygen fed systems. When such variant systems are new on the market, many claims are made that they will perform in some way differently from older systems. Such claims tend to become much more modest as experience of the system grows.

One of the claims often made for new processes is for very low sludge yields because of an enhancement of endogenous respiration. This is clearly very desirable, as it reduces the requirement for sludge treatment plant. It is common for zero sludge yields to be claimed. To the best of my knowledge, no zero-yield claim has ever been substantiated. Sludge yield is mainly a function of sludge residence time. The longer the residence time of the organisms, the greater the degree of endogenous respiration. This factor is independent of whether the microorganisms are in a standard system or some modern variant.

Aerated lagoons are a development of simple pond treatment systems, usually with the inclusion of surface aeration systems. They are essentially identical to an AS system with a high retention time, such as the extended aeration system.

Aerobic sludge digestion is simply the aerobic biological treatment of produced sludges. It uses the enhancement of endogenous respiration to reduce the solids level of the sludges. It is a highly energy intensive process compared with the anaerobic alternative.

ANAEROBIC ATTACHED GROWTH PROCESSES

Unlike the aerobic processes, attached growth processes are relative newcomers in anaerobic treatment.

The simplest and most robust system is the anaerobic filter, a flooded filter, usually operated in upflow, such that microorganisms are retained within the bed. It may however be operated in downflow for high suspended solid applications. The support media are made from materials such as ceramics, minerals, glass, plastics, or wood. The reactor volume required may be 2–5 days' hydraulic retention time, less than the 5–14 days minimum requirement for the conventional continuous stirred tank reactor (CSTR) system. They are mainly used to treat food and distillery effluents, although there is some US experience with high strength organic chemical wastes.

A more recent technology is the expanded bed reactor; with claimed retention times for 90% chemical oxygen demand (COD) removal of 0.5–3 days. These should be considered with some caution, as there are few full-scale examples at present. The far finer media within this type of reactor make it intolerant of high suspended solid concentrations.

ANAEROBIC SUSPENDED GROWTH PROCESSES

The original anaerobic treatment technology—the covered lagoon—is rarely used nowadays because of environmental and efficiency considerations. Retention times may be measured in weeks or months. A variant of this, where the covered lagoon is mixed, is basically operated in a similar way to a CSTR, but with lower capital cost.

The most common type of suspended growth anaerobic reactor is the CSTR. Despite the name, operation need not be continuous, and mixing is usually achieved via gas recirculation. The unit's lack of any internal media or baffles means very high suspended solid levels may be tolerated. COD loadings are modest, and hydraulic retention times are of the order of 5–14 days, so the reactors tend to be relatively large. This is a well-established and robust design.

The upflow anaerobic sludge blanket (UASB) reactor developed more recently is an advanced system. It can give improved performance, but it is relatively difficult to set up and operate. Effluent is injected at the bottom of the reactor, passes through a dense bed of sludge, and through integral three-phase separators before being sent to disposal. No additional mixing is required. Getting the sludge to granulate and settle properly is crucial and inline equalization may also be necessary, as UASB reactors need a fairly constant feed composition (see also Chapter 24: Sludge Treatment Unit Operation Design: Physical Processes and Fig. 24.3).

COMBINED AEROBIC TREATMENT PROCESSES

There are several processes that combine features of AS and filter designs. In summary, these usually consist of an AS plant in series with a biofilter of some sort.

The two processes may be in close association, as in the activated biofilter process, or simply consist of an AS plant polishing the effluent from a "roughing" filter. In the latter case, it is simply a case of integrating the design of the two processes.

The activated biofilter process claims great improvements on conventional filter loadings, with AS being fed into the filter influent to boost available biomass. Examples of the process are however rare.

SECONDARY TREATMENT PLANT DESIGN

Secondary treatment is always biological treatment, simply because there is no cheaper way to remove dissolved and colloidal matter than by getting microorganisms to do the job.

The most popular biological treatment process worldwide is the AS process, invented over a century ago in Manchester, in the United Kingdom, but informed by experiments carried out at the Lawrence Experimental Station in Massachusetts, the United States.

Since the secondary treatment of sewage was invented in the temperate climate of Britain and the Northern United States, it is traditionally an aerobic treatment. It might be argued that, had it been invented in a tropical climate, the technology might have progressed differently, since anaerobic treatment is far more energetically favorable with a warmer effluent.

The AS process involved the aeration of a suspension of microorganisms, but there were already techniques involving the use of organisms attached to substrates, dating all the way back to the bucket-fed sewage farms of the Middle Ages.

Sewage farms are still in use today in arid developing countries, but intensified attached growth processes such as the trickling filter were developed at least as long ago as the AS process.

From a chemical engineer's point of view, the AS process is a CSTR with recycle, and the trickling filter a plug flow reactor (sometimes with recycle).

While this gives a useful insight into the design and operation of these plants, what happens inside them is irreducibly complex. It is practically impossible to design them from first principles, and it is pointless to try. Designers always refer to a set of heuristics developed through the long experience of engineers with the design of municipal facilities. In all cases, the inherent advantages, and disadvantages of attached and suspended growth processes are constant (Table 14.2).

The demands of fashion and "marketing breakthroughs," however, mean that these two basic processes are constantly being repackaged, and sold as if they were something new. The first few installations of these "new" variants ignore the longstanding heuristics, but it then becomes clear that, to paraphrase a famous fictional engineer, "ye cannae ignore the rules of sewage treatment." The traditional heuristics may be conservative, but they are robust.

Table 14.2 Comparison of Attached and Suspended Growth Processes

	Attached Growth	**Suspended Growth**
Energy requirements	Lower	Higher
Operability	Easier	Harder
Maintenance requirements	Lower	Higher
Robustness	Higher	Lower
Fly and odor nuisance potential	Higher	Lower
Footprint	Higher	Lower
Optimum size	Lower	Higher

Even in those very rare situations when there has been a genuine innovation, such as the MBR, it turns out that tradition must guide its design. Early MBR designs with aeration basins designed based on laboratory scale work were woefully inadequate. At the end of the day, an MBR is an AS plant which achieves solids retention with a microfilter in place of the settlement tank. It has taken us since 1962 (the date of its first commercial scale installation) to learn how little was genuinely new in the case of the MBR. This is the reason experienced designers are so conservative.

SUSPENDED GROWTH PROCESSES

There are essentially two aerobic suspended growth processes, namely the AS process and the MBR. Each of these have anaerobic equivalents. There are also anaerobic designs without the solids retention essential to AS and MBR.

Let us first consider the world's most popular (if not always most successful) sewage treatment process: the AS process.

ACTIVATED SLUDGE PHYSICAL FACILITY DESIGN AND SELECTION

The AS process basically comprises mixing incoming effluent with a stream of concentrated microorganisms collected from the outgoing treated effluent, and then aerating the mixture. The mixture is referred to as mixed liquor. Aeration of this mixture allows the microorganisms to feed on the soluble and insoluble organic constituents of the wastewater, leading to their incorporation into the bodies of the organisms, or their breakdown into CO_2, water, and so on. The process has been in use for a long time, and there are many variants on the basics. All of them are however at bottom alike in their mode of action.

The detailed selection of the aeration devices, blowers, pumps, and solids separation devices that make up an AS plant (Fig. 14.3) is usually the responsibility of the contractor who is to build the works.

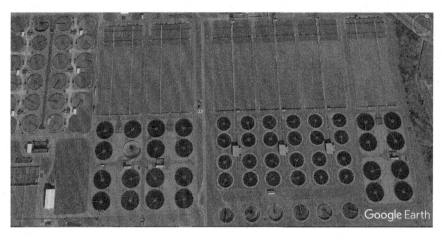

FIGURE 14.3

AS treatment at Minworth Sewage Treatment Works, the United Kingdom.

Courtesy: Google Earth 2017.

Table 14.3 AS Design: Rules of Thumb

Efficiency of aeration	High: fine bubble porous diffusers of all types Medium: jet aeration; surface aerators Low: coarse bubble perforated aerators of all types; propeller aerators
Blower selection	Roots-type positive displacement blowers are used on smaller works, but higher efficiencies are available on centrifugal blowers at around 5000 m³/h free air delivery (FAD)
Primary settlement tank	Size may be estimated by allowing an upflow velocity of 1.5 m/h
Settlement tanks for secondary settlement of activated sludge	Size may be estimated at an upflow velocity of around 0.9 m/h

Table 14.3 gives some rules of thumb that might prove useful in assessing budget and conceptual stages of design.

ACTIVATED SLUDGE PROCESS DESIGN

If you are actually designing the process, further data will be required. Having assessed the load to be treated, as described earlier, many parameters must be determined. These are usually as follows.

Yield (aka sludge production factor)

Yield (Y) is determined using Eq. (14.1).

Sludge production factor:

$$Y = \frac{\text{kg sludge produced}}{\text{kg BOD removed}} \tag{14.1}$$

Table 14.4 gives typical figures for Y for domestic sewage to allow sludge production to be calculated.

Sludge loading

The sludge loading ratio for the aeration tank, also known as the F:M ratio, is determined using Eq. (14.2).

F:M ratio:

$$F : M \text{ ratio} = \frac{\text{Mass of biochemical oxygen demand (BOD) applied per day}}{\text{Mass of mixed liquor suspended solids (MLSS)}} \tag{14.2}$$

Fig. 14.4 illustrates the relationship between sludge yield and F:M ratio.

The F:M ratio should usually be in the range 0.2−0.4. Values of 0.2−0.25 should be used where year-round nitrification of domestic sewage is required. These values may be halved for industrial effluent.

Table 14.4 Values of Sludge Production Factor Y

SS:BOD	≪0.5	0.5	0.6	0.7	0.8	0.9	1.0	1.1
Conventional	1.0	1.1	1.1	1.1	1.2	1.2	1.2	1.2
Nitrifying	0.8	0.9	0.9					
Extended aeration	0.7	0.8	0.8	0.8	0.9	0.9	0.9	0.9

SS, incoming suspended solids; BOD, incoming BOD.

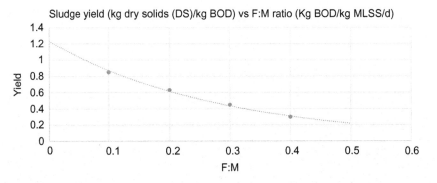

Sludge yield (kg dry solids (DS)/kg BOD) vs F:M ratio (Kg BOD/kg MLSS/d)

FIGURE 14.4

Sludge yield vs F:M ratio.

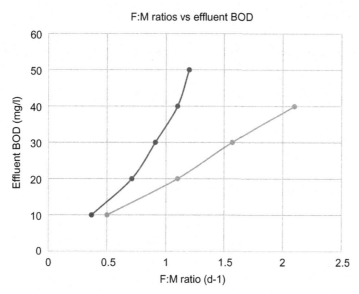

FIGURE 14.5

Relationship between F:M and effluent BOD.

Extended aeration systems have F:M ratios of up to 0.1, nitrifying systems 0.1–0.25, and carbonaceous-only systems 0.25–0.5.

There is an approximate relationship between F:M and effluent BOD for an incoming sewage BOD of 250 mg/L to the aeration tank as shown in Fig. 14.5. The steeper of the two curves is at 12°C, the shallower at 17°C.

Maintaining mixed liquor suspended solid concentrations of the order of 2500–3500 mg/L is usually desirable, mainly because of limitations on final settlement tank design. I have however seen AS plants run at 10–12,000 mg/L MLSS, with an aeration tank which had the appearance of boiling mud. They performed better than one might expect, though they were far from compliant with their specified treated water quality.

Tank volume

Having fixed these parameters, the aeration tank volume required may be calculated using Eq. (14.3).

Tank volume calculation:

$$V(m^3) = \frac{BOD(kg/day) \times 1000}{F : M \times MLSS(mg/L)} \qquad (14.3)$$

Whatever the results of this calculation, the minimum volume used should be equivalent to 8 h hydraulic retention time.

Tank geometry

The dispersion number method (Eq. 14.4) may be used to determine tank geometry. A long aeration tank may be thought of as a series of square tanks N in number.

Dispersion number method:

$$N = 7.4L \ Qs(1 + R)/WH \tag{14.4}$$

where

L = tank length (m)
Qs = average incoming flow rate (m³/s)
R = recycle ratio (RAS flow/Qs)
W = tank width (m)
H = tank depth

L and W should be manipulated such that N lies in the range 8−12. This will give a plug flow geometry favoring good sludge settlement.

Oxygenation capacity

Design oxygenation capacity (the required quantity of oxygen to dissolved per volume of water per hour) varies according to the F:M ratio and nitrification/denitrification requirements. A summary of typical values is given in Table 14.5.

Oxygen requirements

Oxygen requirements vary according to several factors. Oxygen requirements for BOD removal may generally be calculated using Eq. (14.5). Alternative approaches are given in Table 14.6 and Fig. 14.6.

Oxygen requirement calculation:

$$R = aB + bMV \tag{14.5}$$

where

a and b are constants (0.75 and 0.05, respectively, for F:M ratios of 0.1−1.0)
R = daily oxygen requirement (kg/day)
B = BOD removal (kg/day)
M = MLSS (kg/m³)
V = aeration tank volume (m³)

Table 14.5 Typical Total Oxygenation Capacity

Description	F:M/day	Total oxygenation capacity
Carbonaceous only	0.25	1.49 BOD
Nitrifying	0.12	1.75 BOD + 5.4 NH$_3$-N
Nitrifying/denitrification	0.12	1.75 BOD + 3.6 NH$_3$-N
Nitrogen and phosphorus removal	0.12	1.75 BOD + 3.1 NH$_3$-N

Table 14.6 Oxygenation Efficiency for Common Devices

Aeration System	Oxygenation Efficiency (kg/kWh)
Fine bubble diffused air	1.5–3.6
Coarse bubble diffused air	0.9–1.2
Vertical mechanical surface aerator	1.5–2.2
Horizontal mechanical surface aerator	1.2–2.4

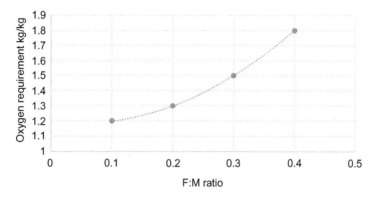

FIGURE 14.6

Oxygen requirements vs F:M ratio.

Where nitrification is to be carried out, 4.3 kg/day of oxygen per kg ammonia oxidized shall be added to the figure calculated by equation 14.5. It is normal to multiply the product of these two calculations by a peaking factor of 1.5 to compensate for variations in loading.

There is however a variation in oxygen requirement proportional to F:M ratio (Fig. 14.6).

When it comes to calculating how much oxygen must be dissolved to meet the oxygen demand of the bugs, things become more complex still.

The normal chemical engineering approach of analyzing a flux of something over a specified area falls down due to an unavoidable lack of knowledge of the true coefficient of gas transfer, and the true area over which gas transfer is taking place. These two terms are consequently in practice rolled up into a single term K_La, the overall mass transfer coefficient, with units of s^{-1}. This is covered in detail in any academic textbook on the subject, but an outline of the key points follows.

K_La may be determined empirically for a system using a clean solution of sodium sulfite, and the results of this test adjusted to account for such factors as the differential solubility of oxygen in wastewater, temperature, system geometry, and wastewater impurities. The last two correction factors are known as α and β.

$\alpha = K_La$ in process water/K_La in clean water and
$\beta =$ saturation concentration in process liquor/saturation concentration in clean water.

α varies between 0.3 and 1.2, and β between 0.7 and 0.98. Values of 0.75 and 0.95, respectively, are most commonly used for design in the absence of laboratory data.

The correction factors are applied under normal operating pressures as given in Eq. (14.6).

Determination of mass transfer coefficient:

$$N = K_La\alpha N_0 \frac{(\beta Cs - C)}{Cs} \tag{14.6}$$

where

$N =$ transfer rate (kg/h)
$N_0 =$ standard transfer rate ($= K_LaCsV/1000$)
$Cs =$ saturation oxygen concentration (mg/L)
$C =$ concentration with respect to time (mg/L)

The power required to achieve this transfer will depend on the aeration system being used. Oxygenation efficiency for several common devices is shown in Table 14.6.

Sludge age and mean cell residence time

The conversion of ammonia (NH_3/NH_4^+) to nitrate (NO_3^-) which occurs during an AS process is directly proportional to sludge age. Sludge ages typically vary in practice from 3 to 30 days, and are calculated as in Eq. (14.7).

Sludge age calculation:

$$\text{Sludge age (days)} = \frac{\text{Total mass of MLSS in aeration tank (kg)}}{\text{Total suspended solids (TSS) entering aeration tank (kg/day)}} \tag{14.7}$$

Mean cell residence time (MCRT) estimates the time the average microorganism will stay in a system, and is preferred by some as a measure of the age of the biomass in a system. As with sludge age, more is better when it comes to degree of nitrification.

MCRT also has units of days, typically varying in practice from 5 to 15 days, and is calculated as follows:

Mean cell residence time:

$$\text{MCRT (days)} = \frac{(\text{MLSS (kg) inaeration} + \text{MLSS (kg) inclarifier})}{(\text{SS wasted (kg/day)} + \text{effluent SS (kg/day)})} \tag{14.8}$$

Returned sludge

A proportion of the secondary sludge must be returned (usually continuously) from the clarifier to the aeration tank to maintain the required MLSS concentration.

The required sludge return rate is normally calculated using one of three methods.

MLSS Method

The MLSS method is based on the SS of the mixed liquor (MLSS) and the settled sludge, and is calculated using Eq. (14.9).

MLSS method:

$$\% \text{ Return sludge} = \frac{(\text{MLSS (mg/L)} \times 100)}{(\text{Return sludge SS (mg/L)})} - \text{MLSS (mg/L)} \qquad (14.9)$$

(To convert to an actual RAS flowrate, multiply the daily flow to aeration by this percentage.)

Settleability Method

The settleability method is based on a 30-min settleability test, and is often less accurate than the MLSS method due to the limitations of the test. It is calculated using Eq. (14.10).

Settleability method:

$$\frac{\% \text{ Return sludge} = \text{settled sludge @ 30 minutes (mL/L)} \times 100}{(1000 - \text{settled sludge @ 30 minutes (mL/L)})} \qquad (14.10)$$

(As mentioned earlier, to convert to an actual RAS flowrate, multiply the daily flow to aeration by this percentage.)

Sludge Volume Index Method

The Sludge Volume Index (SVI) method (Eq. 14.11) is similar to the MLSS method but uses the SVI to estimate the return sludge SS concentration.

SVI method:

$$\text{SVI} = \frac{\text{Settled sludge volume (mL/L)} \times 1000}{\text{SS (mg/L)}} \qquad (14.11)$$

TRICKLING FILTERS

Trickling filters (Fig. 14.7) tend to be designed based on a balance between several parameters. For conventional low-rate filters, the most common parameters used are volumetric hydraulic and organic loading. The method of operation sets these parameters.

These filters can be operated in single filtration mode, with sewage passing through the filter once. Alternatively, for stronger effluents, they may be run in recirculation or in alternating double filtration modes. In recirculation, incoming effluent is added to a constantly recirculating flow through the filter. In

FIGURE 14.7

Trickling filters: Matlock STW.

Courtesy: Google Earth 2017.

Table 14.7 Organic Loading Figures

Single filtration	0.10–0.12 kg BOD/media m³/day
Recirculation	0.15–0.18 kg BOD/media m³/day
Alternating double filtration	0.20 kg BOD/media m³/day
Design hydraulic loadings for this system (sewage m³/media m³/day)	
Single filtration	0.5
Recirculation	1.2
Alternating double filtration	1.8

alternating double filtration, two filters are used in series, with the order being changed from time to time.

Organic loading figures commonly used for bed depths of approximately 2.0 m, with traditional media of size 40–60 mm are given in Table 14.7.

An absolute minimum hydraulic loading of 0.25 m³/m³/day should be maintained.

The associated humus tanks are designed based on a HRT of 5 h at dry weather flow (DWF), and 1.2 m/h upflow velocity.

Higher rate filters tend to be designed based on data obtained from the media manufacturers, and may also require proprietary items such as powered distributors.

HYDRAULIC LOADING RATE

Hydraulic loading rate (HLR) is the ratio of volumetric flowrate to total filtration media volume (both first and second stages in a double filtration configuration).

Table 14.8 Design Parameters: Stone Media: hydraulic loading rate (HLR) at dry weather flow (DWF)

Regime	Media Size (mm)	Effluent Quality (95 %ile)			HLR at DWF
		SS mg/L	BOD mg/L	NH$_3$ mg/L	
Single filtration	40	30	20	6	0.4
Double filtration	65/30	30	20	7	0.7

Table 14.9 Design Parameters: Maximum Organic Loading Rates (MOLR) for Stone Media

Media Size (mm)	MOLR (kg BOD/m^{3}media/day)
30	0.10
40	0.13
50	0.20
65	0.25
100	0.50

It is a process-critical design parameter, and is therefore usually the first thing to be established by the designer, in an iterative process with organic loading rate. The design parameters given in Table 14.8 would be expected to produce a design meeting a Royal Commission (20:30:10 BOD:SS:NH$_3$) effluent quality.

MAXIMUM ORGANIC LOADING RATE

After an initial trial HLR has been chosen, the organic loading rate on the media is checked, based on a range of acceptable media sizes, as the acceptable organic loading rate varies with media size.

If a suitable media size for the type of effluent is not selected, problems such as ponding will result. Care should therefore be taken with the quoted specific surface area of smaller media; the high values often quoted are only applicable if organic loads are not too high. Excessive organic loads on media can block the channels though which effluent flows, so the large specific surface area of small media will be negated and ponding will result.

The maximum loadings given in Table 14.9 should not be exceeded for stone media.

If initial design exceeds maximum loading, the design should be retried with a larger media and consequent variation in HLR until a design which satisfies both HLR and MOL is obtained.

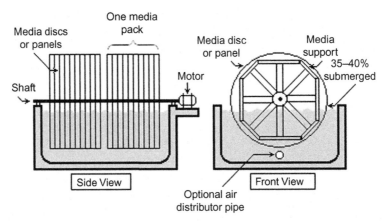

FIGURE 14.8

RBC (schematic).

FIGURE 14.9

A small RBC.

ROTATING BIOLOGICAL CONTACTORS

Rotating biological contactors (RBCs) (Figs. 14.8 and 14.9) tend to be entirely proprietary items. Design parameters tend to be based on loading per plate surface area. Usually, the most important design item is not a process consideration, but rather the shaft and its associated bearings, which are responsible for most problems with RBC operation.

NATURAL SYSTEMS

POND TREATMENT

Pond systems vary from what are almost natural wetland systems, through to what are slightly crudely constructed variants on conventional high-rate treatment systems.

Aerobic systems may be created which require no artificial aeration, with photosynthesis providing the system's oxygen requirements. By simply deepening the system, a lower facultative/anaerobic level is created which provides additional treatment in the absence of oxygen underneath the aerobic surface zone.

Substantial deepening of the ponds results in a pond that is anaerobic for virtually all its depth, as discussed in the suspended growth anaerobic treatment section.

CONSTRUCTED WETLANDS

Although completely natural treatment is the oldest treatment of all, there has been a recent upsurge in interest in natural treatment technologies such as managed or constructed wetlands, or overland systems.

Overland systems were some of the earliest attempts to enhance treatment of sewage prior to discharge. Modern systems have evolved from simple spreading over fields, to address the problems of odor and spreading of debris.

Nowadays, overland systems have constructed slopes and arrangements for collecting surface runoff. Some preliminary treatment is provided to prevent debris spreading and the blocking of spray nozzles. Soil types also tend to be less permeable than in the past.

Slow rate systems use intermittent application to vegetation via furrows or spraying after some preliminary treatment. The intermittent operation keeps conditions aerobic, suppressing odor formation. These options are relatively low rate, and lose a significant proportion of the water directly by evaporation and by transpiration through the plants. Required soil types are of intermediate permeability.

Retaining a higher proportion of the water, for example in aquifer recharge, requires high-rate systems such as the rapid infiltration system. In this, water that has had some preliminary treatment is applied to infiltration basins overlying highly permeable soils, and either allowed to reach groundwater or recovered via an underdrain or well system. It is unusual for there to be vegetation in the infiltration beds. Treatment occurs as the water percolates through the soil, but the rapid infiltration means that these systems have less volumetric treatment capacity than the systems described earlier.

Constructed wetlands are usually made from *Phragmites* or local reeds, grown over beds of gravel, or sometimes soil. There are also systems based on water hyacinths. The system may be filled with water such that the gravel is submerged, or the water surface may be below the gravel surface, which suppresses mosquito nuisance. Treatment is achieved by means of both microorganisms within the water, attached to roots and gravel, and by the reeds and other aquatic plants. The plants provide oxygen to keep the system aerobic. It is occasionally necessary to supplement this with artificial aeration. Odor and mosquito problems may occur if the system is not kept aerobic.

Reed beds and other constructed wetlands are a versatile technology. They can be used for complete treatment, but are most commonly used as a simple reliable tertiary treatment stage, giving a high degree of solids, P (70%) and N removal.

They take up a lot of land, but have few moving parts, making them well suited to provide a final stage of treatment, after package treatment plants such as RBCs, in locations with unreliable power supplies.

They can be sized approximately using the Grant and Moodie equation (Eq. 14.12).

Grant and Moodie Equation:

$$A = 3.5P^{0.35} + 0.6P \tag{14.12}$$

where

A = total first stage area (m^2)
P = population equivalent (number of people served).

In the United Kingdom, Eq. (14.13) has also been used to determine bed area (Cooper et al., 1996).

Alternative bed sizing calculation:

$$Ab = Qd(lnC0 - lnCt)KBOD \tag{14.13}$$

where

Ab = surface area of bed (m^2)
Qd = average daily flow rate of wastewater (m^3/day)
$C0$ = daily average BOD5 of influent (mg/L)
Ct = required daily average BOD5 of effluent
$KBOD$ = rate constant (m/day)

KBOD (reported in Cooper (1999)) has been measured as:

- 0.0813–0.1 for 49 secondary systems in Denmark
- 0.06–0.1 for secondary systems in the United Kingdom
- 0.31 for tertiary systems

AERATED LAGOONS

In the case where solids recycle is carried out, aerated lagoons may be considered as irregularly shaped AS systems. Settlement facilities will therefore need to be provided to remove biomass for recycle in just the same way as in conventional AS systems. These systems tend to be shallow basins like the lagoon in construction, rather than conventional settlement tanks.

Aerated lagoons without solids recycle are also used, and consideration of this lack of recycle must be made in design.

STABILIZATION PONDS

In small communities, effluent may be treated by means of simple ponds. These take a high land area per population equivalent treated, but have very low energy requirements, and usually unsophisticated construction. They vary in the proportions of aerobic and anaerobic treatment intended. Larger plants may have many of the preliminary treatment stages of more sophisticated works.

TERTIARY TREATMENT PLANT DESIGN
THE NEED FOR TERTIARY TREATMENT

The sensitivity of watercourses receiving effluent discharges to residual BOD, suspended solids, and inorganic nutrients from discharges of treated wastewater can be variable. In the worst cases, these constituents may cause uncontrolled microbial growth that stifles all life in the body of water. Manmade organic compounds and viable pathogens may contaminate water abstracted for drinking purposes, where discharge has been to surface or groundwater.

TREATMENT TECHNOLOGIES

Tertiary treatment technologies include the following, which can be used alone or in combination:

- AS
- Biological nitrification
- Biological phosphorus removal
- Carbon adsorption
- Denitrification
- Dosing with metal salts
- Granular media filtration
- Reed beds

Metcalf and Eddy provide a useful indication of performance for these technologies.

TERTIARY TREATMENT PLANT DESIGN

Tertiary treatment processes are more commonly proprietary than secondary treatment processes, usually being newer (or at least new variants on old processes). Secondary treatment was developed in large part to deal with the 1912 Reports of the UK Royal Commission. This stated that treated domestic and industrial effluent treatment for discharge should contain no more than 20 ppm BOD, 30 ppm suspended solids, and 20 ppm ammonia. These reports are the source of the conventional division of treatment into primary and secondary stages, but in 1912 a requirement for tertiary treatment was simply not envisaged.

It is possible to obtain far higher performance from conventional processes than the Royal Commission standard, but effluent treatment processes are being tasked with meeting higher and higher standards of treatment over time, and we are already at a point where tweaks to conventional secondary treatment cannot always meet regulatory requirements.

It is increasingly common to see enhanced requirements for solids removal, intended to maintain the clarity of the body of water.

However, other water quality concerns usually drive tertiary treatment requirements, especially concerns about "eutrophication," in which there is an explosion in the numbers of microorganisms in the body of water receiving treated effluent, supported by nutrients in the effluent. The oxygen demands of these organisms then make the body of water become anaerobic, killing the aerobic organisms which are present.

The two nutrients most commonly targeted are nitrates and phosphates, essential for the growth of the organisms responsible for eutrophication. By restricting these key nutrients, we can prevent eutrophication.

BIOLOGICAL NUTRIENT REMOVAL

It is now common for limits to be set on the concentrations of nitrogen and phosphorus which may be discharged to watercourses.

It is possible to manipulate oxygen and nutrient levels such as to enhance uptake of nutrients by biomass in biological treatment processes.

Biological nitrogen removal

The commonest compound of nitrogen in effluent coming into a sewage works is ammonia. The first stage of removal is nitrification, in which the ammonia is converted to nitrate by biological action. This occurs in biological systems having sufficient aeration capacity and biomass, known as single stage nitrification. Alternatively, additional systems may be provided to carry out the process separately.

The nitrate formed can be converted to nitrogen gas, and completely removed from the system by means of denitrification. It is necessary to exclude oxygen for the process to occur, in what is termed anoxic conditions.

Properly designed conventional AS plants can oxidize ammonia and other nitrogenous compounds to nitrites and nitrates to a high degree to meet stringent ammonia standards of 5 mg/L.

Sometimes this "nitrification" stage is achieved in a separate nitrifying filter, which is commonly considered to be tertiary treatment. As such filters are plug flow reactors, they can achieve higher degrees of treatment than the CSTR of the AS process; 1 mg/L NH_3-N is a common standard.

Nitrates and nitrites can however cause eutrophication so, in sensitive receiving waters, total nitrogen standards can be applied, requiring the removal of nitrates and nitrates as well as ammonia.

One way of doing this is to add a "selector" or "contact" tank upstream of the aeration basin. This is essentially an unaerated tank of 5%−15% of the volume of the aeration basin in which a 100% recycle of treated effluent from the settlement tank is combined with incoming primary treated effluent. This creates an anoxic environment with high BOD and nitrate concentrations which favors those organisms which can obtain their oxygen from nitrate. These organisms "denitrify" the effluent, turning nitrate into gaseous nitrogen which is removed in the aeration basin. The high F:M ratio produced in the selector zone also discourages the growth of poorly settling filamentous organisms.

In the commonplace absence of laboratory test data, allowing a volume based on a HRT of 15 min at DWF will be likely to be adequate for a selector zone. Since the mixing effect of aeration devices will be absent, it should be baffled or mechanically mixed to keep solids in suspension.

This is a well-established process, known to all competent AS process designers, though sophisticated variants on it to be discussed later are proprietary technology.

Dedicated denitrifying filters may alternatively be used to remove nitrates.

It is possible as demonstrated earlier to design additional capacity into conventional AS or attached growth biological treatment plants in order to nitrify ammonia to nitrate. Complete removal of nitrogen requires denitrification of this nitrate to nitrogen. Again, this can be designed into a conventional AS plant as reasonably minor modifications, providing anoxic conditions for the returned sludge addition, and possibly addition of an external carbon source.

Alternatively, the processes may be split, with carbonaceous removal, nitrification and denitrification in separate stages. The separate nitrification reactor may utilize suspended or attached growth of microorganisms.

The 4-stage Bardenpho process, developed in South Africa, is an example of a process where specific areas have been designed to carry out certain processes. Separate reactors carry out carbon removal, nitrification, and denitrification. More simply, an oxidation ditch with areas free of aeration, and controlled influent addition will carry out these stages in a slightly less well-defined way, giving less efficient removal per unit volume than the 4-stage Bardenpho process.

Design of Nitrifying Biological Filters

Nitrifying filters, often called nitrifying tricking filters or NTFs are low rate dedicated processes which oxidize ammoniacal nitrogen (NH_3-N) to nitrate or nitrite, installed downstream of AS or biofiltration plants which do not give sufficient nitrification to meet consent.

The design data given later can be used where the BOD load to the nitrifying filter is less than 30 mg/L, 30 mm stone media (or plastic equivalent) is used, the filter is well ventilated, and there is sufficient alkalinity present. Depths above 4 m are impractical.

Table 14.10 Design Parameters for Nitrifying Trickling Filters

Irrigation Velocity (m/day)	Required Depth of Filter	
	Incoming NH$_3$-N = 20 mg/L	Incoming NH$_3$-N = 15 mg/L
5	3.6	2.7
10	4.2	3.1
20	4.9	3.6
30	5.9	4.4
40	6.8	5.0

Nitrification is temperature- (and therefore often season-) dependent and the design data given are "worst case," for 2 mg/L effluent ammonia as N in temperate climates. These design figures should also give a BOD reduction of around 25%–30%, and SS reduction of 15%–30%, but these reductions should be ignored in any downstream design, as they are not reliable.

Table 14.10 allows required NTF depth to be established for a given incoming ammonia level and irrigation velocity. The design ammonia level should include consideration of recirculation (where used).

Recirculation can be used to control peak influent NH$_3$ concentration. Required recirculation rates can be calculated using Eqs. (14.14a) and (14.4b).

NTF recirculation rates (a) where influent NH$_3$ including recirculation = 20 mg/L:

$$R = \frac{Ni - 18}{20} \tag{14.14a}$$

NTF recirculation rates (b) where influent NH$_3$ including recirculation = 15 mg/L:

$$R = \frac{Ni - 13}{15} \tag{14.14b}$$

where

R = recycle ratio, i.e. $\frac{\text{recycled flow}}{\text{influent flow}}$
Ni = incoming NH$_3$ concentration (mg/L)

The area required may then be calculated using Eq. (14.15).
NTF area calculation:

$$A = \frac{DWF[1 + R] \times [1.25]}{IV} \tag{14.15}$$

where

A = filter plan area (m^2)
R = recycle ratio
DWF = dry weather flow (m^3/day)
[1.25] = average flow/dry weather flow ratio (assumed)
IV = irrigation velocity (m/day)

Biological phosphorus removal

Conventional AS plants remove around 50% of incoming phosphate, making 2 mg/L in the treated effluent readily achievable. Regulators are however setting ever more stringent targets for phosphorus removal. Meeting standards of 0.5−1 mg/L require 90% or more phosphorus removal, and even lower standards, down to 0.1 mg/L, are being discussed.

Certain microorganisms will remove phosphate from solution and store it in crystalline form within themselves under anaerobic (rather than anoxic) conditions, a process known as luxury phosphate uptake. They will release the stored phosphate again in aerobic conditions in the presence of fatty acids.

By manipulating how anaerobic the environment is, along with the concentrations of fatty acids, we can concentrate phosphate in biological or chemical sludge for disposal, or even recover it in a usable form. Plant designers will normally be buying in the treatment technology for these processes, rather than designing it from scratch.

There are several commercial processes that provide the alternating anaerobic and aerobic conditions necessary for enhanced phosphorus uptake. Examples of these are the A/O process, the Phostrip process, and others using a SBR.

The A/O process is mainstream removal, while the Phostrip process uses the addition of fatty acids to produce a concentrated side stream of phosphate. The sequence of operations to obtain phosphate removal with an SBR is fill, anaerobic stir, aerobic stir, anoxic stir, settle, and decant.

Biological combined nitrogen and phosphorus removal

There have been several sophisticated developments (mainly in South African universities) of the AS process. These involve many stages with controlled conditions allowing biological removal of both nitrate and phosphate in a single plant. These are all proprietary processes, so the overall plant designer's responsibilities will lie in specification and accommodation of the requirements of the process supplier.

There are several commercial processes that combine the biological treatment of both nitrogen and phosphorus. The 5-stage Bardenpho and A2/O processes are most common. These are essentially modifications of the 4-stage Bardenpho and A/O processes previously described. The Bardenpho process is modified by the addition of a terminal anaerobic stage, and the A2/O includes an anoxic zone treating recycled mixed liquor. Less common variants are the UCT and VIP processes, essentially modifications of the commoner processes.

NOVEL PROCESSES
BIO-AUGMENTATION

There are many companies offering cultures of microorganisms for enhancement or augmentation of biological treatment plants. However, in almost all situations,

I would consider these cultures to be worthless. Whatever the composition of the effluent, and no matter how unusual the chemicals which it contains, if there are microorganisms which can treat it, they will arise naturally in a well-designed system without the necessity to purchase them. Suppliers of these products will tend to claim that, while the above may be true for all other products, their product is different from the others in some significant way. In my experience, this has never proven to be the case upon rigorous investigation.

The only occasion when products of this type may be useful is when a system's biomass has been entirely lost, and it is not possible to obtain viable biomass from another plant.

DEEP SHAFT PROCESS

ICI first developed this process at their Billingham site in the United Kingdom. The reactor is a very deep (up to 150 m deep) cylindrical shaft fitted with a smaller concentric steel tube. Air is added in such a way as to produce a circulating movement within the reactor.

Primary settlement is not required, and high oxygenation efficiency is obtained. It is however difficult to get the sludge produced to settle, as the effluent is highly saturated with CO_2 during the high-pressure stage at the bottom of the reactor. Instead, it is common to use this natural flotation to remove sludge.

PURE OR ENHANCED OXYGEN PROCESSES

Companies that produce oxygen offer systems using pure oxygen to treat wastewater. It is interesting to note that companies that do not produce oxygen tend not to offer such systems.

While it is true that the higher partial pressures of oxygen achieve using pure oxygen improve oxygenation efficiency, there is no cost saving after the cost of purifying oxygen and of the special oxygen dissolution equipment are taken into consideration.

The only useful applications of this process are where there is a limited space, and/or high strength waste to be treated. The system does sometimes offer a more compact process, albeit at a high price in running costs. It is also claimed that the process is more resistant to high instantaneous loading.

BIOLOGICAL AERATED FLOODED FILTERS

These units combine biological treatment with filtration in such a way as to require no secondary settlement facilities for normal secondary treatment performance. They are therefore space efficient.

The units contain some sort of (usually granular) medium that forms a substrate for the growth of organisms. The whole unit is flooded with water beyond the top of the medium in service. As wastewater passes through, a combination of biological and physical action removes both solids and soluble impurities. The filters need to be backwashed periodically to remove the build-up of undigested solids.

A wide variety of proprietary systems is available, differing mainly in the type of medium and aeration system used.

SEQUENCING BATCH REACTORS

In SBRs, a single tank functions as both aeration tank and settlement tank. The various stages of treatment are separated in time within this tank, rather than in space in different tanks. The system allows a great degree of flexibility, as aerobic, anoxic, and anaerobic conditions can be provided, as well as differing degrees of mixing, and times of static settlement. Multiple tanks in parallel are provided to allow continuous operation despite individual tanks having batch fill and draw operation. Aeration plant must be of a low efficiency type to accommodate sludge build-up on the bottom of the tank during the settlement phase.

The process is as old as the AS process, but has been "rediscovered" recently, and promoted aggressively. There is no evidence that sequencing batch reactors can provide more treatment in less space than conventional continuous systems. Running costs are likely to be higher due to the low aeration efficiency.

MEMBRANE BIOREACTORS

A genuine innovation in wastewater treatment is a rarity. The emergence of MBRs has been brought about by the availability of relatively cheap membranes. Several systems are available, some of which are suitable for mainstream municipal effluent treatment, and some only for specialist industrial use.

The commercially available systems are detailed in Table 14.11.

The membranes used are usually of similar pore size to ultrafiltration membranes. In the case of the gravity head municipal systems, pore sizes are somewhat larger, as available driving force is only a few meters of water, sometimes supplemented by vacuum on the outlet side.

The systems give a greater filtration performance for the discharged water than that given by conventional secondary clarifiers, often offering high degrees of disinfection. Unit sizes are smaller than conventional treatment plant of the same performance (although not as small as originally claimed). Sludge bulking and other clarifier related problems are supposedly eliminated. However, other claims from bench and pilot scale experience, such as very high solid concentrations in the reactor and low to zero sludge production, are not proving true in practice.

However, there are some drawbacks. Aeration devices tend to be of low efficiency types. Fouling of the membranes can be a big problem, and the systems are expensive to purchase and maintain. A general comparison between submerged and side stream units is given in Table 14.12.

Table 14.11 Summary of Commercial Membrane Bioreactor Systems

Market Leaders	Membrane	Pore Size (μm)	Membrane Format	Module Format	Product Name	Packing Density (m²/m³)	Flux (lmh)
Kubota	Cl₂PE	0.4	fs	iMBR	EK515	120	17–24
Mitsubishi	PE	0.4	hf	iMBR	SUN	485	8.5–12
Mitsubishi	PVDF	0.4	hf	iMBR	SADF	approx. 200	30–34
GE-Zenon	PVDF	0.04	hf	iMBR	ZW 500d	304	17–24
2nd Tier							
Koch-Puron	PES	0.05	hf	iMBR	Puron	260	14–26
Siemens-Memcor	PVDF	0.04	hf	iMBR	MemJet B10R	334	17–24
Norit	PVDF	0.03	tub	sMBR	AirLift F4385	308	50–60
Toray	PVDF	0.08	fs	iMBR	Toray	135	29

Reprinted from Pearce, G (2008). Introduction to membranes: MBRs: Manufacturers' comparison: part 1, Filtration + Separation, 45:2 pp 28-31, Copyright 2008 with permission from Elsevier.

Table 14.12 Comparison of Submerged and Side Stream MBR Units

Submerged MBR	Side Stream MBR
Aeration costs high ($\sim$90%)	Aeration costs low ($\sim$20%)
Very low liquid pumping costs (higher if suction pump is used $\sim$28%)	High pumping costs (60%–80%)
Lower flux (larger footprint)	Higher flux (smaller footprint)
Less frequent cleaning required	More frequent cleaning required
Lower operating costs	Higher operating costs
Higher capital costs	Lower capital costs

In addition to these membrane separation bioreactors, there are extractive and membrane aeration types in development.

Anaerobic membrane bioreactors

A novel application of membranes in wastewater treatment is now commercially available as Veolia's Memthane process, which combines AD with cross-flow UF membranes.

System design and optimization

MBRs essentially differ only from conventional AS processes in the separation technology employed, despite claims made based on laboratory scale work. For aerated MBR systems, process design therefore consists of design of an aeration tank as used in an AS plant (other than operating at a slightly higher MLSS) and the design of the cross-flow LP membrane unit which replaces the settling tank.

Academic approaches integrating these two separate designs focus on the system's two aeration requirements—for the biological process, and for membrane cleaning, as follows:

- SADm is the membrane aeration demand per unit membrane area (membrane cleaning linked parameter)
- SADp is the membrane aeration demand per unit permeate flow (biological capacity linked parameter)

A rule-of-thumb design net flux of 20 LMH is suggested for an aerobic MBR's associated membrane separation stage. Both aeration and membrane tanks are made as small as possible (but no smaller, as early designers learned to their cost). Sludge flow through the tank should be 4–5 times permeate flow.

Anaerobic MBRs are similarly AD with a filter. We can size the bioreactor based on a sludge loading of 5–10 kg COD/m^3/day (much lower than initial claims of 30), and 8 day HRT. Reports of 40—50,000 mg/L solid concentrations in AMBR reactors have been made.

Membrane fluxes are lower than for corresponding aerobic versions, but because they are all side stream in configuration, 20–40 LMH has been reported at small scale.

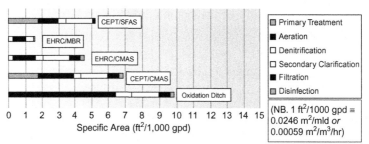

Figure 6.6: Footprint Example – MBR vs Alternate Technologies

CEPT – Chemically Enhanced Primary Treatment
CMAS – Complete Mixing Activated Sludge
EHRC – Enhanced High Rate Clarification
SFAS – Step Feed Activated Sludge

FIGURE 14.10

Footprint example: MBR versus alternate technologies. CEPT Chemically Enhanced Primary Treatment, CMAS Complete Mixing Activated Sludge, EHRC Enhanced High Rate Clarification, SFAS Step Feed Activated Sludge.

Reprinted from Pearce, G (2008) Introduction to membranes: An introduction to membrane bioreactors, Filtration+ Separation, 45:1, pp32-35, Copyright 2008 with permission from Elsevier

Applications of Membrane Bioreactors

MBRs are now a relatively established technology, and where their additional costs are justified by a requirement for their superior performance, they are a valid choice.

They are not however without their operational problems, are quite frequently sold into applications where another technology would be a better fit, and are still developing as a technology.

Where they do score well is where we need a small plant with a high degree of treatment mainly because membranes reduce the footprint required for solids settlement by a large factor (Fig. 14.10), and provide a positive barrier to escape of fine solids, enhancing effluent quality.

UPFLOW ANAEROBIC SLUDGE BLANKET PROCESS

Biological treatment processes were developed in cold weather regions, where aerobic bacteria have higher growth rates than anaerobic. Hence virtually all the main treatment processes are aerobic, and anaerobic processes are mainly used for sludge stabilization. In Holland in the 1970s, research showed that a certain type of anaerobic reactor could be used to treat warm, preferably quite high strength effluents direct, giving stable sludge and biogas in a single stage. Over time this has been adapted to provide a process that can treat municipal effluent in regions where incoming sewage temperatures are greater than 20°C. They are becoming very popular in South America, especially in Brazil, where most new facilities incorporate the process. Recent research in the United Kingdom and Scandinavia shows that the process also works in temperate climates with slightly increased retention times.

It is claimed that plant construction costs are similar to those of primary treatment facilities, and they yield twice the organic matter removal of conventional primary treatment facilities. It has been claimed that South American plants have construction costs of US$30−40 per person, as compared with US$80−100 per person for conventional secondary treatment plant. Sludge yields are far lower, and the plants are mechanically simple.

The units however have a potential for producing H_2S in high sulfate wastewaters, which led to odor problems earlier in development.

Effluent produced is not quite up to aerobic treatment standards, but the process dramatically reduces the load on downstream units. Polishing may be achieved by means of aerobic biological treatment, but may be done by means of physical/chemical treatment such as DAF. Lagoons designed for nutrient removal are most commonly used in Brazil for post treatment.

ANAEROBIC FILTER PROCESS

In anaerobic filters, the biofilm is retained within the reactor on filters made from materials such as ceramics, glass, engineered plastics, or wood. The filter material typically occupies 60%−70% of the reactor volume. The reactor can be operated in upflow or downflow. Downflow is less common, and better suited to effluent containing high suspended solid levels. They require a smaller area than CSTRs or lagoons, because of their low HRT. They have to date mainly been used on effluent from the chemical industry containing easily digested organics. They are simple and robust, but care should be taken to maintain the biofilm in optimal condition.

EXPANDED BED ANAEROBIC REACTOR

Expanded bed reactors are the newest type, developed to further reduce reactor volumes. An inert carrier material supports the microorganisms, which are expanded to 110% by an upflow recirculation of wastewater. The system is intolerant of high suspended solid levels.

A variant developed in the United Kingdom sugar industry eliminates the carrier material, and relies upon a granular sludge like that found in the UASB reactor. This is known as the expanded granular sludge bed design.

TWO AND THREE PHASE ANAEROBIC DIGESTION

As distinct from 2-stage digestion, which usually comprises a heated mixed reactor, and a stabilization tank following, these processes work by splitting the phases involved in digestion in space and time. The first stage of digestion, carried out by a distinct microbial population, is acidogenesis, where the long chain molecules within the incoming feed are broken down into simple fatty acids. The second stage (methanogenesis) takes these fatty acids, and turns them into hydrogen, methane, and carbon dioxide (biogas) and biomass.

It is possible to optimize the stages separately so, for example, acidogenesis could be carried out batchwise, and methanogenesis continuously. Or a mesophilic acid phase could be followed by thermophilic and mesophilic methanogenesis reactors in series, a process known as temperature phased anaerobic digestion or TPAD. A sequencing batch variant of TPAD is now being developed in the United States.

One of the advantages of the batch approach is that both EU "Biosolids in Agriculture" and US "Class A" legislation are now requiring a guaranteed time/temperature holding period for ensuring pathogen kill. An alternative approach, prepasteurization of sludges, is being developed in the United Kingdom. In the United States, prepasteurization processes such as "Ecotherm" are also available. There are also thermal hydrolysis pretreatment processes such as "Cambi" in development, which offer pasteurization as a fringe benefit.

FURTHER READING

Cooper, P. F., et al. (1996). *Reed beds and constructed wetlands for wastewater treatment.* Medmenham, Marlow, UK: WRc Publications.

Cooper, P. F. (1999). A review of the design and performance of vertical-flow and hybrid reed bed treatment systems. *Water Sci. Technol., 40*(3), 1—9.

Metcalf., Eddy, Inc. et al. (2014). *Wastewater engineering: treatment and resource recovery* (5th ed). New York, NY: McGraw-Hill.

Dirty water hydraulics

15

CHAPTER OUTLINE

Introduction 203
Minimum Velocities 204
Open Channels 204
    Straight Channel Headloss 204
    Circular Channel Headloss 206
    Shock Losses in Channels 206
    Peripheral Channels 208
Weirs 211
    Thin Weirs 211
    Broad Weirs 212
Screens 213
    Bar Screens 213
Advanced Open Channel Hydraulics 214
    Critical and Supercritical Flow 214
    Hydraulic Jumps 214
    Choking 215
    Inlet Works 215
    Control Structures 215
    Sump Volumes 216
    Sump Flow Presentation and Baffles 217
Plant Layout Hydraulics 218
Further Reading 218

INTRODUCTION

The variable concentrations of various sizes and types of solids, as well as fats, oils, and greases (FOGs) in dirty water mean that standard hydraulic calculations are likely to be inaccurate. For example, solids can settle out, and FOGs encrust equipment, pipes and channels used to handle dirty water, so even pipe internal diameter is not a constant.

An Applied Guide to Water and Effluent Treatment Plant Design. DOI: https://doi.org/10.1016/B978-0-12-811309-7.00015-1

Professional engineers use well-established heuristics to get around these problems. I am unsure of the original sources of much of what follows in this chapter, but I do know that it is the consensus approach to handling the hydraulic design of the most common types of dirty water handling structures and equipment.

MINIMUM VELOCITIES

Minimum superficial velocities of 1 m/s are recommended for streams having a significant solids or FOG content, to prevent settlement or encrustation.

For pumped wastewater pipes, allow initial estimates of superficial velocities for pumped wastewater pipes of 1.5−2.0 m/s for incoming lines prior to degritting, up to 3 m/s postdegritting, and up to 1 m/s for gravity lines.

Sludge lines should be a minimum of 100 mm NB, with maximum 1.5 m/s superficial velocity.

Initial estimates of superficial velocities for air and odor control lines should be 10 m/s.

OPEN CHANNELS

STRAIGHT CHANNEL HEADLOSS

If an outlet weir is present, for conceptual design purposes, it usually suffices to assume that the depth of flow is set by the depth over this weir. If there is no outlet weir, calculation of depth of flow is more complicated; 100 mm of water depth at all points in the channel should be allowed for conceptual design purposes. Setting the flow velocity at less than or equal to 1.5 m/s at this depth should result in reasonable losses down the channel.

A more accurate general approach based on a specific energy method involves determining a known or control depth, adding shock loss effects, and then friction effects. It should be noted that friction is usually a minor component in sewage treatment plants.

Rectangular channels

For more accurate requirements, we normally use the *Gauckler−Manning−Strickler* (the "Manning") formula (Eq. (15.1)).

Manning formula

$$v = \frac{k}{n} Rh^{2/3} \cdot S^{1/2} \tag{15.1}$$

where

v = average velocity (m/s)
k = conversion factor (1 for SI, 1.49 for US customary units)
n = Manning n value (dimensionless) (see Table 15.1)

Table 15.1 Manning Formula n Values[1]

Material	Manning n
Natural Streams	
Clean and straight	0.030
Major rivers	0.035
Sluggish with deep pools	0.040
Excavated Earth Channels	
Clean	0.022
Gravelly	0.025
Weedy	0.030
Stony, cobbles	0.035
Metals	
Brass	0.011
Cast iron	0.013
Smooth steel	0.012
Corrugated metal	0.022
Floodplains	
Pasture, farmland	0.035
Light brush	0.050
Heavy brush	0.075
Trees	0.15
Nonmetals	
Glass	0.010
Clay tile	0.014
Brickwork	0.015
Asphalt	0.016
Masonry	0.025
Finished concrete	0.012
Unfinished concrete	0.014
Gravel	0.029
Earth	0.025
Planed wood	0.012
Unplanned wood	0.013
Other	
Corrugated polyethylene (PE) with smooth inner walls	0.009–0.015
Corrugated PE with corrugated inner walls	0.018–0.025
Polyvinyl chloride with smooth inner walls	0.009–0.011

Courtesy: LMNO Engineering, Research, and Software, Ltd.

---

[1]https://www.LMNOeng.com/manningn.htm (accessed 5 November 2017).

Rh = hydraulic radius (m)

S = slope of the water surface or the linear hydraulic headloss ((L/L) = hf/L)

The Manning formula covers sloping bottom channels, and those of all shapes. The discharge formula, $Q = AV$, can be used to manipulate the Manning formula by substitution for V. Solving for Q then allows an estimate of the volumetric flow rate (discharge) without knowing the limiting or actual flow velocity. In practice this is best done using MS Excel's capacity for iteration via the "goal seek" command.

The Manning formula carries an assumption of "constant flow," implying that the slope of the water surface and the bottom of the channel are the same (only true in long channels of constant cross-section) though it is often used where this assumption is not strictly true. Table 15.1 gives approximate Manning formula n values for a range of materials.

CIRCULAR CHANNEL HEADLOSS

The Manning formula can also be used with circular channels, but the variability in cross-sectional area with depth of flow makes this considerably more difficult. See "Open Channels" section for another approach.

SHOCK LOSSES IN CHANNELS

Shock losses from bends, contractions, and expansions may be calculated as follows:

Entry B1 > B2	Approximate headloss (m) = 0.015 $(v1^2 - v2^2)$
Entry B1 < B2	Approximate headloss (m) = 0.026 $(v1^2 - v2^2)$
Bend	Approximate headloss (m) = $0.015v^2$

Friction losses can be added using Eq. (15.2).

Friction losses

$$\text{Headloss due to friction, } hf = 2.55 \times 10^{-4} \times \left(Q^2/B^3 y^3\right) \times (2y + B)L \qquad (15.2)$$

where

Q = volumetric flow rate
B = channel width (m)
$B1$ = upstream of $B2$
$v1$ = velocity at $B1$ (m/s)
y = calculated depth
L = length of channel

Outfalls, entries, and exits

Outfalls

Fig. 15.1 shows a free channel outfall. Submerged outfalls should normally be avoided.

Flumes

Fig. 15.2 shows a flume design.

Alternatively the methods in *BS/ISO 4359:2013* can be used.

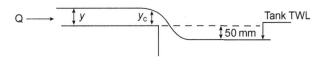

FIGURE 15.1

Free channel outfall.

Normal depth y, $(m) = \left(\dfrac{Q}{1.71B}\right)^{2/3}$

Critical depth y_c, $(m) = \left(\dfrac{Q^3}{B^2 g}\right)^{1/3}$

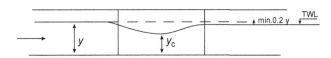

FIGURE 15.2

Flume.

Q = volumetric flow rate

B = channel width (depth y)

b = width in narrowest part of flume where depth is y_c

Normal depth y, $(m) = \left(\dfrac{Q}{1.71B}\right)^{2/3}$

Critical depth y_c, $(m) = \left(\dfrac{Q^3}{b^2 g}\right)^{1/3}$

Penstocks

We can divide penstocks into two types, namely weir penstocks and conventional penstocks.

Weir penstocks behave like sharp-edged weirs. Flow and head over the weir can be estimated using Eq. (15.3).

Weir penstocks

$$Q = 1.73 \times WH^{1.5} \tag{15.3}$$

where

Q = volumetric flow rate
W = penstock opening width
H = head of water over weir

The volume of discharge of a conventional penstock can be estimated using Eq. (15.4).

Conventional penstocks

$$Q = 0.7 \times WH \times (2gh)^{0.5} \tag{15.4}$$

where

W = penstock opening width
H = head of water over weir
h = upstream head
g = acceleration due to gravity

PERIPHERAL CHANNELS

The profile of a peripheral channel (Fig. 15.3) can be predicted from Eq. (15.5).

Peripheral channels

$$y_c = \left(\frac{q^2}{B^2 g}\right)^{1/3} \tag{15.5}$$

where

$q = Q/2$
$y = 1.5 y_c$
$y_{max} = 2 y_c$

The total flow Q is effectively split into two as shown in Fig. 15.3.

Aeration tanks/collection channels

Aeration tanks and collection channels are designed as per peripheral channels though q will vary dependent on the location of the collection chamber, and will not necessarily be $Q/2$.

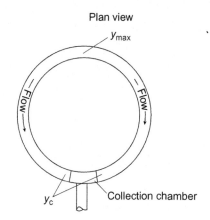

Plan view

FIGURE 15.3

Diagram of a peripheral channel.

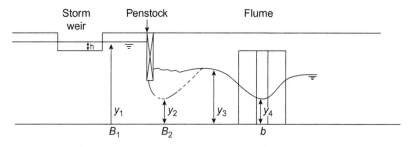

FIGURE 15.4

Hydraulic design of storm overflow.

y_2 = depth after penstock where supercritical flow is present

y_4 = depth in the flume where critical flow is present

b = width of flume at y_4

B_2 = width of channel at y_2

Storm overflows

A storm overflow (Fig. 15.4) is conventionally an interacting system of overflow weir, conventional penstock, and flume. Supercritical flow is generated through the penstock which may generate a hydraulic jump downstream.

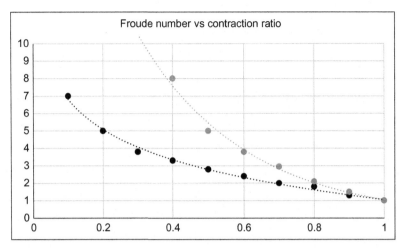

FIGURE 15.5

Choking and contraction ratio.

The essential feature of storm overflow design is ensuring a high likelihood of choking conditions, to prevent flow shooting through the flume by manipulating the upstream and downstream channel widths.

To design the storm overflow:

- Choose a trial value of b
- Find contraction ratio $= b/B_2$
- Find $y_3 = \left(\frac{Q}{1.71B}\right)^{2/3}$
- Find $V = \frac{Q}{B_2 y_3}$
- Find Froude number at $y_3 = \dfrac{v}{\left(g y_3\right)^{0.5}}$
- Check for the Froude number for likelihood of choking in Fig. 15.5 (Values lying in the area below the blue curve show choking is certain. Values falling in the area between the two curves indicate possible choking (by hydraulic jump). Values in the area above the orange curve show choking is unlikely.)
- If at least possible, OK
- If not, vary b and repeat
- Once OK, determine y_2 by trial and error from $y_2^3 - y_1 y_2^3 + 0.051 \frac{Q^2}{B_2^2} = 0$
- (y_2 is approximately $\frac{0.226Q}{B_2(y_1^{0.5})}$)
- Penstock opening is approximately $1.67 y_2$
- Calculate flow over storm weir Qs and select a desired head over weir h
- Calculate weir length using $L = \frac{Q_s}{1.84 h^{1.5}}$
- If L is too high, (> 10 m) select a larger h and redo

Check choking is certain or possible using Fig. 15.5. If not vary contraction ratio to make choking at least possible.

WEIRS

Calculation of depth of flow over a weir is mostly dependent on the shape of the weir. My own experience on large water features indicates that 6 l/s/m be allowed for broad-crested weirs, and 4 l/s/m for sharp-crested weirs. These flows gave depths of the order of 10−15 mm over the weir.

If it is desired to estimate depth of flow over a weir more accurately, several equations are available to determine depth of flow for several different weir cross-sections. Note that these standard calculations may not apply to weirs over 10 m long. Advice should be sought from an expert when considering weirs of over 10 m in length.

Weirs are generally divided into two classes, being "thin"/"knife edge" or "broad." This distinction may be defined by means of whether the nappe is free or adheres as it leaves the weir edge. This in turn is a function of the relationship between the head of water passing over the weir, and the breadth of the weir. Here, broad weirs are defined as being those where weir breadth is greater than three times the head upstream of the weir.

THIN WEIRS

A knife-edge or thin weir (Fig. 15.6) will have a flow rate at a given depth over the weir according to the formula in Eq. (15.6) if approach velocity is less than 0.6 m/s.

Thin weir flow rate calculation

$$Q = 1.84 \times (W - 0.2h) \times h^{1.5} \tag{15.6}$$

where

W = weir length (m)
h = head upstream of the weir (m)
Q = flow rate (m^3/s)

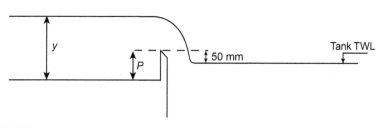

FIGURE 15.6

"Thin" weir.

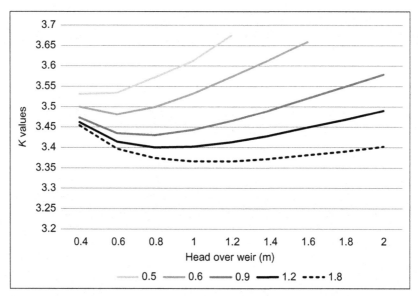

FIGURE 15.7

K values for sharp-edged weirs in Bazin's equation.

For deeper channels, and higher approach velocities, Bazin's formula ($Q = KbH^{3/2}$) may be used. Fig. 15.7 shows the *K* values for sharp-edged weirs in Bazin's formula at the height in m from channel bottom to crest of weir with which each line corresponds.

BROAD WEIRS

Determination of flow over broad-crested weirs is more difficult than for knife-edge weirs, and there are several complicating factors, such as the degree of detachment of the nappe. A clinging nappe gives a higher discharge flow for a given head than a free nappe.

The flow at a given weir breadth and head may be calculated using Eq. (15.7).

Broad weir flow calculation

$$Q = C \times W \times h^{1.5} \tag{15.7}$$

where

W = weir length (m)
C varies from 1.4 to 2.1 per weir shape and discharge condition. A value of 1.6 may be taken for estimation purposes

Steps

Steps can be treated as a type of broad weir, with flow changing from subcritical to critical a little back from the edge of the step, a free nappe, and a fully developed hydraulic jump downstream.

The depth of flow over a freely discharging step like this can be estimated using Eq. (15.8).

Estimating depth of flow over a step

$$h = 0.72 \times \left(\frac{QW^2}{g}\right)^{1/3} \tag{15.8}$$

SCREENS

Manufacturers of specific screens will be able to provide details of the required installation details of their kit, and headloss curves in these standard installations. The most basic type of screen is covered in the following section as an exemplar.

BAR SCREENS

Velocities

Approach velocity should be >0.4 m/s, velocity at average flow through screen no more than 0.6 m/s, and at peak flow no more than 0.9 m/s.

Rough head losses

For a clean bar screen, we can predict approximate headloss through the screen using Eq. (15.9).

Estimating headloss through clean screens

$$hf = \frac{1}{0.7} \times \left(\frac{(v2^2 - v1^2)}{2g}\right) \tag{15.9}$$

where

hf = headloss (m)
$v1$ = approach velocity (m/s)
$v2$ = velocity through bar opening (m/s)
g = acceleration from gravity
0.7 = headloss coefficient

Screens which are clogged with rags, etc. can have far higher head losses.

More accurate head losses

A commonly used equation which allows allowance to be made for bar shape in partially clogged bar screens is given at Eq. (15.10).

Table 15.2 Shape Factors for Cross-Sections in Headloss Calculations

Sharp-edged rectangle	2.42
Rectangular with semicircular side upstream	1.83
Circular	1.79
Rectangular, semicircular up and downstream	1.67
Teardrop	0.76

Estimating headloss through partially blocked screens

$$hf = \beta \left(\frac{w}{b}\right)^{4/3} \frac{v_a^2}{2g} \sin \theta \qquad (15.10)$$

where

hf = headloss across the screen (m)
β = shape factor
w = maximum bar width (mm)
b = minimum aperture width (mm)
v_a = velocity in the approach channel (m/s)
g = acceleration due to gravity
θ = angle of inclination of the bars to the horizontal (commonly 30°)

Shape factors for bar cross-sections are described in Table 15.2.

ADVANCED OPEN CHANNEL HYDRAULICS
CRITICAL AND SUPERCRITICAL FLOW

Critical velocity in water flow is any faster than the maximum wave speed, so it is analogous to supersonic velocity in a gas. A Froude number (*Fr*) of greater than 1 is associated with supercritical flow.

HYDRAULIC JUMPS

Where we have a transition from supercritical to subcritical flow, we see a hydraulic jump. A jump is a turbulent region where shallow supercritical flows decelerate rapidly to deeper subcritical flows (as for example when flow passes over a weir or approaches a flume).

- a strong jump occurs when $Fr > 9$
- a steady jump occurs when $4.5 < Fr < 9$
- an oscillating jump occurs when $2.5 < Fr < 4.5$

- a weak jump occurs when $1.7 < Fr < 2.5$
- an undular jump occurs when $1 < Fr < 1.7$

Hydraulic jumps are extremely effective energy dissipators. The stronger the jump, the higher the energy loss: strong jumps can incur losses of 85% + .

CHOKING

Lateral contractions in open channels reduce channel width and increase discharge per unit width. Contracting flows beyond a certain limit will result in a phenomenon called choking. Under choking conditions, flows pass through critical depths in the contracted channel reach.

INLET WORKS

The inlet works of a sewage treatment plant have three main components, namely screening, grit removal, and storm tanks. In addition to these, there are commonly may be flow control and measurement structures, like flumes, flow distribution chambers, etc.

CONTROL STRUCTURES

Water/sewage control chambers

To split flow evenly a design containing several weir penstocks (Fig. 15.8) of width L is recommended. The penstocks should be able to achieve a complete flow shutoff as well as modulating duty.

Top Water Level (TWL) is determined using Eq. (15.11).

Penstock head calculation

$$H = \left(\frac{q}{1.84W}\right)^{2/3} \tag{15.11}$$

where

q = flow per penstock
W = clear width of penstock
H = head over weir

Activated sludge control chamber

Activated sludge control chambers (Fig. 15.9) are commonly used to combine sludge flows from several clarifiers, and are therefore in essence a reversal of flow direction through a chamber as in the last section, with maximum TWL in the divided chambers. Determination of this max TWL is as in the last section. Weir penstocks in RAS applications are normally actuated to allow control of RAS flows.

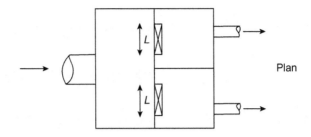

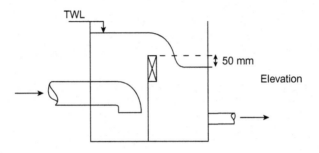

FIGURE 15.8

Weiring penstock.

Maximum TWL should be set 50 mm below clarifier TWL- pipeline losses to chamber, and downstream TWLs should be calculated from this.

RAS is reliably similar enough in density and viscosity to water that no special allowances need to be made.

SUMP VOLUMES

Submersible or drywell pumps fed from sumps are a common feature of effluent treatment plants. The next section covers sizing and key design features of sumps.

The working volume between pump start and stop levels in cubic meters can be calculated using Eq. (15.12).

Sump working volume calculation

$$V = 0.9 \times \frac{Q}{Sh} \tag{15.12}$$

where

Q = pump capacity (l/s)
Sh = permissible number of starts per hour (allow 10 if no manufacturers information)

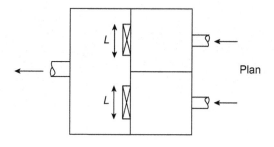

Plan

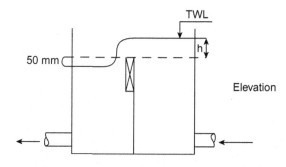

Elevation

FIGURE 15.9

Activated sludge control chamber.

The depth below sewer invert of sump TWL can be calculated using Eq. (15.13).

Depth below sewer invert calculation

$$D = \frac{V}{F} + 0.5 \qquad (15.13)$$

where

F = CSA of sump (m^2)
V = working volume (m^3)

SUMP FLOW PRESENTATION AND BAFFLES

Incoming sewers and channels

Incoming sewers and channels should be placed so as to distribute flow evenly and symmetrically around the pump. Large freefalls should be avoided as they can lead to odor and other problems.

Ideally the approach flow should be straight onto the pump, and there should be a uniform approach channel/sewer cross-section.

An inlet baffle may be provided below the incoming sewer to correct any deviation from these ideal conditions.

Benching

Sumps should have benching which diverts flow toward pump suctions and prevents build-up of solids. The slope of such benching should be not less than 45° to the horizontal.

Stalls for individual pumps can be used, but this is not likely to be the most economic option.

Bellmouths

To avoid vortex formation, and other effects that reduce pump performance, the minimum distance from the suction bellmouth to the sump floor should be greater than 80 mm, or $D/2$, where D is the bellmouth diameter.

Bellmouths should be between 2D and 3D apart edge to edge.

PLANT LAYOUT HYDRAULICS

In sewage treatment plants:

- Use gravity flow through works as much as possible
- Use open channels and chambers to control flows as much as possible, especially early in the process
- Avoid weirs in positions which may rag up
- Weiring penstocks used should not have a central spindle
- Use weiring penstocks in preference to conventional penstocks as flow control devices
- Design weirs and flumes so as not to drown
- Design tanks and channels to be drained down in a reasonable time
- Design in sufficient standby capacity to allow for one unit out of service
- Avoid low superficial velocities on outlet channels, which can lead to algal/fungal growth and knock-on consent problems
- Do not provide permanent ladders in pumping sumps
- Sumps should be covered by open mesh or enclosed for odor control
- Removable covers for pump removal should be kept as small as possible
- Provide a readily accessible means of isolation of the flow into the sump
- Drain the valve chamber into the pumping sump

FURTHER READING

Judd, S. (2013). *Watermaths: Process fundamentals for the design and operation of water and wastewater treatment technologies.* UK: Judd & Judd Ltd.
Degrémont Suez. (2007). *Water treatment handbook* (7th ed.). Cachan, France: Lavoisier.

Industrial effluent treatment engineering

4

INTRODUCTION: AN OVERVIEW OF INDUSTRIAL EFFLUENT TREATMENT

Industrial effluents discharged to sewer form an important part of the load on the average sewage treatment plant. It is, however, common for industrial facilities to carry out some or all the treatment of their effluent on-site.

Treatment of industrial effluent involves combining techniques from sewage treatment with those used in potable water treatment, as well as several specialist techniques. The general principle is still the same: a stage-by-stage reduction in contaminants of decreasing particle size.

Cost and robustness are very important in industrial effluent treatment. Nonspecificity of processes is key to robustness: designs need to remove entire classes of contaminant, rather than specific ones. The key to both low cost and robustness is usually avoiding novelty.

The most common treatments are therefore coagulation/flocculation followed by settlement, filtration, or flotation. These processes offer very good nonspecific removal of solids, fats/oils/greases, colloidal matter, and several substances (such as dyestuffs) that bind to the surface of the floc particles.

Membrane treatments are becoming more popular as membranes become cheaper, but they are still generally far more expensive and less robust than traditional processes, so they have a long way to go before they will be the designer's first choice options.

Industrial effluent characterization and treatment objectives

16

CHAPTER OUTLINE

General 221
Calculating Costs of Industrial Effluent Treatment 222
Industrial Wastewater Composition 222
Industrial Effluent Case Studies 223
    Case Study: Paper Mill Effluents...223
    Case Study: Plating Effluents 224
    Case Study: Petrochemical Facility Wastewaters...225
    Case Study: High Sulfate Effluents 225
    Case Study: Vegetable Processing Effluents 227
Problems of Industrial Effluent Treatment 228
    Batching..228
    Toxic Shocks 228
    Nutrient Balance 228
    Sludge Consistency 229
    Changes in Main Process 229
Further Reading 229

GENERAL

Industrial wastewater may be reused after treatment, though it is most commonly discharged to sewer or environment.

If it is reused, the specification for the treated water will be derived from the needs of its user. Treated effluent is most commonly used for low grade activities such as site wash-down or (especially in the Arabian Gulf, where businesses have a legal requirement) irrigation. I have however seen it treated to the point where it was used to wash salad leaves for human consumption, though this was in my view an uneconomic public relations exercise.

Discharge to sewer will be regulated by the owner of the sewer, and discharge to environment by the environmental regulator.

Knowing how much it will cost to have effluent treated by a third-party undertaker (where available) is useful because it allows designers to determine if it will be economic to treat onsite rather than exporting. In the United Kingdom, trade effluent charges for sewer discharge are priced using the Mogden formula (see Eq. (16.1)), which allows a rational determination of the most economically favorable degree of treatment for an effluent. Since the formula is based upon the cost of treatment, it may be that also provide an approximate guide for other developed nations, but costs of water treatment in less developed countries are (perhaps surprisingly) far higher.

CALCULATING COSTS OF INDUSTRIAL EFFLUENT TREATMENT

In the United Kingdom, costs for treatment by municipal undertakers can be predicted using the Mogden Formula (Eq. (16.1)).

Mogden Formula

$$\text{Cost}, C = R + [(V + Bv) \text{ or } M] + B(Ot/Os) + S(St/Ss) \tag{16.1}$$

where

R = Reception charge
V = Primary treatment charge (also referred to as P)
M = Treatment charge where effluent goes to a sea outfall
Bv = Biological treatment charge (also referred to as $B1$ and Vb)
Vm = Preliminary treatment charge for discharge to outfalls
B = Biological oxidation charge (also referred to as $B2$)
Ot = Measured chemical oxygen demand (COD) of effluent
St = Measured total suspended solids (SSs) of effluent
Os = Standard COD of crude sewage
Ss = Standard SSs concentration in crude sewage
S = Sludge treatment charge

UK readers can download a Mogden Formula calculator free of charge from the Waste and Resources Action programme (WRAP) (see Further Reading).

INDUSTRIAL WASTEWATER COMPOSITION

Industrial wastewater is primarily categorized in the same way as sewage. Measurements of parameters like biochemical oxygen demand (BOD), COD, pH, and alkalinity can allow the designer to evaluate possible process options, though they may be misleading. Many industrial wastewaters contain substances which cause problems for treatment plants at relatively low concentrations.

A more extensive analysis of wastewater informed by the chemicals used on site, and a knowledge of those which cause problems for effluent treatment plants is therefore usually required. Key chemicals of concern are detergents, fats, oils and greases, heavy metals, and biocides. Variations in temperature and pH can be far greater than in municipal plants, and the designer needs good information on these.

Flow and load variations are very important to the industrial effluent treatment plant designer, and these can often be reduced by resource efficiency audit based on a mass balance over the main plant. It is therefore always best if possible to undertake water and waste minimization in the main process on site prior to designing an end of pipe effluent treatment plant.

Advice on how to do this can be found in the WRAP publication "Saving Money Through Resource Efficiency: Reducing Water Use" (see Further Reading). The effort taken to do this will be a sound investment, and can radically reduce the capital and running cost of the effluent treatment plant. The traditional approach of considering only the main process, and ignoring the cost of effluent treatment, has not been financially viable for many years now in developed countries.

It is commonly the case that the designer's starting point will have insufficient information on all the above parameters. The resource efficiency study will therefore bring the added benefit of generating a working estimate by mass balance. The designer will usually however be relying to some extent upon informed guesswork based upon knowledge of the industry sector. The luxury of being able to carry out lab or pilot plant work to inform design is, in my experience, every rare.

INDUSTRIAL EFFLUENT CASE STUDIES

The following are examples of treatment processes from different industry sectors, to illustrate several typical types of effluent and their associated treatment processes.

CASE STUDY: PAPER MILL EFFLUENTS

While different kinds of paper, card, and board produce effluent with different characteristics, all produce effluent with a mixture of coarse particles and finer fibers of raw material, cellulose, and lignin-derived organics.

Coarse materials are removed by screening, and the screened effluent is treated with coagulant and flocculant to remove finer fibers, as well as some of the soluble and colloidal organics. Traditionally, a settling tank was used to remove the flocculated solids, but dissolved air flotation (DAF) often works far better, as the material tends to be lighter than water.

FIGURE 16.1

Paper mill effluent treatment plant.

Courtesy: Google Earth 2017.

Whether DAF or a settling tank is used, the flocculated solids tend to form a "matted" sludge, so it is essential to use a very heavy-duty scraper mechanism and to withdraw sludge continuously, usually recycling it to the feed to the mill.

Traditionally, paper mills were located on rivers, and the partially treated effluent produced by the above processes was discharged directly to environment. This is unlikely to be allowed anywhere in the world nowadays.

At a paper mill effluent treatment plant (Fig. 16.1), biological treatment (mostly activated sludge) normally follows screening and DAF. Nutrient supplementation (at a minimum, N and P) and pH adjustment may well be required. Aeration systems should be designed to handle peak loads, and provision made for recirculation of treated effluent.

CASE STUDY: PLATING EFFLUENTS

Chromium plating requires that the metal to be plated is immersed in a sodium cyanide/caustic soda solution as a pretreatment, after which it is rinsed free of alkaline cyanide in running water, producing a rinse tank overflow contaminated with alkali and cyanide.

The pretreated parts are then plated in a chromic acid bath and subsequently rinsed free of surplus plating solution in running water.

Thus two waste streams are produced: an alkaline cyanide effluent and an acidic chromate effluent, each containing traces of the base metals which were immersed in them. These effluents must be treated separately because, if they were combined, they would release cyanide gas. They are treated in opposite ways, one under acidic reducing conditions, the other under alkaline oxidizing conditions, in both cases under the control of redox probes.

The chromic acid (Cr^6) is pretreated by reduction to chromate (Cr^3) using sulfur dioxide gas or sodium bisulfite along with sulfuric acid to give an operating pH below 2.4, with a hydraulic retention time (HRT) of around 20 min.

The cyanide effluent is oxidized with alkaline chlorine to nitrogen and carbon dioxide, using sufficient added caustic soda to give pH 9.5–10, again with a HRT of around 20 min.

The products of these two reactions are then mixed to allow them to largely mutually neutralize, with a final trim in pH to the optimum for precipitation of chromium and other metallic contaminants as hydroxides (see Fig. 18.2) in a settlement tank.

CASE STUDY: PETROCHEMICAL FACILITY WASTEWATERS

Crude oil and condensate refineries generate a large amount of wastewater that has both process and nonprocess origins. Depending on the type of crude oil, composition of condensate, and treatment processes, the characteristics of refinery wastewater vary according to a complex pattern. The design and operation of modern refinery wastewater treatment plants are challenging and are essentially technology driven.

The quantity of wastewaters generated and their characteristics depend on the process configuration. As a general guide, approximately 3.5–5 m^3 of wastewater per ton of crude are generated when cooling water is recycled. Refineries generate polluted wastewaters, containing BOD and COD levels of approximately 150–250 and 300–600 mg/L, respectively; phenol levels of 20–200 mg/L; oil levels of 100–300 mg/L in desalter water and up to 5000 mg/L in tank bottoms; benzene levels of 1–100 mg/L; benzo(a)pyrene levels of less than 1–100 mg/L; heavy metals levels of 0.1–100 mg/L for chromium and 0.2–10 mg/L for lead; plus other pollutants.

Refineries also generate solid wastes and sludges (ranging from 3 to 5 kg per ton of crude processed), 80% of which may be considered hazardous because of the presence of toxic organics and heavy metals.

Given the complex and diverse nature of refinery wastewater pollutants, a combination of treatment methods is often the norm before discharge. Therefore separation of wastewater streams, such as storm water, cooling water, process water, sanitary, sewage, etc., is essential for minimizing treatment requirements. Fig. 16.2 shows the different stages of treatment at a typical plant.

CASE STUDY: HIGH SULFATE EFFLUENTS

Sulfuric acid and its salts, commonly used in industry, produce an effluent which can directly or indirectly attack the concrete used in some sewers and most effluent treatment plants.

FIGURE 16.2

Petrochemical wastewater treatment plant.

Courtesy: Google Earth 2017.

The pharmaceutical industry, e.g., commonly produces effluent high in sodium sulfate as well as lower concentrations of many other chemicals. Pharmaceutical production is almost always done batchwise, with the consequent requirement for flow and load balancing covered elsewhere.

Sulfates may be removed to some extent by precipitation as the calcium salt, but this is not as straightforward as it might appear. Calcium sulfate is unusual in having reduced solubility at higher temperatures, a higher solubility when sodium ions are present, a lower solubility when chloride ions are present, and both neutralization and precipitation can be interfered with by many other ions. It also has a higher solubility under practically attainable conditions than some regulators will permit to be discharged to sewer, let alone direct to environment.

One approach is to dose a balanced effluent with hydrochloric acid to increase the chloride content, then add lime to neutralize and add calcium ions. Alternatively, calcium chloride can be used to provide both chloride and calcium, followed by lime neutralization. I have come across other approaches using sophisticated chemistry, ultrasonic reactors, and so on, but I would not recommend them.

After treatment, there will be 2%–5% w/w suspended solid calcium sulfate in the effluent. Dense mineral particles like these settle very readily, producing a hard-to-pump 10%–20% dry solids (DS) non-Newtonian sludge, unless care is taken to keep solids concentrations down to 5% by rapid removal of settled sludges. These sludges can be pumped via a stirred buffer tank to a filter press or suchlike, to produce a 30%–50% solids cake. The filtrate from the press should be blended with the feed to the treatment plant for recycling.

FIGURE 16.3

Effluent treatment at a vegetable processing plant.

Courtesy: Google Earth 2017.

Such a treatment process might not be able to reduce sulfate levels below 1–2000 ppm. If a more stringent standard is required, expensive processes such as reverse osmosis might be necessary, unless the regulator can be convinced that the cost of these is excessive.

CASE STUDY: VEGETABLE PROCESSING EFFLUENTS

Effluents from vegetable preparation tend to come in two kinds. There are high solids/low BOD effluents from vegetable washing operations, and lower SS/higher BOD effluents arising when rinsed vegetables are cut and cooked. One site may carry out both operations, producing an effluent with varying proportions of the two effluent types. A typical effluent treatment plant for a vegetable processor is shown at Fig. 16.3.

A common difficulty for the effluent treatment plant designer in this sector is that vegetables are harvested and processed seasonally. It may be that many different vegetables are handled in successive campaigns by a single plant, and these vegetables are handled differently at different times of the year. Flow and load balancing is required, usually producing an effluent with a BOD in the range 100–2000 mg/L.

It is easy to assume that rejected vegetables, and associated soil, sugars, and starches form the bulk of the pollutants. However, syrups, oil, and other materials used in processing can add a lot of BOD and unexpected floating matter. For example, food processors may spray parsnips and potatoes with oil to sell in ready-to-roast form at Christmas. Careful questioning about seasonal/year-round use of additives is thus advised.

The treatment of such effluents commonly starts with a rundown screen for removal of coarse vegetable matter. Next there will be flow and composition balancing in a mixed, often aerated tank, with the facility for removal of settled solids. pH adjustment can be done in this tank, but it is better to do it inline on the feed to the plant from this tank.

Where there is a reasonably constant flow of effluent, such effluents are very well suited to biological treatment, usually aerobic, though some supplementary N and P addition may be required.

There can however be problems with maintaining an adequate level of biomass in a biological treatment plant all year round if there are significant periods when there is no load on the plant. In such situations, anaerobic treatment may be preferred (especially if COD is at least 2000) as it more likely to be responsive after extended low load conditions.

As far as aerobic treatment is concerned, recirculating high-rate filters have been reasonably commonly used as roughing treatment, followed by activated sludge in what is sometimes called the contact process. This should produce Royal Commission standard effluent if properly designed.

PROBLEMS OF INDUSTRIAL EFFLUENT TREATMENT
BATCHING

Many processes, especially in the food, pharmaceutical and fine chemical sectors, commonly use batch production processes. Effluents from batch processes are produced intermittently at high flowrates. This can be dealt with by the process designer either providing a large inline buffer capacity prior to the treatment plant, or by treating the wastes on a batch basis as they are produced.

TOXIC SHOCKS

There may be wide variations in production of certain substances, processes that are carried out infrequently, or emergency situations that lead to occasional high concentrations of toxic materials being sent to the effluent treatment plant. These may not be diluted by other arisings in the way they would at a municipal facility.

Either the works can be designed with online or offline buffering capacity to handle these situations, the process selected can be designed to handle the full loading, or alternative arrangements can be made for dealing with the consequent toxic shock.

NUTRIENT BALANCE

While industrial effluents may have very high BOD levels, nutrients may be limiting for biological growth. Appropriate levels of N, P, S, and so on as detailed in earlier sections need to be present for biological processes to work. There are several proprietary formulations available, but it is doubtful whether most of them offer significant advantages over commercially available agricultural fertilizers.

SLUDGE CONSISTENCY

If there are problems at municipal sewage treatment plants with obtaining consistent sludges, these may be multiplied several fold at industrial plants. There may be a greater proportion of chemical sludges and there may be a greater number of different sludge types. There will almost certainly be less sludge inventory to smooth out any inconsistencies. Design of consolidation and thickening facilities to cope with greater inconsistency may be required.

CHANGES IN MAIN PROCESS

Not only may the main process or processes be highly variable on a hourly and daily basis, but the processes carried out might also vary on a seasonal or campaign basis. The effluent produced may therefore vary very widely in composition and strength throughout the year. Such changes may be on a timescale too long to handle by means of buffering capacity. It will therefore be advisable to consider operation of the effluent treatment plant as part of the preparations for changes to be made in the main processes. In the case of campaign manufacture, where there may be unique events and contaminants, pilot trials may be advisable.

FURTHER READING

UK, Waste & Resources Action Programme (WRAP). (2013). *Saving money through resource efficiency: Reducing water use.* Available at <http://www.wrap.org.uk/sites/files/wrap/WRAP_Saving_Money_Through_Resource_Efficiency_Reducing_Water_Use_0.pdf> Accessed 09.10.17.

UK, Waste & Resources Action Programme (WRAP). (2017). *Mogden Formula tool.* Available at <http://www.wrap.org.uk/content/mogden-formula-tool-0> Accessed 23.02.17.

Industrial effluent treatment unit operation design: physical processes

17

CHAPTER OUTLINE

Introduction 231
Gravity Oil/Water Separators 231
Coalescing Media 233
   Hydrophobic Media: Packed Beds 233
   Hydrophilic Media: Nutshell Filters 233
   Removal of Toxic and Refractory Compounds 234
Adsorption 234
Membrane Technologies: Oily Water 235
Membrane Technologies: Removal of Dissolved Inorganics 236
Further Reading 236

INTRODUCTION

Many of the processes used in industrial effluent treatment are the same as those used in municipal clean or dirty water treatment. Space constraints on industrial sites tend to mean that high-intensity processes like dissolved air flotation (DAF) are favored over low-intensity alternatives such as sedimentation, though design quality is often low in industrial effluent treatment.

The main difference, other than the use of specific chemicals to remove or enhance treatability of specific chemicals present at high concentrations in the effluent, is the use of various kinds of separators for oils and other light nonaqueous phase liquids (LNAPLs).

These separators are the main physical process used in industrial effluent treatment which is not used in municipal treatment.

GRAVITY OIL/WATER SEPARATORS

Removal of LNAPLs from mixtures is a fairly common requirement in the treatment of industrial effluents. LNAPLs may be hydrocarbons, or they may be biologically derived fats, oils, and greases. The units that carry out this process are known as oil−water separators or interceptors (see Fig. 17.1).

An Applied Guide to Water and Effluent Treatment Plant Design. DOI: https://doi.org/10.1016/B978-0-12-811309-7.00017-5

FIGURE 17.1

Oil–water separator (*green* tank (dark gray in print version) right of picture).

Courtesy: Expertise Ltd.

The most commonly accepted standard for the design of Oil Water Separators is the API 421 Monograph "Design and Operation of Oil-Water Separators" (see Further Reading).

The standard covers conventional rectangular channel units and parallel plate separators. It is the parallel plate separators that are most commonly associated with this standard.

The standard gives an equation (Eq. (17.1)) which can be rearranged to show that required area for separation.

Separation Area Calculation Derived From API 421

$$AH = \frac{Qm}{0.00386\left(\frac{(Sw - So)}{\mu}\right)} \tag{17.1}$$

where

Qm = design flow in cubic feet per minute

AH = horizontal separator area in square feet

Sw = specific gravity of the wastewater's aqueous phase (dimensionless)

So = specific gravity of the wastewater's oil phase (dimensionless)

μ = wastewater's absolute (dynamic) viscosity in poise

The product of this equation represents the required area of tank surface, plus the projected plate area required to effect separation of 60-μm oil globules.

The monograph also gives design details, and the results of much practical experience of operation of these separators. It is therefore recommended that anyone considering designing one of these units obtains the monograph, and discusses their application with suppliers of the plate packs to be used in the design.

If we require a higher performance than API 421 demands, we can use Stokes Law to determine the required surface area, based on the rise rate of the smallest droplet we wish to remove. We can then see how many plates are required using the "projected area" method.

This method involves assigning an effective surface area for settlement of each plate in the stack of its projection onto the base of the tank. The plates should be separated by 19−37 mm, and the plate inclination angle should be 30−45 degrees from vertical.

COALESCING MEDIA

We can enhance the removal of very fine oil droplets with media which capture droplets impinging on their surface. If the media have a hydrophobic surface (as is the case with many plastics) the oil sticks to it and, over time, large droplets are formed which float away. If we use more hydrophilic media, (such as crushed nutshell) we can wash accumulated oil off the media after a collection period.

HYDROPHOBIC MEDIA: PACKED BEDS

Passing oily wastewaters with low suspended solids through a bed of random hydrophobic packing, such as 25 mm polypropylene pall rings or saddles, enhances the collection of fine droplets of oil by coalescence. The absence of any backwashing facility makes this unsuitable for water with significant solids content, which is why nutshell filters (see "Hydrophilic Media: Nutshell Filters" section) are more popular for produced water and other oil and gas industry wastewaters.

HYDROPHILIC MEDIA: NUTSHELL FILTERS

Depth filters filled with graded crushed nutshells (usually walnut or pecan) have a long record of accomplishment as the final stage of oily water treatment. They can produce a very high-quality water, suitable for reinjection when used with produced water.

There are now five generations of designs. Each has dealt in successively more sophisticated ways with the main problem of these filters, namely, the separation of the collected oil from water and nutshells. For the fifth-generation designs, maximum feed oil levels of 100 ppm and surface loadings of 13−20 gpm/ft^2 are recommended for 98% removal of NAPL and SS $> 2 \, \mu m$, though process engineers usually only specify such filters, with the detailed design being completed by their proprietary suppliers. I strongly recommend this approach, having seen in my expert witness practice the results of unwise attempts to design these units from first principles.

REMOVAL OF TOXIC AND REFRACTORY COMPOUNDS

There are two types of refractory organic compounds which escape conventional treatment. Firstly, there are the substances which are of genuine concern because they have been shown to be of risk to people or the environment. Then there are the enormous number of substances which have not been shown to have any harmful effects, and are often present at concentrations far below those shown to have any detectable effects in scientific tests.

The first class of substances, and any others which we might have rational reasons to suspect might be classed among them, must be reliably removed from wastewater to a level below that which causes harmful effects. The second class of substances should, at least in my opinion, be of no concern to engineers, although they may well have political mileage and be potentially fruitful for academic research.

There are many specific techniques used in industrial effluent treatment to remove specific compounds, but in general, the essence of an effective technique in effluent treatment is nonspecificity. The techniques used therefore tend to fall into two categories: adsorption and oxidation.

Adsorption is usually achieved by means of granular activated carbon (GAC), although powdered activated carbon (PAC) is also used. In applications where something is known about the composition of the effluent to be treated, isotherms are available for specific chemicals (binary mixtures might also be analyzed using isotherms). In the more common case of unknown or more complex mixtures, empty bed contact times, hydraulic loadings, and depths of carbon are employed in designing nonspecific absorption columns. The units may be downflow, and therefore incorporate solids removal, or alternatively upflow.

ADSORPTION

The most common adsorption process in industrial effluent treatment is GAC treatment, though other proprietary adsorbents are also used.

Activated carbon is produced by charring coconut or other nutshells, coal, or wood in a controlled manner. This process results in a very open-textured carbon material, full of macroscopic and microscopic channels throughout its bulk. The action of the carbon in removing dissolved substances from solution, and attaching them to its surface happens mostly within the microscopic channels which make up the bulk of its surface area.

The material is used either as a fine powder (PAC) or alternatively in a granular form (GAC). PAC tends to be added directly, and GAC held within a vessel, and contacted with a flow of wastewater passed through the vessel. The retention of GAC means that it may be recovered, and regenerated for further use, whereas PAC used is lost, usually in sludge production. Onsite regeneration is however rare. Instead, at a site level, "spent" GAC is normally replaced with virgin material.

The adsorption process is one of nonselective adsorption of all substances that are adsorbed by GAC, until the available binding sites are exhausted, and contaminants break through the end of the bed. The "time to breakthrough" is the most important design parameter, and is theoretically determined by means of adsorption isotherm data available from GAC suppliers.

An additional design consideration in industrial applications may be the lack of availability of isotherms reflecting the absorption characteristics of activated carbons for the specific contaminants present. As discussed previously, empty bed contact time and superficial velocity form the main design considerations.

Adsorption can be combined with a gravity OWS, and coalescing media, as in the design in Fig. 17.2.

In Fig. 17.2, the inclined plate pack is on the left, the coalescing media is in the solid *red* (gray in print version) container, and on the far right, bags of GAC remove the final traces of oil to allow direct discharge to environment.

MEMBRANE TECHNOLOGIES: OILY WATER

Treatment of oily waters arising at petrochemical refineries has traditionally been performed using a mixture of physical, chemical, and biological processes. API oil–water separators and DAF are used to reduce oil levels, and then the

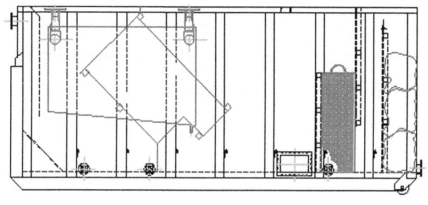

FIGURE 17.2

Cross section through a combined adsorption and gravity OWS.

Courtesy: Expertise Ltd.

wastewater is aerated along with activated sludge, the treated effluent settled, and sand filters used to polish the final effluent.

There can be advantages to replacing the sand filters with membranes. More radically, both the final settlement tank and the sand filters can be replaced by membranes.

The water produced by this process is reliably clean enough to be fed to RO membranes for recycle to process as high-purity water, suitable as boiler feed.

Careful pretreatment of oily waters going into membrane bioreactors (MBRs) is required. There are several case studies of inadequate treatment of the feed to MBRs, resulting in highly variable outlet chemical oxygen demand levels and in the production of an oily foam requiring expensive tankering away.

One such site trialed the spinning membrane system which is currently being promoted by several lab-based investigators, but found its power cost made it uneconomic compared with sponge-ball cleaned tubular membranes as a pretreatment.

MEMBRANE TECHNOLOGIES: REMOVAL OF DISSOLVED INORGANICS

There must be a good reason to remove dissolved organics. The required techniques tend to be expensive to install and/or run. Most are essentially techniques borrowed from drinking water production and usually require extensive pretreatment.

Membrane processes have different names, but are characterized by the size of the holes in the membrane. Tight membranes (with small holes) such as those used for reverse osmosis have very high head losses (perhaps 15 bar), but they can remove molecules selectively. Ultrafiltration membranes have larger holes, and remove colloidal matter, and large organic molecules only. Head losses are of the order of perhaps 1 bar.

Membranes may be combined with electrochemical processes, as in electrodialysis, with electrical flux separating ions across a semipermeable membrane. These processes however tend to be unreliable, especially in effluent treatment applications.

FURTHER READING

American Petroleum Institute. (1990). *API 421 (withdrawn) Management of water discharges: Design and operation of oil-water separators.* Washington, DC: API.

Peeters, J., & Theodoulou, S. (2005). *Membrane Technology Treating Oily Wastewater for Reuse. Corrosion.* Houston, TX: NACE International.

Industrial effluent treatment unit operation design: chemical processes

18

CHAPTER OUTLINE

Introduction 237
Neutralization 238
Metals Removal 239
    Precipitation as Hydroxides 239
    Precipitation as Sulfides 239
    Precipitation by Reduction 240
Sulfate Removal...240
Cyanide Removal 240
Organics Removal 241
Other Chemical Processes 241
    Oxidants 241
Disinfection 242
    Chemical Disinfection...242
    Dechlorination...242
Further Reading 242

INTRODUCTION

Chemical dosing is very important in industrial effluent treatment processes.

The simplest form of chemical dosing is the addition of acids and alkalis. Such dosing is performed to moderate pH to meet discharge consents, to optimize pH dependent separation processes, or for corrosion or scaling control.

As in municipal water treatment, coagulant chemicals are commonly added to industrial wastewater, and variables such as pH adjusted in such a way as to minimize the solubility of the product of the addition. This usually means either adsorption of pollutants onto the surface of the coagulant, or entrapment of pollutant particles within the matrix of precipitated coagulant.

Industrial effluent treatment differs from municipal in that chemicals may also be added to directly react with the pollutant, to destroy it or convert it into a more readily removed form. (There is an exception to this rule: ferric dosing for phosphate removal in municipal effluent treatment performs this function as well.)

An Applied Guide to Water and Effluent Treatment Plant Design. DOI: https://doi.org/10.1016/B978-0-12-811309-7.00018-7
237

NEUTRALIZATION

Industrial effluents may require correction to near-neutral pH prior to discharge (pH 5–9 is a common range). They may require tighter pH control to control solubility of key contaminants, or control reaction conditions. Some engineers still think that pH control is best carried out with stirred tanks and/or batch dosing rather than in-line to provide buffering and prevent excessive pH swings, but control of pH correction is in my opinion best achieved by means of pH probes with associated controllers, which will in turn operate dosing pumps to add acid or alkali to a low retention time mixer in proportion to the pH shift required. Modern equipment has no problem controlling pH to within 0.1 pH units.

I do however commonly see systems with actuated valves used for additions of acid and alkali, high retention time, poorly agitated mixing vessels, and other poor design features (see Fig. 18.1 for an example). There are, regrettably, many unskilled designers in the industrial effluent treatment sector. I discuss design issues further in Chapter 28, Classic mistakes in water and effluent treatment

FIGURE 18.1

Poor chemical dosing arrangements in a cyanide production plant.

plant design and operation. As Fig. 18.1 shows, standards of maintenance may also be less than ideal.

In addition, pH control can be used to prevent scaling or erosion of pipework. The determination of the requirement for dosing of this nature is carried out using the Langelier (or similar) Index. Alternatively, rectification of excessively scaling or aggressive waters can be carried out using lime addition, sometimes in conjunction with other chemical dosing.

METALS REMOVAL
PRECIPITATION AS HYDROXIDES

Many metal ions can be removed by filtration after precipitation as hydroxides by manipulating pH (Fig. 18.2).

The main challenge for the industrial effluent treatment plant designer is that there are commonly a number of different metals present each with a different minimum solubility pH value. The operating pH must therefore represent a compromise between these ideal pH values.

PRECIPITATION AS SULFIDES

If metals cannot be removed as hydroxides, the more expensive and dangerous approach of precipitating them as their sulfides can be used instead (Fig. 18.3).

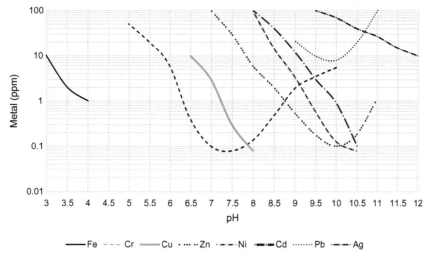

FIGURE 18.2

Metal hydroxide solubility versus pH.

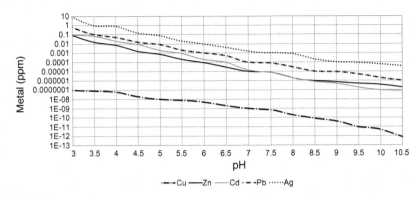

FIGURE 18.3

Metal sulfide solubility versus pH.

Hydrogen sulfide dosing is used in conjunction with pH control. However, the use of H_2S (as toxic as hydrogen cyanide) means that special care must be given to safety aspects of the design.

Complexing agents may also be used instead of sulfides

PRECIPITATION BY REDUCTION

Cr^{6+} can be converted to the far less soluble Cr^{3+} by reduction with sulfite.

SULFATE REMOVAL

Sulfate can be precipitated as the calcium salt to some extent, but it is reasonably common for this process to be insufficient to remove enough sulfate to satisfy consent to discharge to public sewer.

I have come across several other novel techniques for sulfate removal. These may reportedly work well in the lab, and appear to have translated into a few full-scale installations, but they do not in my experience perform well in practice. Great caution should be exercised in considering the use of alternative processes, especially any which claim to remove more sulfate than would be expected from the solubility of calcium sulfate.

Sulfate can only be practically, reliably removed to hundreds of ppm levels using expensive processes such as ion exchange or reverse osmosis.

CYANIDE REMOVAL

Cyanides can be removed from wastewaters using oxidizing agents. The most commonly used approach is alkaline chlorination using NaOH and Cl_2 at pH 11, though hydrogen peroxide or SO_2/air (both with copper II catalyst), and complexation with iron II sulfate have also been used.

ORGANICS REMOVAL

Biological processes are usually the cheapest way to remove biodegradable organic materials, but not all organic substances are biologically treatable. These hard to treat compounds are called refractory.

There are many specific techniques used in industrial effluent treatment to remove specific compounds, but the essence of a robust technique in effluent treatment is nonspecificity. The techniques used therefore tend to be either adsorption (see Chapter 17: Industrial effluent treatment unit operation design: physical processes) or oxidation (see "Other Chemical Processes" section).

OTHER CHEMICAL PROCESSES

Many of the chemicals discussed in the preceding sections have alternative uses, e.g., in controlling sludge bulking by filamentous organisms.

Control of pH by addition of lime or other alkaline substances may be undertaken to prevent sludges becoming obnoxious, or to give optimal conditions for biological growth.

Iron compounds may be used to bind H_2S preventing odors and corrosion. There are also other chemicals to inhibit H_2S formation, micronutrient preparations, and so on, which may be of greater or lesser value.

Ion exchange can specifically remove ions, selected by means of resin specification. The ions adsorbed are released during regeneration of the resin. The resins tend to have problems dealing with dirty water, with high levels of suspended solids, organics, certain metals, and so on. Pretreatment to remove these substances is usual.

There are several additional processes coming into use in tertiary treatment applications, such as the use of oxidants to remove reactive dyes and other refractory organics.

OXIDANTS

As well as the reasonably commonplace case of chemical oxidation of cyanide using alkaline chlorine, advanced oxidation processes are less commonly used to remove recalcitrant organics. These use strong oxidizing agents such as ozone, or UV-activated peroxide to cleave the strong bonds of organic ring structures which are often responsible for the lack of biodegradability of these substances.

These strong oxidant agents are expensive and will preferentially oxidize any less recalcitrant compounds present in the effluent, so they may need extensive pretreatment. Any chemical oxygen demand (COD) other than the recalcitrants will mop up oxidant, so as much COD as possible will need to be reliably removed by less expensive means before the advanced oxidation stage.

DISINFECTION
CHEMICAL DISINFECTION

As noted in Chapter 13, chemical disinfection is less popular nowadays, due to the likelihood of leaving residual disinfectant chemicals, or their reaction products in the wastewater. Historically, almost all known disinfectants have been used at some point.

Chlorine remains the most commonly used, because of its cheapness, ready availability, and ease of control of residual.

DECHLORINATION

Chlorination is perhaps the most controllable of the chemical disinfection techniques, as the addition of sulfur dioxide may be easily controlled to give zero chlorine and chlorine product residuals.

The control of this addition by means of combined chlorine residual monitoring leads to very good control of release of chlorine and its compounds, if monitoring and mixing processes are well designed and reliable.

An alternative technique for removal of chlorine in free and combined forms is granular activated carbon contact. This is more expensive than SO_2 dosing, but is 100% reliable if the bed is not exhausted.

FURTHER READING

Forster, C. F. (2003). *Wastewater treatment and technology*. London: ICE Publishing.

Mattock, G. (Ed.), (1978). *New processes of wastewater treatment and recovery*. Chichester: Ellis-Horwood.

Nemerow, N. L., & Dasgupta, A. (1991). *Industrial and hazardous waste treatment (Environmental engineering series)*. London: Van Nostrand Reinhold International.

US Industrial Environmental Research Laboratory. (1980). *Summary Report: Control and treatment technology for the metal finishing industry, sulfide precipitation*. Cincinnati, OH: US Environmental Protection Agency.

Industrial effluent treatment unit operation design: biological processes

19

CHAPTER OUTLINE

Introduction 243
Inhibitory Chemicals 244
    Aerobic Inhibitors..244
    Anaerobic Inhibitors 244
    Prediction of Heavy Metal Inhibition of Digestion Process 244
Nutrient Requirements..246
Microbiological Requirements 246
Aerobic Treatment Design Parameters..247
Anaerobic Treatment Design Parameters 247
Further Reading 248

INTRODUCTION

Biological processes can treat a surprising number of chemical residues in wastewater which might not be thought of as viable "food" for microorganisms.

The ratio of chemical oxygen demand (COD) to biochemical oxygen demand (BOD) is often used to indicate how treatable an effluent is by biological means. COD:BOD ratios much above 3 are usually taken as an indication that biological treatment is not likely to be a viable option.

Where industrial effluents are biologically treatable, there will often be a marked difference in performance between anaerobic and aerobic approaches.

As anaerobic digestion (AD) operates at around 35°C, heating requirements mean that a COD of above 2000 is usually required for AD to be economically favored in a temperate climate, though COD figures of one-tenth of that may be viable in tropical climates.

The specific chemicals which inhibit aerobic and anaerobic treatment processes are to some extent different, as are the nutrient requirements.

INHIBITORY CHEMICALS

AEROBIC INHIBITORS

Table 19.1 summarizes compounds which have been reported as inhibitors of aerobic biological treatment.

See Further Reading for references to the inhibitory concentrations of these compounds in activated sludge, trickling filters, and nitrification processes.

ANAEROBIC INHIBITORS

It is probable that many of the above compounds are also inhibitory to anaerobic treatment, though they may be more or less toxic than they are to aerobic treatment. The only compounds specifically cited in literature as inhibitory to anaerobic treatment are summarized in Table 19.2.

Light also inhibits anaerobic digestion, as indeed does overfeeding.

PREDICTION OF HEAVY METAL INHIBITION OF DIGESTION PROCESS

Laboratory experiments have shown a relationship between heavy metal concentrations and a 20% inhibition of digestion as set out in Table 19.3.

The effect of multiple heavy metals can be accounted for using Eq. (19.1) (Fig. 19.1).

Inhibitory Effect of Heavy Metals on Digestion

$$K\left(\frac{\text{meq}}{\text{kg}}\right) = \frac{\frac{[\text{Zn}]}{32.7} + \frac{[\text{Ni}]}{29.4} + \frac{[\text{Pb}]}{103.6} + \frac{[\text{Cd}]}{56.2} + (\text{Cu})/47.4}{\text{Sludge solids concentration(kg/L)}} \tag{19.1}$$

where figures in square brackets are concentrations in mg/L.

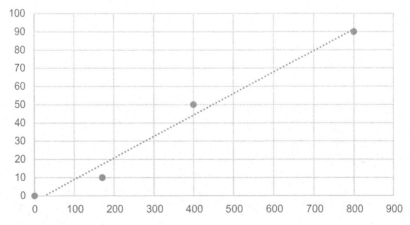

FIGURE 19.1

Probability of AD failure versus K (see equation 19.1).

Table 19.1 Aerobic Inhibitors

1,2-Dichlorobenzene	Furans
1,2-Dichlorobenzene	Green liquor (paper mills)
1,2-Diphenylhydrazine	Hexachlorobenzene
1,3-Dichlorobenzene	Hydrogen sulfide
1,4-Dichlorobenzene	Iodine
1,4-Dichlorobenzene	Isothiocyanates
2,4,6-Trichlorophenol	Lead
2,4-Dichlorophenol	Methyl Ethyl Ketone
2,4-Dimethylphenol	Mercaptans
2,4-Dinitrotoluene	Mercury
2-Chlorophenol	Methyl chloride
Acetylenic compounds	Naphthalene
Acrylonitrile	Nickel
Alkyl amines	Nitrobenzene
Ammonia	Oxygen scavenging compounds
Anthracene	Ozone
Antibiotics	Pentachlorophenol
Arsenic	Peracetic acid
Benzene	Peroxyacetic acid
Bisphenols	Phenanthrene
Black liquor (paper mills)	Phenol
Bromine compounds	Phenolics
Cadmium	Pyridine and derivatives
Carbamates	Quaternary ammonium compounds
Carbon tetrachloride	Silver
Chlorate	Sodium hypochlorite
Chlorinated phenolics	Sulfate
Chlorine dioxide	Sulfide
Chlorite	Sulfide scavengers with an amine group
Chlorobenzene	Surfactants: (cationic, anionic, nonionic)
Chloroform	TCA (Trichloroacetate)
Chromium	TCE (Trichloroethylene)
Copper	Terpenes
Cyanide	Tetrachloroethylene
Diesel range organics	Thiocyanate
Dioxins	Toluene
Dipropylene glycol monomethyl ether	Trichloroethylene
DPTA (Pentetic acid)	Trichlorofluoromethane
EDTA (Edetic acid)	Xylene
Ethylbenzene	Zinc
Filming amines	

Table 19.2 Anaerobic Inhibitors

Sodium	Potassium	Calcium
Magnesium	Aluminum	Ammonia
Chromium	Iron	Cobalt
Copper	Zinc	Cadmium
Nickel	Long-chain fatty acids	Alkyl benzenes
Halogenated benzenes	Nitrobenzenes	Phenol and alkyl phenols
Halogenated phenols	Nitrophenols	Alkanes
Halogenated aliphatics	Alcohols	Halogenated alcohols
Aldehydes	Ethers	Ketones
Acrylates	Carboxylic acids	Amines
Nitriles	Amides	Pyridine and its derivatives
Surfactants	And detergents	Oxygen
Herbicides	Antibiotics	

Table 19.3 Inhibitory Effect of Heavy Metals

Metal	Batch Digesters: Concentration (mg/kg Dry Solids)	Typical Concentration in Digested Sludges (mg/kg Dry Solids)
Nickel	2000	30–140
Cadmium	2200	7–50
Copper	2700	200–800
Zinc	3400	500–3000

NUTRIENT REQUIREMENTS

The commonly held benchmark is that aerobic processes require a BOD:N:P ratio of 100:5:1. It is worth noting that this can be met with standard agricultural fertilizers. Sophisticated nutrient mixes (with or without added proprietary bugs) are part of the industry of "marketing breakthroughs" which surrounds industrial effluent treatment.

MICROBIOLOGICAL REQUIREMENTS

As well as sophisticated nutrients with added bugs, it is possible to buy the bugs with or without super nutrients.

My first job after university was working for a company which sold this type of bug. I discovered there that the bugs in question were obtained by sampling a treatment plant which worked well, and then growing those of the bugs present which also grew well in the lab. A small amount of the resulting culture was sprayed into a carrier such as wheat bran, or mixed with a solution which would protect the spores of the bugs which were grown up.

There are usually no live bugs in such products. The best case scenario is that the product being sold contains traces of the spores of the bugs which grew in culture inoculated with bugs from a working plant treating a similar wastewater to the one requiring treatment. However, not all bugs grow in culture, and not all cultured bugs form spores. The best case will still be inferior to obtaining some sludge from a working plant, but even this may prove ineffective, due to the nature of biological processes.

The bugs which will grow in a plant are those which are best suited to the ecological niche which the plant represents. This ecological niche is competitive, so the best adapted bugs will outcompete those which are less well adapted. Bugs can multiply exponentially in favorable environments, and die in unfavorable environments. Even adding a ton of live bugs every day may not be enough if you have not created a proper home for them.

AEROBIC TREATMENT DESIGN PARAMETERS

Equalization of industrial wastewaters in a tank with a 24-h hydraulic retention time (HRT) is recommended prior to biological treatment. The equalization tank may be mixed or plug flow, and may be aerated to prevent septicity.

This general specification is however not universal. In the pesticide industry, neutralization is usually undertaken in basins with HRT in the region of 6 min, and 14 W/m^3 mixing requirements.

Biological treatment takes place in aerated lagoons with HRTs from 2 to 95 days, in activated sludge plants with HRTs from 7 to 79 h, or (less popularly) in trickling or high-rate filters. BOD removals from 87% to 98% and COD removals from 61% to 90% are obtained in these plants.

ANAEROBIC TREATMENT DESIGN PARAMETERS

Wastes which have proven economically treatable by AD include:

1-Octanol
Acetylsalicylic acid
Acrylic acid
Benzoic acid
Brewery effluent
Butylbenzylphthalate
Confectionery

(Continued)

Continued

Dairy waste
m-Cresol
Nitrophenol
Paper and pulp
Pharmaceutical
Phenol
Polyethylene glycol
Potato waste
Rendering
Slaughterhouse
Sugar beet
Terephthalic acid

However, it should be noted that novel types of AD, such as upflow anaerobic sludge blanket, are in my experience often not particularly robust, especially in the medium to long term. Lab studies, and even pilot plant work, may be misleading.

I was once instructed to investigate problems at one of the plants which had originally featured as a government-sponsored case study promoting AD for industrial effluent treatment. The plant had initially worked very well, but several years down the line it was no longer working at all. The plant was in a sugar mill, and the lime used in sugar processing had produced inert lime granules in the reactor with little biological activity. Conversely, I have also heard of many sites where granule formation is hard to maintain.

FURTHER READING

Anthony, R. M., & Briemburst, L. H. (1981). Determining maximum influent concentrations of priority pollutants for treatment plants. *Journal Water Pollution Control Federation, 53*(10), 1457–1468.

Chen, Y., Cheng, J. J., & Creamer, K. S. (2008). Inhibition of anaerobic digestion process: A review. *Bioresource Technology, 99*(10), 4044–4064.

Jenkins, D. I., & Associates. (1984). *Impact of toxics on treatment: Literature review.*

Russell, L. L., Cain, C. B., & Jenkins, D. I. (1984). Impacts of priority pollutants on publicly owned treated works processes: A literature review. In *1984 Purdue industrial waste conference.*

US EPA. (1986). *Interferences at publicly owned treatment works.* Cincinnati, OH: US Environmental Protection Agency.

Industrial effluent treatment hydraulics

20

CHAPTER OUTLINE

Introduction 249
Hydraulics and Layout 249
Hydraulic Design 250
Flow Balancing 250
Sludge Handling 251

INTRODUCTION

In my experience, the only difference between the hydraulics of municipal and industrial effluent treatment plants is that industrial effluent treatment plants are so often designed as afterthoughts, fed in short sharp batches by the main process into which most of the design effort went.

In this chapter, I have attempted to offer some basic guidance on how to approach a design, for those conscientious designers who consider all aspects of the plant during the design process.

The designer should always remember that effluent treatment plants (ETPs) are commonly seen as at best a nuisance by those who operate the main process. The designer should assume that any chemical used on the site will find its way to the ETP, and all means of transferring flow to the ETP will be run at their maximum capacity.

Although businesses which supply industrial effluent treatment plant have some of the highest profit margins of anyone in the water industry, many such plants which I encounter are cheap and poorly constructed, based on best outdated technology, and designed with little ingenuity. In my experience the hydraulic aspect of these designs tends to be the weakest area of all.

HYDRAULICS AND LAYOUT

Ideally the plant should be located at a low point on the site, to allow gravity feed from the main process. Channels may be favored for effluent collection, but care should be taken in a situation where heavier-than-air flammable vapors might collect in such a channel, allowing the possibility of rapid spread of fire around a site.

In my experience a very common approach is to feed the plant from several large pits or sumps around the site, emptied over a short period of time by large pumps. However, this approach is never optimal. The requirements of operation and commissioning for the ability to empty a tank quickly may make a pit a good idea, allowing a hydraulic buffer between the tank and the ETP, but the point of such a tank is not that it be emptied quickly. The pumps and pipework used to empty the pit would ideally be sized to restrict flow out of the pit. Experience suggests that they tend to be run at their maximum rated capacity, to maximize flexibility for the main process, or just to allow operators to attend to something else.

HYDRAULIC DESIGN

Although, in the preceding section, I recommended deliberately restricting flow from catch pits into the plant, I would nonetheless recommend that all flow restrictions should be deliberate. The strong tendency to design industrial effluent treatment plants to be as cheap as possible should be resisted. Any successful main plant will tend, over time, to produce more effluent. An allowance for this should be built into the hydraulic design.

The normal approaches to hydraulic design used by water treatment plant design specialists should be followed. A common mistake made by designers from other sectors trying their hand at effluent treatment plant design is to use active flow control, such as modulating valves, where specialists would use passive flow splitting chambers, and to use pipes where specialists would use channels.

It is always better to hire a genuine specialist, though there are many more amateurs operating in industrial effluent treatment than there are in the municipal market. The relatively small size of the projects, and their one-off nature makes them unattractive to the larger established water treatment contractors.

FLOW BALANCING

An industrial effluent treatment plant should normally be designed to handle the highest average daily flows experienced over the course of a year. The peak design throughput of the plant should be the average hourly flow on this peak day. A 24-h HRT in-line flow balancing tank upstream of the plant will allow accommodation of any instantaneous flows which are higher than the average.

The alternative (sizing the plant to handle instantaneous peak flow) will be likely to cost far more money, and be less controllable and flexible, due to turndown ratio restrictions and feed flow and composition variability.

SLUDGE HANDLING

The particles in industrial effluent may be far denser than those found in municipal water and wastewater treatment plants. Industrial effluents can consequently produce very dense, hard to pump sludges.

The high degree of variability of many industrial effluents can also mean that the density, viscosity, temperature, and chemical composition of sludges can vary dramatically over time.

This needs to be considered when designing hoppers, tanks, and other sludge holding facilities, along with the type and ratings of pumps, pipes, valves, and so on.

Rodding points or similar provision for solids settlement are recommended wherever wastewater or sludges are to be held and at bends in sludge pipework. Pipe runs between rodding points should be restricted to a length which can be rodded-through effectively.

Sludge treatment engineering

5

INTRODUCTION: AN OVERVIEW OF SLUDGE TREATMENT

The treatment of domestic and industrial clean and wastewaters in engineered plants produces (as a waste product) mixtures of water and solids, which are known as sludge.

The treatment of the organic component of waters by biological means generates sludges, comprising some of the original wastewater solids, plus significant quantities of the organisms used in the treatment processes.

Very significant quantities of sludge are produced in nations with a high degree of treatment of wastewaters. The United Kingdom, e.g., would fill to a depth of 1 m in a year with sewage sludge if it were not disposed of.

Sludges are treated by a process of dewatering, followed by destruction of the residual solid matter. Resource recovery is then carried out, most commonly using anaerobic digestion, which yields both a biogas fuel, and a digestate that can be used as a fertilizer.

Throughout these processes, consideration must also be given to the nuisance factors associated with sludge, in particular odor and pathogen content.

The methods for disposal of treated sewage sludge are varied, but none is universally popular. The main techniques used are:

- Landfill
- Incineration
- Use as fertilizer
- Sea disposal
- Land application

Each of these techniques has its advantages and disadvantages.

There are also experimental techniques such as gasification and pyrolysis, which are not yet market ready (some of which will never be market ready).

Sludge characterization and treatment objectives

21

CHAPTER OUTLINE

Treatment Objectives...255
   Volume Reduction 255
   Resource Recovery 256
   Other Objectives..256
Sludge Characteristics 257
   Wastewater Sludges...257
   Potable Water Sludges 259
Disposal of Sludge 261
Beneficial Uses of Sludge 262
Solids Destruction..262
Further Reading 263

TREATMENT OBJECTIVES

The main problem with sludge is its sheer mass, the great majority of which is water. Most water and effluent treatment sludges are around 99% water.

The primary objectives of sludge treatment are therefore mass/volume reduction by removal of water and, where possible, resource recovery.

VOLUME REDUCTION

Sludge varies in its proportions of solid materials and water. In a municipal plant its concentration generally varies between 0.5% and 10% dry solids as originally yielded. It is uneconomic to transport sludges with a lower dry solids content for long distances, so sludges are usually thickened to a higher dry solids content (perhaps 10% dry solids) prior to transportation off site.

Thickened sludges may also be dewatered to produce a still thicker material by removal of water. Thickened sludges will usually have a solid or semisolid appearance. The oldest still-current technology for sludge dewatering is the plate filter press, which was invented in 1853.

Dewatered sludges may be dried to drive out residual moisture, producing a solid material (Fig. 21.1). This is nearly always by means of thermal drying. The

FIGURE 21.1

Dried sewage sludge[1].

Courtesy: Hannes Grobe.

material produced may be classified to reduce the range of particle sizes, and/or pelletized to produce consistently sized pellets of material. Almost all thermal drying processes need to use heat to evaporate water from the sludge, as well as electrical energy to move sludge so that it may be dried.

RESOURCE RECOVERY

Anaerobic digestion is a process whereby organic material is broken down by microorganisms in the presence of hydrogen to produce a combustible methane-rich gas known as biogas. This biogas can then be used as fuel, while the remaining matter, known as digestate, can be used as a fertilizer. Anaerobic digestion is not a new technology, but has grown in use since the late 20th century and is now the most widely used process in the treatment of sewage sludge.

OTHER OBJECTIVES

Secondary objectives center on nuisance reduction. Sludges can be highly odorous, and they can contain pathogens.

---

[1]Licensed under CC BY-SA 2.5: https://creativecommons.org/licenses/by-sa/2.5/

SLUDGE CHARACTERISTICS
WASTEWATER SLUDGES

Primary sewage sludges are generally more offensive than secondary sludges. As a general rule, the greater the degree of treatment, the less offensive sludges are (when fresh). Due to the nature of the various treatment processes, it is also usually the case that sludge solids are denser and sludges have higher solid contents for lesser degrees of treatment. Primary sludges may have solids with densities of 1400 kg/m³, secondary sludges around 1300 kg/m³, and tertiary filter backwashings 1200 kg/m³. The exception to the rule is trickling filter humus at around 1450 kg/m³.

Sludges from chemically dosed primary settlement tanks may have far higher densities than these, with high doses of ferric and lime to primary tanks potentially yielding sludge solids densities of 2200 kg/m³.

Historically, the heavy metal content of sludges also used to be of great interest, as it used to form the basis of allowable loadings for land spreading before the spreading of untreated sludges was banned in the EU. The various heavy metal concentrations were converted by loading factors into a nominal equivalent amount of zinc for this procedure.

Often the most important properties of sludge are hydraulic. The ability of sludges to be pumped, dewatered, and so on depends often on the nature of the sludge. Non-Newtonian properties are commonplace with thicker sludges.

Sewage sludge yields

Sludges may be yielded from each of the stages of treatment. Primary sludges are usually produced at 2%−10% solids; secondary sludges at perhaps 0.5% solids. Even the most concentrated of these is 90% water. Primary sludges tend to be more offensive than secondary sludges when fresh, although secondary sludges are susceptible to putrefaction.

In addition to these common sludges, additional processes such as nutrient removal, chemical treatment, digestion, and so on may produce sludges with different characteristics. Scum from primary treatment is also usually treated alongside sludges. There may also be import of sludges from septage, industrial sites, and so on.

Approximate sludge dry solids mass values are given in Table 21.1 for use in conceptual design, in the common case where no lab results are available

In addition, the following inputs should be taken into account:

- Works liquors: recycled flows, most significantly from sludge dewatering
- Trade effluent: estimated as described earlier
- Imported sludge: estimated or known from survey

More accurate estimates of sludge volumes per head of population per day can be calculated using the values in Table 21.2.

Table 21.1 Rough Estimate Mass of Dry Solids

Sludge Type	Mass of Dry Solids (g/h per day)
a. *Primary sludge*	50
b. *Secondary sludge*	
Activated Sludge (AS)	
Settled sewage, nitrifying	35
Settled sewage, high rate	40
Unsettled sewage, nitrifying	45
Unsettled sewage, high rate	65
Filter Sludge	
Low rate, nitrifying	20
High rate	40
c. *Tertiary sludge*	1
d. *Cosettled sludge*	
Primary and nitrifying AS	85
Primary and low-rate filter	70
e. *Anaerobically digested sludge*	
Primary	32
Cosettled primary and nitrifying AS	54
Primary and low-rate filter	45

Table 21.2 More Detailed Estimate of Dry Solids Mass

Sludge Type	Dry Solids (%)	Sludge Yield (kg Sludge/kg BOD per day)	Sludge Volume (m³/day)
Primary Sludge	8.0		$0.76 \times S \times DWF$
Secondary Sludge			
Surplus Activated Sludge			
Conventional carbonaceous	0.8	1.0	$0.13 \times BOD$
Low rate, nitrifying	0.8	0.8	$0.10 \times BOD$
Extended aeration, nitrifying with sludge stabilization	0.8	0.7	$0.09 \times BOD$
Wasted Filter Sludge			
Low rate, nitrifying	2.0	0.5	$0.03 \times BOD$
High rate	2.0	1.0	$0.05 \times BOD$
Rotating Biological Contactor (RBC)	2.0	0.5	$0.03 \times BOD$
Cosettled Sludges			
Primary and low-rate filter	6.0		$1.0 \times S \times DWF$
Primary and nitrifying AS	3.0		$2.0 \times S \times DWF$

BOD, *BOD into biological treatment (kg/day)*; DWF, *dry weather flow (m³/s)*; S, *crude sewage SS concentration (ppm)*.

Table 21.3 Sludge Yield From Domestic Sewage by SS/BOD Ratio

SS/BOD ratio	0.5	0.7	0.9	1.1
Conventional	1.1	1.15	1.2	1.2
Nitrifying	0.9			
Extended aeration	0.8	0.85	0.9	0.92

Table 21.4 Sludge Production Rates

Sludge Loading Rate	Rate of Sludge Production (% Age)
0.10	5
0.15	9
0.20	13
0.25	18
0.30	23
0.45	33
0.50	45

As shown in Table 21.3, sludge production may also be estimated based upon the usual ratio of suspended solid:biochemical oxygen demand (SS:BOD) at the aeration tank, assuming a settled sludge volume index (SSVI) of 125.

These sludges will require containment and treatment. Their high water content means that dewatering/volume reduction will form an important part of these treatments, in order to minimize the size of downstream treatment units. The water removed from the sludge will usually be quite strong in terms of chemical oxygen demand, and often offensive in character.

Yields of primary sludge are often estimated based on removal of 50% of incoming SS at 5% dry solids. Yields of secondary sludges vary with the degree of treatment.

Taking the figures given for sludge loading rate given in the earlier chart, the rates of sludge production as a percentage of total existing mixed liquor solids are given in Table 21.4.

Design of secondary sludge treatment facilities is usually based on a yield of the above sludge quantities at 0.5% DS.

Yields of other sludges vary per the exact process being used. Yields from anaerobic digestion will be covered in a later section.

POTABLE WATER SLUDGES

The (unfortunately long out of print) WrC Technical Report TR189 gives very detailed practical advice on waterworks sludge characterization and treatment, which is as far as I know still current and certainly a sound basis from which to work. Some useful facts from this publication are summarized as follows:

- The solids in waterworks sludge have a specific gravity of around 2.5. Their contribution to density becomes significant at >1% DS sludge. 1 tonne of wet sludge occupies $(1 - 0.6S)$ m^3, where S is the fractional solids content
- Solids content is a poor predictor of handleability in waterworks sludge. 25% dry solids sludges can be a pourable fluid, whilst 19% DS filter cakes can be a hard solid
- Incoming suspended solids concentration to a waterworks can be estimated as 2 × turbidity in Formazin Turbidity Units (FTU)
- Sludges are usually yielded from upflow clarifiers at about 0.2% DS, or as low as 0.1% with low incoming color and turbidity
- Filter wash waters require 2−4 h of static settlement, and can yield 1% DS sludges

Potable water sludge thickening

TR189 states that continuous polymer thickening can produce sludges with 5%− 20% DS from any practical solids concentration feed, with recovered water turbidity of 4−8 FTU.

More precisely, it predicts the following performance for a surface loading on the thickener of 1.5 m/h on the following basis:

Thickened sludge solids = (Total sludge solids in raw water (mg/L))/(0.25 × Coagulant hydroxide dose (mg/L)) + (0.02 × Color removed (°H))

Water treatment sludge yield

The yield of sludge from drinking water treatment can be estimated using by the equation taken from TR189:

Sludge solids produced in color removal (ppm) = 0.2 × Color removed (°H)

A more sophisticated and comprehensive equation based on this is:

$$Total\ solids\ load\ (mg/L) = 2.9 \times [Al_3^+] + 1.9 \times [Fe_3^+]$$
$$+ 1.6 \times [Mn_2^+] + 0.2 \times Color\ (°H) + [SS]$$

All terms in square brackets are in mg/L, and include any dosed chemicals. The highest expected concentrations of all parameters should be used in estimation.

The design of sludge treatment facilities is usually based upon solids being initially recovered as sludges with solids contents of 0.1%−0.2% DS for upflow clarifiers, 1% from filter backwash recovery tanks, and 5% for dissolved air flotation.

DISPOSAL OF SLUDGE

The disposal of produced sludge is potentially a major operating expense for sewage treatment operators. The cheapest and most environmentally beneficial option is to spread it on agricultural or arboricultural land.

There are however several potential problems with spreading to land. From a purely scientific point of view, the capacity of the land to handle residual pollutants, most notably heavy metals, is limited. Limits on application based on controlling the maximum loading of heavy metals onto agricultural land are therefore necessary. More contentiously, there are public concerns about other pollutants which have not, to date, been shown to harm the environment, human, or animal health.

While sludge is produced all year round, spreading can only be beneficial to crops at certain phases of their growing cycle. There are public concerns about odors arising during and after sludge spreading, which have been addressed by requiring subsurface application. There are also public concerns about the possibility of contagion from sludge spreading.

In the United Kingdom, these concerns have been addressed voluntarily by an agreement between Water UK, the British Retail Consortium (BRC), Department of Environment Transport and Regions (DETR), Food Standards Agency (FSA), Environment Agency (EA), and the Food and Drink Federation (FDF), known as the Safe Sludge Matrix. This essentially requires sludges to have been disinfected to a high standard before spreading to land. Such disinfected sludges are known as *enhanced treated sludge*.

Under the Safe Sludge Matrix, the use of untreated sewage sludge on land used to grow food crops was phased out from 1999; from land used to grow industrial crops that also have a food use from 2001; and on land used to grow nonfood crops from 2005.

There are many methods used to destroy or deactivate microorganisms in sludge to achieve disinfection, such as application of heat, chemicals, or radiation.

Engineers design disinfection processes by means of graphs (known as kill curves) which show the proportional reduction in numbers of viable organisms when exposed to varying strengths of a disinfectant treatment over increasing periods of time (see Fig. 4.11). These curves all have the same general appearance, which we can express in words as "the greater the strength of a disinfectant treatment and the longer it lasts, the lower the proportion of the original number of live microorganisms we can detect." A factor of 10 reduction in microorganism numbers is known as a 1-log reduction.

The sludge matrix defines "enhanced treated sludges" as being "free from Salmonella and ... treated to ensure that 99.9999% pathogens have been destroyed" (at least a 6-log reduction). This is essentially a requirement for disinfection.

The standard does not require a simple reduction in numbers of all types of organisms present, but requires a high degree of destruction of pathogens in general and Salmonella in particular.

A sustained temperature of at least 80°C for 10 min should produce enhanced treated sludges in conventional thermal dryers. Specific tests should be applied to any proposed new processes (see Further Reading).

BENEFICIAL USES OF SLUDGE

The historically most important disposal routes for sludge were land application and sea dumping. Only the first of these is still permitted in the EU (following treatment), and the second is becoming less popular.

Attempts have been made to produce marketable products from sludge, from simple garden fertilizers through building blocks to jewelry.

SOLIDS DESTRUCTION

The most common solids destruction approach is anaerobic digestion, with land spreading or incineration of the residual sludge.

Incineration is often the most logical route for destruction of digested sewage sludge, and can recover useful energy in the process. In Japan, extensive research to use the ash produced by this process has led to the development of sewage bricks, tiles, slag, aggregate, cement, and pumice.

There is however widespread opposition to incinerators in many countries. Although much of the opposition is highly illogical, public fears about incineration influence the viability of plans to install incinerators.

Thermal hydrolysis, pyrolysis, and gasification may be more publicly acceptable than incineration (Fig. 21.2).

The various forms of thermal hydrolysis are however still not that close to market ready practical options, and pilot scale experiments suggest that some of them are too hard to control to ever be practical at full scale.

In Australia and elsewhere, worm cultures are being used to stabilize sludge. The product of vermistabilization exceeds the New South Wales EPA "grade A" requirements for pathogen kill.

FIGURE 21.2

Plant for thermal hydrolysis of sewage sludge.

FURTHER READING

European Commission Directorate-General Environment. (2001). *Evaluation of sludge treatments for pathogen reduction.* Available at <ec.europa.eu/environment/archives/waste/sludge/pdf/sludge_eval.pdf> Accessed 10.10.17.

Sludge treatment unit operation design: physical processes

22

CHAPTER OUTLINE

Sludge and Scum Pumping 265
Grinding 266
Degritting 266
Blending..266
Thermal Treatments 267
Volume Reduction Processes..267
    Thickening...267
    Dewatering 271
    Novel Processes 272
Further Reading 273

SLUDGE AND SCUM PUMPING

Sewage sludges have several characteristics that make the selection of pump type crucial. They may exhibit non-Newtonian properties (i.e., shear stress and shear rate are nonlinear) and tend, for all but the thinnest sludges, to have viscosities more than that of water. They are furthermore continuously variable in their properties. They may be shear-sensitive (such that they are reversibly or irreversibly different after exposure to shear). They are often abrasive in character, and contain a range of solid particles.

In general, low-shear, positive displacement pumps are therefore required for most sludge pumping duties. Certain types of centrifugal pumps may be used, with modifications to accommodate the above problems, but they require very careful design, as their range of conditions handled will be narrow. These may be used for activated sludge handling, due to the low solids content, and large volumes to be pumped.

Probably the commonest type of pump used for thicker sludges is the progressing cavity type. These may be preceded by grinders or bridge breakers integrated within the body of the unit.

Diaphragm pumps have a more limited use, as they are inefficient, and are limited in both delivery head and volume.

Piston or ram pumps may be used for high pressure applications like long-distance sludge pumping, or the filling of filter presses.

Recycle of activated sludge in smaller plants is often done using airlift pumps (see Fig. 12.1). These are designed as described in Chapter 12, Dirty water unit operation design: physical processes.

On larger plants, the scum component of mixed sludges may coat the inside of the pipelines with grease. Facilities for melting this out may have to be provided.

It is generally accepted that sludge lines should not be less than 50 mm nominal bore (NB), irrespective of the volumes of sludge to be pumped. On large plants, this minimum size may be as high as 150 mm NB. Superficial velocities of <1.5 m/s are usually desirable for pumped lines, and far lower for gravity lines, perhaps 0.5 m/s, though solid settlement can be a concern.

Some rules of thumb for digester sludge pipework are as follows:

- The sludge feed to a digester should be via a drop leg with a ventilated upper end to prevent gas traps and syphoning
- Sludge pipework is to be lagged and arranged in such a way as to prevent any gas traps

GRINDING

Sludges may be subjected to grinding (e.g., where there is a high concentration of animal hairs at abattoirs or tanneries, and the destination of the sludge is drying and pelletization). Grinding is however declining in popularity due to the high maintenance requirements of even the most modern grinding units.

DEGRITTING

Where wastewater degritting is ineffective, sludges may contain grit, and it may therefore be necessary to degrit the sludge. Centrifugal units known as cyclone degritters may be used for this duty.

BLENDING

Blended sludges have reduced variability, which hopefully improves the performance of downstream processes.

Sludge blending may be carried out in-process. For example, secondary sludges may be cosettled with primary sludges, or chemical sludges cosettled with secondary sludges.

Anaerobic digesters were also once used as a sort of sludge blending operation, but it is more common nowadays to feed digesters with a blended, standardized sludge to optimize efficiency. This may involve side stream thickening of some of the sludges to allow a consistent solids content to be blended. Alternatively, inline, or in-tank mixers may be used to blend sludges off-line.

Sludge storage facilities may form part of this blending operation, smoothing variations in sludge production rates to allow a consistent blended sludge to be produced. On smaller works, sludge storage within process units may be sufficient but on larger works, tanks—often with mixing facilities—may be provided.

Care must be taken to maintain a low sludge inventory to avoid septicity, which causes problems with downstream processing. If it is not possible to keep sludge fresh, chemical septicity control may be required. Sludge processing facilities usually have odor control facilities designed in.

THERMAL TREATMENTS

Heat treatment of sewage sludges often has a beneficial effect on downstream processes, including dewatering and oil removal. Dewatering of heat-treated sludges without chemical addition is possible.

There are several proprietary systems available for drying and/or size reduction of sludges. The rotary and multiple hearth types are most common, although several dryer technologies used in other sectors have been tried for sewage sludge drying.

Incineration is a cost-effective way of disposing of sludges with a significant organic component, such as sewage sludges and oily refinery sludges. It may even prove profitable. It is however often unpopular with local communities in general and with green pressure groups in particular (however much effort is put into pollution abatement by the designer).

Little needs to be said here about incineration, as incinerators are procured as a complete plant from a specialist supplier.

More speculatively, pyrolysis can produce a synthetic gas (syngas) fuel from sewage sludge by thermal decomposition at high temperature and pressure in the absence of oxygen. This may be more publicly acceptable than incineration, even though the produced fuel is ultimately burned, because the process also produces biochar, which locks up carbon in a form resistant to environmental degradation, with the possibility of anti-global-warming benefits. However, pyrolysis is not presently a commercially available/viable process.

VOLUME REDUCTION PROCESSES
THICKENING

It is conventional to divide volume reduction and dewatering into several stages.

The first of these stages is thickening or consolidation in tanks. This process may be carried out in simple gravity consolidation tanks or stirred tanks, by flotation, centrifugation, or belt thickeners. Addition of chemical agents to

FIGURE 22.1

Sludge thickening tanks.

promote flocculation is usual. The required doses of polymer increase through the list given above as a rule.

Gravity thickeners may be a simple tank with multiple supernatant draw offs up the height of the tank as in Figs. 22.1 and 22.2 or units similar in design in many respects to sedimentation tanks (although usually much taller in UK design practice) as below. Gravity thickening works best with sludges with high solids density, such as primary sludge.

Raked gravity thickener designs for waterworks sludge are described in detail in WRc plc's publications TR 189 (and, more recently, in TT016—see Further Reading). Surface loadings of 1.5 m/h are recommended, in cylindrical steel walled tanks on a concrete base with diameters at least as great as depth, 1.5–15 m in diameter, and working depths up to 3 m.

These tanks are fed from a mixed balancing tank designed to average out loads and volumetric flows of feed over the day. The feed is passed through an orifice plate on its way to the thickener to give a short period of high intensity mixing, at the point where polymer is added. In the absence of trial data, velocity through the orifice should be around 5 m/s for alum sludge, and 11 m/s for iron sludge. Downstream of the addition point, velocity gradient should be less than 300 m/s (Table 22.1).

Thickener diameter can be calculated using Eq. (22.1).

Thickener Diameter Calculation

$$D = \left(\frac{4}{\pi} \times \frac{\text{Feed pumping rate(m}^3/\text{h})}{\text{Thickening rate(m/h)}} \right)^{0.5} \tag{22.1}$$

FIGURE 22.2

Gravity sludge thickener[1].

Courtesy: High Contrast.

Table 22.1 Flow in Smooth Straight Pipes With Average Velocity Gradients of 300 m/s

Bore	mm	25	50	75	100	150	200	250	300
Bore	in.	1	2	3	4	6	8	10	12
Flow	m³/h	1	5	12.5	25	64	125	210	322
Flow	gal/h	216	1080	2760	5500	14,000	27,500	46,000	71,000

Courtesy: WRc plc, TR 189.

Thickener working depth can be calculated using Eqs. (22.2a and b).

Thickener Working Depth Calculations (a) Supernatant Depth; (b) Total Depth

$$\text{(a) Supernatant depth(m)} = 0.225 + 0.05D$$
$$\text{(b) Total depth(m)} = 1.35 + 0.1D \tag{22.2}$$

A conical sludge hopper should be provided in the base of the tank to hold 3 days' production of thickened sludge. The hopper should have an included angle of no greater than 60 degrees, and no space where sludge can stagnate. The thickened sludge pump should be rated at least four times that of the thickener feed rate.

In my experience, the most common error in the use of WRc TR 189 design for continuous thickeners is to overlook the fact that TR 189 assumes a consistent dose of polymer per kilogram of incoming sludge solids which, in turn, relies upon a consistent solids level in the feed to the thickener. The feed tank or sump must therefore be appropriately sized and mixed to buffer variations in flow and solids concentration.

For example, blending 5% DS DAF (dissolved air flotation) sludge with 0.5% DS settled backwash sludge at a tenth of its solids concentration might require a sludge buffer tank with 1−2 days hydraulic retention time.

The treatment of unblended DAF sludges using TR 189 thickeners leads to a solid-limited design, with thickener size determined by rake capacity. Outer echelon scrapers can be omitted from the rake to utilize the minimum diameter provided, giving a hydraulic loading less than 1.5 m/h.

Sludge rake capacity design should be based on four times daily sludge production rate. The discharge pump capacity must exceed the rake capacity. Hydraulic design should be based on the rated capacity of the thickener feed pumps. Discharge arrangements should be kept simple, especially for thicker sludges.

Centrifugation of sludges has only become very popular in recent years with the development of organic polymers. These have made it possible to obtain very high solids content from decanter centrifuges. These continuous centrifuges are superseding the basket types formerly used, which required batch operation. Decanter centrifuges (Fig. 22.3) can be used for thickening without polymer addition in some cases, but it is usually possible to achieve up to 20% DS sludges by the addition of polymer.

FIGURE 22.3

Decanter centrifuge.

Courtesy: Expertise Ltd.

Gravity belt thickeners are entirely proprietary items, dewatering sludges on a moving belt. They require high doses of polymer, and sometimes filter aid, to give similar performance to centrifuges.

DAF designs are very variable between suppliers, and may be circular, rectangular, and have a variety of arrangements to mix sludge and air bubbles. As the process is one of flotation, there is slightly better performance with sludges with lower solids density such as waste activated sludge. The design solids loading rate for sludge thickening DAF should consider the required quality of both thickened sludge and subnatant. To obtain thickened sludge at 5% DS and subnatant <50 mg/L SS, $4-6$ kg SS/m^2 per hour is appropriate when polymer is not used, and $6-10$ kg SS/m^2 per hour when polymer is used.

There are other systems in use, such as rotary drum thickeners and vacuum filters, but their use is less common than the units described earlier.

DEWATERING

The development of organic polymers in the past few decades has changed the technology used for sludge dewatering quite radically.

Where vacuum filtration and batch centrifugation traditionally required heavy manpower inputs (and, in the case of the vacuum filters, high doses of filter aids), decanter centrifuges and belt filter presses can offer completely automated processes, with a single chemical used at low concentration.

Belt filter presses are presently state of the art dewatering equipment in municipal sludge treatment works.

They have a relatively low capital cost at small facilities, and have a lower energy consumption and potential for noise and odor nuisance than centrifuges. They are however not without disadvantages. Centrifuges offer 15% lower capital cost, and half the footprint of belt presses.

Efficient dosing and mixing, use of flocculants, and uniform distribution of flocculated sludge over the belt is essential to the belt press operation. These are the factors which limit belt press size to $200-300$ kg/m^2/h.

As with DAF (but unlike centrifuges), variations in sludge SS are not well tolerated by belt presses. Feed DS and flowrate have to be balanced, necessitating continuous measurement of incoming SS levels.

Clarified filtrate might be expected to have a DS content of $1-2$ g/L, similar to that for centrifuges (but not DAF). However, solids washout from the belt can make this a lot higher, and the 1 m^3/h of washwater per m^3 of feed of which has to be added to wash the belt ultimately adds to the recycle loads.

The WRc publications (see Further Reading) contain detailed design information for centrifuges and belt presses for waterworks sludge. Despite the shortcomings of belt presses, the present trend is away from centrifuges and toward belt presses, after many centrifuge installations failed to match their expected performance over the long term and proved labor and energy intensive.

FIGURE 22.4

Plate filter press.

Plate and frame filter presses (Fig. 22.4) still produce the highest solids content product in many cases, and have the benefit of simple, reliable construction. Their operation has become increasingly automated in recent years.

On smaller plants, or where there is a lot of (preferably isolated) land available and ambient temperatures are consistently high, sludge-drying lagoons are still a viable option. Otherwise, constructed sludge-drying beds may be used.

NOVEL PROCESSES

Ultrasonic sludge conditioning

In sludge, various substances and agents collect in the form of aggregates and flakes, including bacteria, viruses, cellulose, and starch. Ultrasonic energy causes these aggregates to decompose. This alters the constituent structure of the sludge and allows the water to be separated more easily. While this is a new technology, a German supplier of sonication plant makes the following claims:

- Fewer flocculants are needed to reduce the moisture content of the sewage sludge
- Ultrasound attacks the cell walls of bacteria, causing the release of enzymes which accelerate the breakdown of organic material
- The yield of gas is markedly increased, and the overall solid content of the fully decomposed sludge is reduced

This process is reportedly being used with success at some wastewater purification plants. At one of these, where both domestic and industrial outflows are processed (35,000 PE scale), a large-scale installation has been working since 1999. It is claimed that the biogas yield there has increased by more than 25%. The new process is claimed to lower operational costs, too. The effects reportedly became more evident in a full-scale installation at a large treatment plant serving the equivalent of 280,000 inhabitants. On the basis of these experiences, the potential savings could amount to $2.50 per inhabitant per year.

The application of split flow disintegration of organic suspensions using highly intensive ultrasound in various stages of wastewater treatment and sewage sludge treatment has been reported to be an efficient way of solving problems in sewage plants. As with so many new technologies, time will tell if this proves a practical and cost-effective technology.

FURTHER READING

Warden, J. H. (1983). *Water Research Centre Technical Report: Sludge treatment plant for waterworks, TR189*. Marlow: Water Research Centre.

WRc. (1997). *Application guide to waterworks sludge treatment and disposal, TT016*. Swindon: WRc plc.

Sludge treatment unit operation design: chemical processes

23

CHAPTER OUTLINE

Introduction 275
Stabilization 275
Conditioning 276

INTRODUCTION

As with the other areas of water process design, there are three kinds of sludges to consider: sewage sludges (high organics), waterworks sludges (low organics), and industrial effluent treatment sludges, which might be similar to one of the two preceding types, or entirely different (e.g., they might be very oily).

Chemical treatment of sewage sludges may aim to stabilize or deodorize them to prevent odor nuisance, or pathogenicity.

While waterworks sludges may have sufficient organic content for septicity to be a possibility, with the potential for taste or odor problems when water is recycled from sludge dewatering, this is not usually a problem in practice.

STABILIZATION

Sewage sludges can be stabilized to prevent them "going off," resulting in odor nuisance. This often also has the benefit of reducing pathogen count. Lime and heat treatment are the most common techniques.

Lime or some lime-containing waste material may be added to sludge, with the aim of maintaining a pH of 12 or above for 2 h, with residual pH above 11 for a few days. This may be undertaken before or after dewatering. It is not a permanent solution, and does not kill all organisms present. The lime addition may have the effect of improving dewaterability by some processes if carried out prior to dewatering.

Carrying out quicklime addition after dewatering has the secondary benefit of giving a useful temperature increase, which destroys parasite eggs.

An Applied Guide to Water and Effluent Treatment Plant Design. DOI: https://doi.org/10.1016/B978-0-12-811309-7.00023-0

CONDITIONING

The main purpose of sludge conditioning is to improve dewaterability of sludges. Chemicals used to this end include inorganic coagulants, organic polymers, filter aids, and lime. Chemical dose requirements increase approximately as solids density of sludges decrease. Septicity of sludges also increases their chemical dosing requirements.

For all three types of sludge, it is usual to add the chemical agents in a zone of very high shear, followed by sufficient residence time within a low shear area to allow floc to build prior to dewatering equipment.

The most commonly used sludge conditioning chemicals are organic polymers, specifically polyacrylamides. Concerns over the carcinogenicity of acrylamide means that low-monomer grades of polyacrylamide are used in drinking water sludge treatment, and calculations are performed to ensure that levels of monomer recycled to process are always below maximum safe limits.

Sludge treatment unit operation design: biological processes

24

CHAPTER OUTLINE

Anaerobic Sludge Digestion..277
Types of Anaerobic Digestion..278
  Factors Affecting Anaerobic Digestion 278
  Anaerobic Digester Design Criteria 279
Novel Types of Anaerobic Reactor...282
  Anaerobic Filter Reactors 282
  Anaerobic Contact Reactor 283
  Upflow Anaerobic Sludge Blanket 284
  Fluidized (Expanded) Bed Reactor 285
Aerobic Sludge Digestion 286
  Composting..287
  Disinfection 287
Reference...287

ANAEROBIC SLUDGE DIGESTION

Sludge treatment and disposal can account for half of total sewage treatment costs. The commonest form of biological sludge treatment on large works is anaerobic digestion (AD), in which sludge is held in a mixed tank in the absence of oxygen.

Anaerobic sludge digestion is one of the oldest forms of effluent treatment, dating back to the 19th century. About half the population in England is served by sewage works using anaerobic digestion for sludge treatment.

Traditionally, the process was unheated, operating at ambient temperatures, but there are only a few such plants operating nowadays in the United Kingdom, as the process takes 2−3 months at UK temperatures.

Instead of these traditional low temperature digesters operating in the cryophilic range (around 10−15°C), heated digesters are far more commonly used in two temperature ranges. These are the mesophilic range, around 35°C, and the thermophilic range, around 55°C. While metabolism proceeds much faster at 55°C, the range of acceptable temperatures either side of the optimum is much narrower. The mesophilic range is the most commonly used, largely for this reason.

Anaerobically digested sludge is normally black, with an inoffensive odor and texture.

TYPES OF ANAEROBIC DIGESTION

- cryophilic (10−15°C), mesophilic (25−45°C), or thermophilic (50−60°C)
- wet (5−15% dry matter in the digester) or dry (over 15% dry matter in the digester)
- continuous flow or batch
- single, double, or multiple digesters
- vertical tank or horizontal plug flow

FACTORS AFFECTING ANAEROBIC DIGESTION

Alkalinity

The various stages of digestion produce or consume fatty acids (see Chapter 3: Biology) and if there is insufficient alkalinity present, the lack of buffering capacity makes for a fluctuating pH which affects digestion efficiency. The fatty acid/alkalinity ratio w/w should ideally be around 0.1, or at least below 0.35.

Composition of Raw Sludge

Nitrogen must be present at a level of at least 2.5% w/w of dry organic matter, and phosphorus at 0.5% to support the organisms responsible for AD. While municipal sewage sludge will exceed these requirements, industrial effluent sludges may not.

Inhibitory Substances

There are many chemicals which can inhibit anaerobic digestion, or even cause it to fail entirely if present in a high enough concentration, but the main problems in practice are with heavy metals (see Table 19.3 and Eq. 19.1), halogenated hydrocarbons, and anionic detergents.

Method of Addition of Raw Sludge

Ideally, raw sludge should be fed to the digester continuously to avoid fluctuations in gas production, but this is often impractical. Feed is usually intermittent in practice, perhaps occurring for 20 min every hour.

Mixing

Mixing of the digester contents homogenizes them with respect to their composition, and temperature, increasing digestion efficiency and preventing the formation of scum and grit layers, as well as enabling the release of gas from in lower regions of the digester.

pH Value

Methanogens are pH sensitive, so digesters are best maintained at a pH value of 6.8−7.5.

Primary versus Secondary Sludges

Secondary sludges are less amenable to digestion that primary sludge because they have a lower solids content, and a lower volatile matter (VM): total suspended solids ratio.

Sludge Type and Concentration

Digesters work best if fed with a consistent type (or a consistent blend of types) of sludge at a consistent dry solids (DS) content of around 8%−12%. Measures can be taken to produce such a standard blend of sludges from the various kinds arising on site and/or being imported by, for example, thickening part of the flow and blending it back with raw sludges to produce the digester feed.

Temperature

Variations of $\pm 2°C$ do not affect mesophilic processes, but they can affect thermophilic processes, the control requirements associated with this sensitivity being a key reason for its lower popularity.

ANAEROBIC DIGESTER DESIGN CRITERIA

Table 24.1 summarizes general design criteria for traditional mesophilic anaerobic digestion processes in cylindrical tanks.

Table 24.1 Generalized Mesophilic Digester Design Criteria

Floor slope	0−60°
Gas calorific value	22.5−26 MJ/m³
Gas production	0.9−1.2 m³/kg VM removed
H:D ratio	1:1−3:1
Input sludge	2%−12% DS
Operating pressure	100−300 mm WG
pH	6.8−7.5
Retention time	12−30 days (+10% for grit build-up)
Solids loading	1.5−2.1 kg SS/m³/day
Temperature	35°C
Total solids removal	30%−35% of total input solids
VM Loading	1.5 kg VM/m³/day (acceptable range 0.2−2.8)
VM Removal	35%−50%

Types of Heating

Digesters can be heated by means of a heating jacket or internal heat exchangers, or directly by steam injection or submerged flame. One dedicated external heat exchanger per digester, (commonly double pipe or concentric plate types) is the modern norm.

Sludge Mixing

Digesters are usually mixed via unconfined gas mixing, though a central screw mixing pump, circulation pumps, or internal gas lift pumps are also used.

Unconfined gas mixing systems tend to use 25 mm NB nitrile lined neoprene pipework dropping from a roof manifold to individual leaf spring diffusers. The pipework needs inline backflow protection from nonreturn valves and antisyphon legs to protect the gas compressor, as leaf spring diffusers do not provide effective backflow protection.

Rules of Thumb

Gas Flowrate

The total gas flowrate required for mixing one digester can be calculated using Eq. (24.1).

Total digester gas flowrate calculation

$$Q = \frac{0.077 \times \text{volume of digester}}{\ln(P(\text{abs}))} \tag{24.1}$$

where

$P(\text{abs})$ = absolute pressure of the gas.

Assume that gas flow through each diffuser = 4 L/s (at STP) to work out how many diffusers are needed.

Gas Production

Gas production can be determined using Eq. (24.2).

Gas production

Gas production(m³/day) = (sludge feed volume) × (dry solids content) ×
(feed sludge volatile matter) × (volatile matter removal) × (gas
production/PE/day) × 1000 kg/m³ (24.2)

where

Gas production/PE/d = 1 ft³(0.03m³)/population equivalent/day
m³/day gas production = 0.3 × daily kg DS feed (kg/day)
1m³/m³ volume of digester capacity/day (assuming 15 days HRT digester, feed 5% DS).

This simplifies to:

> Feed volume(m^3/day) $\times$ % dry solids in feed $\times$ 3(m^3/day)(assuming feed 75% VM, 40% VM removal, 1m^3 of biogas produced/kg volatile matter removed).

Alkalinity and Anaerobic Digestion

The production of fatty acids during the acetogenic stage and the production of carbon dioxide as a biogas component make the pH and amount of alkalinity present in the sludge a design consideration. Alkalinity and pH may need to be adjusted to make the process work well (see McCarty, 1964 for more details).

Gas Handling Equipment and Pipework Design

Gas handling equipment including compressors, diffusers, and flare stacks need to be suitably sized for instantaneous gas production rates (allow twice average):

- gas pipework will generally be constructed of 316SS/ABS/GMS/MDPE
- pipework needs to slope to condensate traps at ground level, because as gas cools outside the digester, water condenses
- gas superficial velocity in pipework needs to be <3.5 m/s to prevent water entrainment
- gas pipework must include flame traps at any potential ignition source
- compressors and gas pipework need to be outdoors and above ground
- buildings must be well ventilated
- gas holders need to have a volume of at least 1/30 of total digester volume

Rules of Thumb: Boiler Design

- Water inlet temperature: 71°C
- water outlet temperature: 82°C
- overrate the boiler shell by at least 30% with respect to natural gas, to allow for the larger volume of flue gases produced by biogas

Zoning Data

Lower Explosive Limit (LEL)	5% CH$_4$/air
Upper Explosive Limit (UEL)	15% CH$_4$/air

Volatile Matter Destruction

The "Van Kleek" method (Eq. 24.3) may be used to determine volatile matter (VM) destruction.

Van Kleek method

$$\%VM \text{ removed} = \left(1 - \frac{((100 - A) \times B)}{((100 - B) \times A)}\right) \times 100 \qquad (24.3)$$

This method assumes that only VM is removed during digestion (i.e., there is no significant solids loss by settlement).

Alternatively, the O'Shaughnessey equation (Eq. 24.4), which is essentially a rearrangement of the Van Kleek Method, may be used.

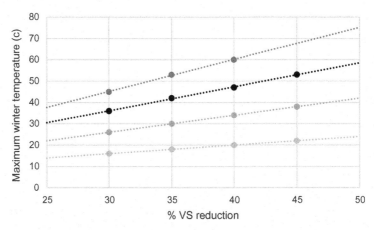

FIGURE 24.1

Maximum digester temperature in winter versus volatile solids reduction for different raw sludge TS concentrations in a 50,000 population equivalent AD plant.

O'Shaughnessey equation

$$\%\text{VM removed} = \frac{100(A - B)}{A(100 - B)} \times 100 \tag{24.4}$$

where

A = % dry matter of raw sludge VM
B = % dry matter of digested sludge VM

Digester Temperature

Digester temperature may vary through the year due to weather conditions (see Fig. 24.1). In temperate climates, winter maximum temperature determines minimum solids reduction performance.

The lines refer from top to bottom to 8%, 6%, 4%, and 2% DS sludges, respectively.

NOVEL TYPES OF ANAEROBIC REACTOR

Table 24.2 summarizes the more novel types of anaerobic digester.

ANAEROBIC FILTER REACTORS

In an anaerobic filter reactor (Fig. 24.2), organisms are established on a packing medium. It may be operated in upflow or downflow mode. The packing medium provides a mechanism for separating the solids and gas produced from the liquor.

Table 24.2 Novel Types of Anaerobic Digester

Type	Typical Retention Time (days)	Typical COD Removal Efficiency (%)	Typical Organic Loading Range (kg COD/m³/day)
Anaerobic filter	1.5–5	70–80	0.5–15
Contact reactor	5–8	70–80	2–5
Covered lagoon	3–5	Up to 90	1–5
Upflow anaerobic sludge blanket	0.3–5	70–90	5–15
Fluidized/expanded bed reactor	0.5–3	70–90	5–30

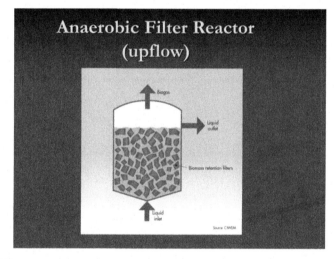

FIGURE 24.2

Anaerobic filter reactor.

Courtesy: CIWEM.

Operation in downflow mode is better for liquors with high suspended solids concentration.

Design loadings are 0.56–15 kg COD/m³/day for pharmaceutical waste and paper mill foul condensate, respectively.

ANAEROBIC CONTACT REACTOR

In an anaerobic contact reactor (Fig. 24.3), organisms are maintained in suspension with liquor for treatment. Organisms are then separated from the liquor prior

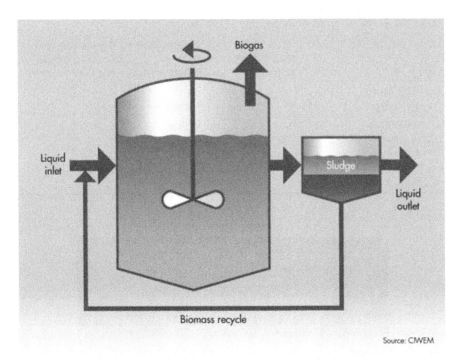

Source: CIWEM

FIGURE 24.3

Anaerobic contact reactor.

Courtesy: CIWEM.

to discharge of effluent, and the organisms are recycled to the reactor. A degasifier between the reactor and clarifier prevents floating sludge.

This design is useful for treating high solid wastes.

Design loadings are 0.1–3.5 kg BOD/m^3/day for keiring and slaughter house wastes, respectively.

UPFLOW ANAEROBIC SLUDGE BLANKET

In a upflow anaerobic sludge blanket (UASB) (Fig. 24.4), the formation and maintenance of granules is very important for good treatment. Good granule formation requires neutral pH, so pH adjustment prior to the reactor may be required. See also Chapter 14, Dirty water unit operation design: biological processes.

An upflow velocity of 0.6–0.9 m/h is required to keep sludge in suspension but not wash it from the system. Liquid residence times of about 3–8 h can be achieved with high strength wastewaters.

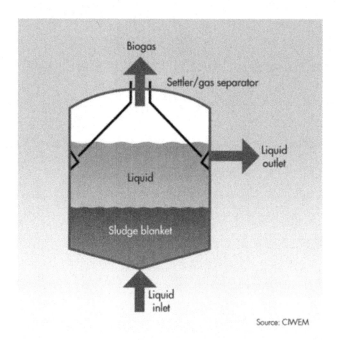

Source: CIWEM

FIGURE 24.4

Upflow anaerobic sludge blanket.

Courtesy: CIWEM.

Design loadings are $10-95\,kg\;COD/m^3/day$ for sugar beet and brewery wastes, respectively.

However, there are some issues with treatment of sugar beet effluent via UASB, such as:

- calcium build-up leading to granular sludge deposition
- calcium build-up leading to scaling of pumps and lines
- acidification outside reactor
- insufficient biomass inventory

UASB is a sensitive technology and can be expensive to design and maintain.

FLUIDIZED (EXPANDED) BED REACTOR

In a fluidized/expanded bed reactor (Fig. 24.5), micro-organisms develop on a bed of sand/inert material. The influent is pumped up through the bed, expanding it by about 10%. High concentrations of biomass are maintained.

The effluent is recycled to mix with the feed, depending upon influent strength and the required fluidization velocity.

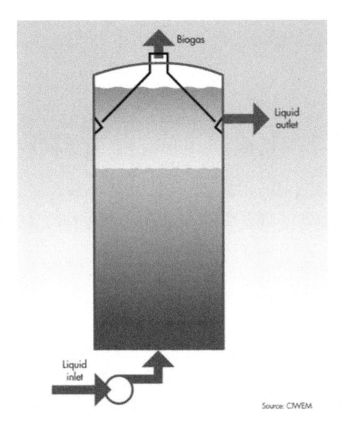

Source: CIWEM

FIGURE 24.5

Fluidized (expanded) bed reactor.

Courtesy: CIWEM.

High rates of removal can be achieved but the process is intolerant of wastes with a high suspended solids content.

Design loadings are $0.8-48$ kg $COD/m^3/day$ for synthetic and paper mill foul condensate, respectively, with typical hydraulic retention times (HRTs) of $0.33-8.4$ day.

AEROBIC SLUDGE DIGESTION

It is possible to treat sludges that are not too obnoxious by means of aeration, effectively a continuing activated sludge process. There are variants with pure oxygen feed, high and low temperature and dosing of pH correction chemicals, but all offer solid reductions rates similar to those of anaerobic digesters.

However in all cases, where used as the sole digestion process, aerobic digestion is more expensive to operate than anaerobic digestion. While some benefits are claimed for its use as pretreatment prior to anaerobic digestion, these are almost always outweighed by its problems with dewatering and high temperature (60–70°C) requirements, so it generally has limited application in the United Kingdom.

COMPOSTING

Composting is becoming increasingly popular. Usually some material is required to lower the overall water content of the material being composted. Previously something like straw was used, but more recently, the use of the solid fraction of municipal waste has led to a dual benefit of reducing straw imports and degrading of both components.

Aerobic composting is more common. Maintenance of the entire system in an aerobic state avoids the production of nuisance odors. This is often carried out by the use of specialist machinery. Composting facilities are often covered, allowing control of water content and the containment of any odors formed.

DISINFECTION

Disinfection of sewage sludges is usually achieved by means of chemical addition, heating, or storage.

Long-term storage is often attached to digestion facilities, most commonly in the United Kingdom in tanks which resemble those used to make up the digester, sometimes called "kill tanks." Traditionally, simple lagoons were used, but these are declining in use, probably mostly on safety grounds.

The most common chemical used for sludge disinfection is lime, as described previously. Chlorine has also been used, but the rapid inactivation of chlorine in the presence of organic matter, and undesirable reaction by-products make its use less common nowadays.

Heating may take many forms but, in all cases, a combination of residence time and temperature determines degree of microorganism kill. The commonest forms of this are pasteurization, composting, incineration, pyrolysis, and drying.

REFERENCE

McCarty, P. L. (1964). Anaerobic waste treatment fundamentals, parts 1–4. *Public Work J.*, *95*(9), p 107–112; *95*(10), p 123–126; *95*(11), p 91–94; *95*(12), p 95–99.

Sludge treatment hydraulics

25

CHAPTER OUTLINE

Introduction 289
Rigorous Analysis 290
Quick and Dirty Approaches 290
   Sludge Volume Estimation 290
   Headlosses for Sludges 291
References 291
Further Reading 291

INTRODUCTION

Sludge hydraulics are challenging, because sludges are often variable in composition and consistency over time. "First principles" approaches to hydraulic calculations are thus frustrated by complexity.

Sludges may (continuously or intermittently) have variable density and viscosity; they may be non-Newtonian fluids; they may shear thin or shear thicken. They may undergo irreversible and undesirable changes to properties such as particle size under shear. It may be important not to kill the microorganisms present by rough handling. If they have significant FOG content, they may coat pipes, reducing hydraulic diameter. Similarly, solids settlement from sludges may reduce pipe diameter over time. The greatest challenges for the theoretician are that the various types of sludge produced on a plant will differ from each other, or even from themselves over time. Furthermore, mixtures of sludges in variable proportions, with different degrees of homogeneity, may be produced and passed on to further treatment.

Sludges are often pumped using positive displacement pumps, so flows are pulsating, and the relationship between flow and pressure is different from that applicable to the more familiar rotodynamic pump. The use of positive displacement pumps means that a certain degree of leeway on achieving given flowrates is built in, but power usage estimates will be inaccurate if viscosity has been significantly underestimated. Positive displacement pumps that are significantly underrated with respect to head requirement can also be subject to excessive maintenance requirements and premature failure.

An Applied Guide to Water and Effluent Treatment Plant Design. DOI: https://doi.org/10.1016/B978-0-12-811309-7.00025-4

RIGOROUS ANALYSIS

The flow behavior of sludges and slurries is complex, and the best references on the subject for wastewater treatment plant designers are hard to obtain. Hathoot (2004) has authored a series of papers on the subject, aimed at providing practical methodologies.

For an approach intended to be used in software, Little (1998) is useful. He has also authored a series of papers on the subject.

QUICK AND DIRTY APPROACHES

If (as is commonly the case) these rigorous approaches are not followed, engineers need to bear in mind that their rough headloss estimates based on multiplying clean water headloss by 1.1 or so can be carrying a significant margin of error, as Fig 25.1 shows. Such approaches are merely guidance for the designer. Even the more rigorous approaches given earlier are only heuristic approximations from a practical point of view.

The general rule is that the higher the solids content, the higher the probability that the sludge will exhibit some non-Newtonian behavior.

SLUDGE VOLUME ESTIMATION

- Clarifier sludge average flow = 0.5%−1% of clarifier incoming flow
- Instantaneous flowrate = 7 × average
- For thickener, instantaneous sludge flowrate is 2−3 × average

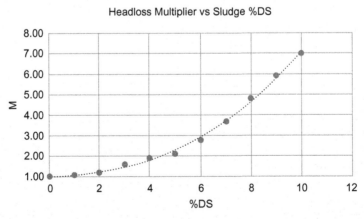

FIGURE 25.1

Sludge headloss multiplier.

HEADLOSSES FOR SLUDGES

We can apply a multiplication factor (see Fig. 25.1) to estimate how much higher headloss is for sludges up to 10% DS compared with water, using Eq. (25.1).

Sludge headloss multiplication factor

$$h_c = M \times \frac{V^2}{2g} \tag{25.1}$$

- For thickened sludges from 2% to 3% DS, head losses Ω1½ times water

REFERENCES

Hathoot, H. M. (2004). Minimum-cost design of pipelines transporting non-Newtonian fluids. *Alexandrian Eng. J.*, *43*(3), 375−382.

Little, S.N. (1998). The flow behaviour of non-Newtonian sludges. PhD Thesis, University of Greenwich, Available at: https://core.ac.uk/download/pdf/9657297.pdf [accessed 09.03.17].

FURTHER READING

Bartlett, R. E. (1979). *Public health engineering: sewerage* (2nd ed). London, UK: Taylor & Francis.

Durand, A., Guerrero, C., & Ronces, E. (2002). Optimize pipeline design for non-Newtonian fluids. *Chem. Eng. Progr.*, March 2002, pp 62−69.

SECTION

Miscellany

6

INTRODUCTION

This section has been used to collect topics which do not fit neatly into the structure of this book, but have nevertheless proven important and useful in my professional practice.

Ancillary processes

26

CHAPTER OUTLINE

Introduction 295
Noise Control 295
Volatile Organic Compounds and Odor Control 296
    Measurement and Setting of Acceptable Concentrations 297
    Covers and Ventilation 297
    Removal of Volatile Organic Compounds and Odors 297
Fly Control 298
Reference 299

INTRODUCTION

The treatment of sewage, industrial effluent, or even drinking water has the potential to create environmental pollution or nuisance.

Environmental pollution generally can be defined objectively as the release of certain substances to environment at levels which cause measurable harm to the environment. The definition of what constitutes a nuisance is often more troublesome as it is a highly subjective concept. What represents a nuisance level of odor or noise varies from person to person, and over time for individuals. There can be a significant psychological component to the threshold. It is however clear that if there are no people, there is no nuisance.

Layout considerations, management plans, and/or abatement measures are commonly required to mitigate the potential of certain processes or equipment items to create nuisance in the forms of noise, odor, vermin, and flies. Alternatively, certain processes can be avoided altogether if it is not practical to control their potential for nuisance complaints from staff or neighbors.

NOISE CONTROL

Placing noisy equipment such as aeration blowers well away from the site boundary (especially any with residential neighbors) is often the cheapest way to avoid nuisance complaints.

An Applied Guide to Water and Effluent Treatment Plant Design. DOI: https://doi.org/10.1016/B978-0-12-811309-7.00026-6

Most kinds of equipment can be shrouded in acoustic enclosures, though these add considerable cost, and may make cooling and maintenance problematic.

VOLATILE ORGANIC COMPOUNDS AND ODOR CONTROL

When designing a new plant, the avoidance of production of odors by excessive agitation, or by allowing septicity to develop should form part of the consideration of all stages of the design of facilities.

There are incredibly complex models available to work out an odor footprint, but I personally never bother with more than the Danish Environmental Protection Agency's equation. This states that the abatement zone surrounding a low-lying odor source can be estimated from Eq. (26.1).

Estimation of odor abatement zone for low sources

$$L = 1.6(RC_{50})^{0.6} \tag{26.1}$$

where

R = air volume from outlet (dry air) (Nm^3/s)
C_{50} = odorant emission concentration (OU/Nm^3)

I use the maximum expected odor yield over 15 min for this. You may not have an estimate in OU/m^3, as it is far more common to have an estimated H$_2$S concentration. In this case, allow 0.0005ppm H$_2$S per OU. This can be used to generate a potential abatement zone as illustrated in Fig. 26.1. If this contains sensitive receptors, abatement is recommended.

For high level sources, the determination of odor abatement zones is more complex; see Danish Environmental Protection Agency (2002).

Any areas whose potential for odor nuisance cannot be designed out should be placed as far away from residential areas (and other sensitive receptors) as possible.

There are however some works which, through climate and/or historical design of sewer or plant, have odor problems. These problems may also arise as development brings housing or industry closer to units previously separated from the general population.

The solutions to these problems may be operational, or it may become necessary to add chemical dosing, tank covers, or complete collection and treatment systems for odorous air.

The release of volatile organic compounds (VOCs) from wastewater treatment facilities has also become of increasing concern in recent years. Many of the solutions are similar to those used for the avoidance of odor nuisance. In addition, the control of organic solvent release to sewers can be carried out at source.

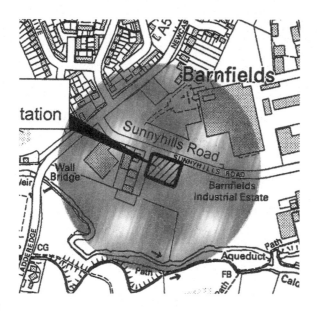

FIGURE 26.1

Example of an odor abatement zone.

Courtesy: Expertise Ltd.

MEASUREMENT AND SETTING OF ACCEPTABLE CONCENTRATIONS

Two main criteria used to set acceptable levels of VOCs and odor are as follows:

- avoidance of the lower explosive limit (LEL), in the case of VOCs and
- avoidance of toxic concentrations of H_2S
- maintenance of acceptable aesthetics, in the case of odors

COVERS AND VENTILATION

Covers should not be air tight but should allow ingress of air, and the system should be maintained under negative pressure using fans rated based on a specified number of air changes per hour, proportional to the likelihood of excessive concentrations of odorous air or VOCs.

REMOVAL OF VOLATILE ORGANIC COMPOUNDS AND ODORS

Generally, both VOCs and other odorous compounds can be removed by oxidation, by adsorption, or by scrubbing. The best technique to use is to some extent dependent on the chemical compounds responsible for the odor.

Oxidation may be biological, chemical, or thermal, and may be wet or dry. It works for all odorous compounds, though in some cases, the products of chemical

FIGURE 26.2

Filter fly[1] (not life size).

Courtesy: Sanjay Acharya.

oxidation are themselves noxious (e.g., the oxidation of H_2S produces SO_2) and the oxidation stage must be followed by scrubbing.

Adsorption is usually carried out using packed beds of granular activated carbon (GAC), though other adsorbents are available. GAC adsorption works best for hydrophobic compounds such as aromatic VOCs and with a dry, or at least noncondensing gas stream. Liquid condensing onto the GAC surface tends to reduce its contaminant removal efficiency dramatically.

Scrubbing usually involves contact with recirculating sprays of aqueous solutions of caustic, bleach, or acid. Sodium hydroxide (NaOH) solution works best with acidic compounds such as free fatty acids, nitrogen and sulfur oxides. Acids work best with alkaline odorants such as amines. Scrubbing with bleach solution is a wet oxidation process. Multiple stages of scrubbing may be used to remove various types of contaminant.

FLY CONTROL

While there are few brand new installations of this type, traditional low-rate tricking filters are notorious for producing clouds of "filter flies" (Fig. 26.2) (aka moth flies or drain flies), for which they are an ideal breeding environment.

[1]Licensed under CC BY-SA 3.0: https://creativecommons.org/licenses/by-sa/3.0/deed.en.

Although these flies do not bite, they reliably annoy any residents subjected to them in high numbers, and they are very likely to be unsanitary.

These flies are controlled by fine netting placed on the filter surface and/or by applications of VectoBac, a biological control based on *Bacillus thurigiensis* toxin. However, there is really no way to prevent filter flies reliably and absolutely from breeding on filters, and conventional insecticides upset the ecosystem of the filter. It is best to either choose an alternative process, or to make sure such filters are a very long way from any residential properties.

Activated sludge plants are also reportedly a source of fly nuisance in the form of mosquitoes in the Southern United Kingdom. VectoBac is also used to control mosquito larvae.

REFERENCE

Danish Environmental Protection Agency. (2002). Environmental Guidelines no. 9, 2002 Industrial odor control. Available at: https://www2.mst.dk/Udgiv/publications/2002/87-7972-297-0/html/indhold_eng.htm [accessed 10.10.17].

Water and wastewater treatment plant layout

27

CHAPTER OUTLINE

Introduction 301
General Principles 302
Factors Affecting Layout 303
    Site Selection 305
    Plant Layout and Safety 307
    Plant Layout and Cost...308
    Plant Layout and Aesthetics 309
Matching Design Rigor with Stage of Design..310
    Conceptual Layout...310
Combined Application of Methods: Base Case..312
    Conceptual/FEED Layout Methodology..312
    Detailed Layout Methodology...313
    "For Construction" Layout Methodology..314
Further Reading 315

INTRODUCTION

It has been estimated that 70% of the cost of a plant is affected by the layout, and its safety and robustness are if anything even more strongly dependent on good layout. Nowadays, however, layout is no longer taught widely in universities. I updated and revised Mecklenbergh's Process Plant Layout textbook in 2016, and I took the opportunity to include material on water and effluent treatment plants, some of which I summarize in this chapter.

There are several reasons for the disappearance of plant layout from today's university curricula and it is arguably part of a wider loss of drawing and development of visual/spatial skills from chemical engineering courses. However, the main reason academics no longer teach layout and drawing is because these skills are the province of the practitioner rather than the academic. Plant layout requires flexing mental muscles which universities usually do not exercise.

An Applied Guide to Water and Effluent Treatment Plant Design. DOI: https://doi.org/10.1016/B978-0-12-811309-7.00027-8

GENERAL PRINCIPLES

We need to consider layout in sufficient detail from the very start of the design process. Even at the earliest stages of design, attempting to lay plant out will unveil practical difficulties.

Layout is not just a question of making the plant look aesthetically pleasing from the air (or, more usefully, from ground-level public vantage points). The relative positions of items and access routes are crucial for plant operability and maintainability, and there are more detailed site-specific considerations.

The three key elements that must be balanced in plant layout are cost, safety, and robustness. Thus, wide plant spacings for safety increase cost but may interfere with process robustness. Minor process changes may have major layout or cost consequences. Cost restrictions, on the other hand, may compromise safety and good layout.

The layout must enable the process to function well (e.g., consideration must be given to gravity flow, multiphase flow, and net positive suction head (NPSH)). Equipment locations should not allow a hazard in one area to impinge on others, and all equipment must be safely accessible for operations and maintenance.

As far as is practical, high cost structures should be minimized, high cost connections kept short, and all connection routing planned to minimize all connection lengths.

Layout issues at all design stages are always related to the allocation of space between conflicting requirements. In general, the object most important to process function should have the first claim on the space. Other objects then fit in the remaining free space, again with the next most important object being allocated first claim on the remaining space. The constraints always conflict, and the art of layout lies in balancing the constraints to achieve an operable, safe, and economic layout.

Layout most obviously affects capital cost, since land and civil works can account for 70% of the capital cost. Operating costs are also affected, most obviously through the influence of pumping or material transfer cost and heat losses, but more subtly in increased operator workload caused by poor layout.

The layout must minimize the consequences of a process accident and must also ensure that safe access is provided for operation and maintenance. Increasing space between items will reduce the risk of domino effects from explosion, fire, or toxic hazards, but it is likely that lack of space means that complete mitigation of fire, explosion, and toxic risk will be unachievable by separation alone.

Layout starts by considering the process design issue of how the equipment items function as a unit and how individual items relate to each other. Such relationships may sometimes be identified by a study of the PFD or P&ID, but not all will be obvious. This is where a GA drawing, experience and judgment become vital to find and balance the physical relationships in the layout.

There are many other factors to consider. Equipment needs to be separated in such a way that it can be safely accessed for maintenance, for other safety reasons (such as zoning potentially explosive areas), and to avoid unhelpful interactions.

Exposure of staff to process materials needs to be minimized. Access to areas handling corrosive or toxic fluids may need to be restricted. This may require the use of remotely operated valves and instruments located outside the restricted area.

In a real-world scenario, we will have a site or sites to fit our plant on to. Different technologies will have a different "footprint." They will have a required range of workable heights and overall area. They may lend themselves better to long thin layouts or more compact plants.

There may be a choice to be made between, for example, permanently installed lifting beams, davits and so on, more temporary provision, or leaving the eventual owner to make their own arrangements.

There are no right answers to these questions, but professional designers will need to be able to demonstrate that they have exercised due diligence. In the United Kingdom, the minimum extent of this due diligence is specified in the Construction Design and Management regulations.

Layout does not require complex chemical engineering calculations but it does require an intuitive understanding of what makes a plant work, commonly known as professional judgment but perhaps more accurately described as "engineering common sense." These qualities cannot be taught formally, but must be acquired through practice. It is, however, possible to start the learning process in an academic setting by design practice, though judgment will mostly be developed in professional practice.

Mastery comes only from learning from experienced practitioners and by testing one's own ideas rigorously. The best way to do this is by listening to those who build, operate, and maintain the plants you design over time, and adopting their good ideas.

The most economical (and easiest to understand/explain to operators) way to lay out a plant is usually for the process train to proceed on the ground as it proceeds on the P&ID, with feed coming in on one side, and product out of the far side of the site or plot. There may, however, be arguments for grouping certain unit operations together in a way which does not correspond with the P&ID order if the site is on multiple plots with any degree of separation.

Land is cheap in some places and expensive in others. Tall plants are generally more favored away from human habitation, but in some regions residents will tolerate a large ugly plant if it provides local employment. Such places might also be willing to allow the location of relatively dangerous processes close to human habitation, as we were before we could afford to be as fussy as we are now.

FACTORS AFFECTING LAYOUT

Layout is, in short, the task of fitting the plant into the minimum practical or available space so that each plant item is positioned to balance the following competing demands:

Cost

- The capital and operating costs must be affordable (e.g., placing heavy equipment on good loadbearing ground)
- The plant must be capable of producing product to specification with the practical minimum levels of operation, control, and management
- Regular maintenance operations should be capable of being performed as quickly and easily as is practical. Units should be capable of being dismantled in situ and / or removed for repair
- The plant must be arranged to promote reasonably rapid, safe, and economical construction, taking into consideration staging of construction/length of delivery period

Health/safety/environment

- Safe and sufficient outgoing access for operators, and possibly incoming access for fire fighters should be designed in
- Operating the plant must not impose unacceptable risks to the plant, its operators, the plant's surroundings, or the general population
- Operating the plant must not impose excessive physical or mental demands on the operators. Manual valves and instruments, for example, need to be easily accessible to operators
- Design out any knock-on effects from fire, explosion, or toxic release
- Avoid off-site impact of noise, odor, or visual intrusion, mainly by moving such plant away from boundary and communities, and presenting attractive offices and landscaping to public view
- The plant, whether enclosed in buildings or outdoors, should not be ugly through uncaring design, but should blend with its surroundings and should appear as the harmonious result of a well-organized, careful design project reflecting credit on the designer and plant operator. See Fig. 27.1 for an example of a sewage treatment works which has been blended into local parkland by the use of a grassed roof structure

Robustness

- The plant and its subcomponents must be arranged so as to operate and make its product(s) as specified
- Process requirements must be accommodated (e.g., arranging plant to give gravity flow)
- The plant must be designed to operate at the planned availability and should not be subject to unforeseen stoppages through equipment failure or malfunction
- Consider how the plant might be expanded in future, and allow space and connections to do so easily
- Consider access to commissioning resources in layout

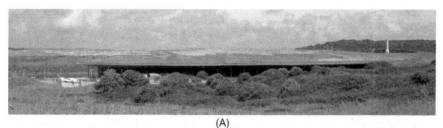

FIGURE 27.1

Blending a sewage treatment works with surrounding parkland (Peacehaven, UK) (A) as seen by the public and (B) aerial view.

(B) Courtesy Google 2016.

SITE SELECTION

The site layout must provide a safe, stable platform for production over a period which may be measured in decades. It is essential to define early in the design process if the site is to be used by the designer or others for a single plant or if several plants are to be installed either now or in the future.

If future plants are planned, some assessment of future development is needed. It might be that space needs to be reserved, road networks planned, and major utility distribution expanded to serve the new plants. When a site specification is drawn up, the site layout aims to make the best use of all features of the site and its environs, for example,

- site topography
- ground characteristics
- natural watercourses and drainage
- climate
- external facilities: water, gas, electricity supplies

- effluent disposal services
- transport of people and goods

Due care and attention also needs to be given to the effects of the plant on the site surroundings, especially:

- housing
- hospitals, schools, leisure centers
- other plants
- forests, vegetation
- wildlife
- rivers and groundwater
- air quality

If we have several candidate sites which we need to choose between, the following factors should be considered at a minimum:

- desired layout of the proposed complex
- cost, shape, size, and contours of land/degree of levelling and filling needed
- loadbearing and chemical properties of soil
- drainage: natural drainage, natural water table, and flooding history
- wind: direction of prevailing winds and aspect, maximum wind velocity history
- seismic activity
- legacies of industrial activity such as mine workings, chemical dumps, and in-ground services
- ease of obtaining planning permission
- interactions with both present and planned future nature of adjacent land and activities

Many of these are environmental factors. We can group them into three categories: natural, manmade, and legislative.

Natural environment

Weather conditions vary greatly from place to place and seasonally. Could our pipes freeze, or could foreseeable high ambient temperatures soften them? How much rainwater might we need to handle? What earthquake and wind loadings do we need to specify? All need to be considered from the start. Well-designed plants are site-specific: plant designs cannot be cut and pasted from one location to another.

Manmade environment

At the detailed design stage, we will need to consider the possibility of effects on and interaction with surrounding plants, installations, commercial, and residential properties. In the United Kingdom, this is likely to involve interaction with regulators such as DEFRA, the Environment Agency, and planning authorities.

Regulatory environment

Potential effluents, emissions, and nuisances (gaseous, solid, and liquid including noise and odor) and any abatement measures need to be considered at the earliest stage. It should never be assumed that regulatory authorities will allow any release to the environment, or that sewage undertakers will allow any discharge to sewer.

As well as the question of simple permission, there will be the question of emission quality, which will set the size and cost of onsite treatment, or the ongoing bill for third party treatment. Not all effluents can be economically treated on site, and third party costs can be very significant, so this needs to be considered at the earliest stages of design. When estimating trade effluent costs, the Mogden Formula (see Eq. 16.1) can be a useful tool.

It should be borne in mind that the regulations which cover releases to environment tend to become more stringent over time. Future-proofing should therefore be considered. If you are not going to include additional plant, you need to at least allow space for such plant.

PLANT LAYOUT AND SAFETY

The IChemE book "Process Plant Design and Operation" (see Further Reading) contains the following suggestions about safety implications of plant layout:

> *Incompatible systems should be separated from each other; humans from toxic fluids; corrosive chemicals from low grade pipework and equipment; large volumes of flammable fluids from each other and from sources of ignition; utilities should be separated from process units; pumps and other potential sources of liquid leakage should where possible not be located below other equipment to minimize the chance of a pool fire. (This is particularly important with fin fan coolers where air movement may fan the flames.)*

HSE guidance on plant layout to meet the Control of Major Accident Hazards (COMAH) Regulations (see Further Reading) states that the most important aspects of plant layout affecting safety are for our designs to:

> *...prevent, limit and/or mitigate escalation of adjacent events (domino); Ensure safety within on-site occupied buildings; Control access of unauthorized personnel and Facilitate access for emergency services.*

So, the advice is consistent. A well-run Hazard and Operability Study (HAZOP) should pick up any issues in this area, but designers should not go in to a HAZOP with ill-considered layouts which will require extensive modification. HAZOP is not supposed to involve redesign. Safety studies should be carried out as part of the design exercise, as described in the next chapter.

As well as in these guiding principles, there are many detailed considerations, such as siting dangerous materials as far as practical from populations and control rooms, considering plan operability and maintainability, safe access and loading/unloading of deliveries and collections.

Regulatory authorities may want to place restrictions on the design with respect to siting, distance of certain structures from people, height of certain structures, size of inventory of specified substances, releases to environment, and so on. These issues need to be considered as soon as possible in the design process. Certain types of plants, especially those likely to be covered in Europe by COMAH legislation, will need special attention.

There are many codes of practice and guidance notes with respect to plant layout as well as many specific codes for issues such as installations handling liquid chlorine. A good place to start with these, especially for UK readers, is the HSE website (see Further Reading). The Dow/Mond indices are also explained there—these can be used at an early design stage as rules of thumb to give outline guidance on equipment spacing.

The general principles which designers most need to bear in mind are inherent safety and risk assessment.

In my book on layout, I give the following principles for separation of source and target:

- Large concentrations of people both on and off site must be separated from hazardous plants
- Ignition sources must be separated from sources of leakage of flammable materials
- Firebreaks are wanted, and are often best provided by a grid-iron site layout
- Tall equipment should not be capable of falling on other plant or buildings
- Drains should not spread hazards
- Large storage areas should be separated from process plant
- Central and emergency services should be in safe areas

I also offer tables of suggested separation distances in the appendixes of that book, which are of course for guidance only, and should be subject to the application of engineering judgment.

PLANT LAYOUT AND COST

As far as capital cost is concerned, the further apart things are, the more piping and cable is required to connect them. Items which are further apart also require more steelwork and concrete, land and buildings to support and contain them. More compact processes often cost more than less space-intensive plant, but land costs money too.

With respect to operating costs, items which are further apart have higher fluid transfer head losses, higher cable power losses, and higher heat losses from hot and cold services. It takes more time for operators to move from one part of a larger plant to the next.

There will be a balance to be struck between capital and operating costs, as designers need to think of how every item of equipment will be maintained, how motors and other replaceable items will be removed and brought back, and how vehicles required by operational and maintenance activities will access the plant during commissioning and maintenance activities as well as during normal operation.

Some examples of cost saving guidelines are as follows:

- Buildings should be as few and as small as practical
- Gravity flow is preferred and, failing that, pump NPSH should be minimized
- The number and size of pipes and connections should be the minimum practical
- Save space and structures wherever possible (consider safety!):
  - Group equipment where practical and safe
  - Make multiple uses of structures, buildings, and foundations
  - Make full use of the height available
  - Consider column locations when laying out equipment in a building/ choosing building type
- Do not bury services under buildings
- Storage tanks can be made of welded steel, bolted glass on steel panels, site cast concrete, preformed concrete, plastics, and composite materials. All materials should be considered, and have implications for layout with respect to construction and maintenance access
- Ground conditions can affect the economics of concrete tankage—they might most economically be fully in-ground, fully out, or intermediate depending on the water table and how hard it is to dig, etc
- Ground conditions can also affect the civil engineering costs of locating heavy structures on a site. Place them where the ground is good
- Structure loading can interact with ground conditions—heavy structures can have a low loading by being shorter and wider. This can be in tension with the process design if a tall thin structure is required

Having considered the most practical issues, we need to remember that we may also need to make our plant aesthetically acceptable.

PLANT LAYOUT AND AESTHETICS

Aesthetics can be very important to engineering designers, but it is absent from engineering curricula. Architects can help us with this: Fig. 27.2 depicts one of the many examples worldwide of architecturally rich sewage pumping stations.

Aesthetics will usually be another issue settled by negotiation. Cost-conscious engineers may prefer buildings to be unadorned metal sheds, while architects, planners, and the general public may want them to be things of beauty (or remain unbuilt). The cost implications of building something to the standards required

FIGURE 27.2

"Temple of Storms" Pumping Station[1].

Courtesy Reading Tom.

where we attract public resistance might make it more economical to build it where communities care more about jobs.

In addition to helping with aesthetics, architects may also help us to produce a plant which is a more pleasant and efficient place of work. Staff morale is important—some even think an attractive layout improves performance.

MATCHING DESIGN RIGOR WITH STAGE OF DESIGN
CONCEPTUAL LAYOUT

Conceptual design aims to do enough work on layout to ensure that the proposed process can fit on the site available, and to identify any layout problems which need to be addressed in more detailed designs.

From a regulatory point of view, we must establish whether enhanced regulatory environments like COMAH or IPPC are going to apply to the plant, and design the layout to suit. There are several issues which must be considered at an early stage.

---

[1]Licensed under CC BY 2.0: https://creativecommons.org/licenses/by/2.0/deed.en.

Prevailing wind

Relative upwind/downwind positions can matter, for example, pressure vessels should not be downwind of vessels of flammable fluids, toxics should not be stored upwind of offices/personnel, and cooling towers need to be as far away as possible from anything which would interfere with airflow, and so arranged as not to interfere with each other.

Indoor or outdoor?

It will be necessary to decide whether the plant will be indoors or outdoors. Good lighting, ventilation and air conditioning, protection for excessive noise, glare, dust, odor, and heat help staff to do their work well, so indoors is often the most comfortable for operators. Indoors is more secret, easier to keep clean, and indoor equipment usually has a lower IP rating, making it cheaper to buy. Outdoor plant is, however, usually best for toxic and flammable vapor dispersal, and buildings are expensive.

Construction, commissioning, and maintenance

Designers need to consider all stages of the plant's life, and specifically vehicle access and space for removal and laydown of plant and subcomponents.

There may also be a requirement for access by tankers and temporary services for maintenance, commissioning, and turnaround activities. This should be identified and accommodated early in design.

Maintenance activities also need to be considered. Instruments and plant requiring regular attendance and maintenance should be reachable by short, simple routes from the control room. The design needs to consider accessing different levels in the plant via ladder or staircase and how to get tools to the working level.

Where equipment is to be maintained in situ, space needs to be left for the people and tools to reach equipment for inspection and repair, the lifting gear required, and laydown for parts.

Where it is to be maintained in the workshop, space should be allowed for people and tools to reach equipment for inspection, disconnection and reconnection, the lifting gear required for removal and replacement, and loading/unloading onto transport to workshop or offsite.

In either case, the working area should be designed to be safe with respect to access, lifting, confined space entry, electrical and process isolation, draining, washing, and so on.

Items requiring regular access for operation, maintenance, or inspection should not be in confined spaces or otherwise inaccessible.

Materials storage and transport

The size of onsite materials storage and transport facilities will be determined in the first instance by practical issues of import and export from site of chemicals and waste material.

The process's requirement for storage and transport facilities will need to be moderated by statutory and commercial standards and codes of practice, as well as planning authority requirements. Commercially available software can assist designers by overlaying a vehicle turning circle over the road layout to allow checks that the roads are suitable for the proposed use.

Materials storage and transport considerations will form part of the hazard assessment process, and will probably need to be at least justified and possibly reconsidered as part of planning and permitting applications.

The provision of utilities such as compressed air and process water needs to be considered at the earliest stage, as their design needs to be well integrated with that of the whole plant.

Emergency provision

Although water does not burn, some contaminants in wastewater can give rise to flammable environments, and anaerobic digestion can lead to large inventories of flammable biogas. In summary, roads accessible by fire trucks need to be at least 4 m wide, suitable for 20 ton vehicles with turning circles of at least 21 m. Of note, 5 m of hardstanding needs to be provided 5–10 m from buildings and items which might require firefighting.

Security

As soon as equipment is on site, a site fence will be needed to protect company and staff property from theft, and prevent unauthorized access for safety reasons. In some cases, it may be necessary to design out features which will shelter protestors at the site entrance, or make plant buildings amenable to occupation.

Central services

Admin, welfare facilities, laboratories, workshops, stores, and emergency services need to be well sited, and should feature on designs from an early stage.

Earthworks

Earthworks may be required to shield tanks of dangerous materials or controls from the site fence, as well as for aesthetic reasons.

COMBINED APPLICATION OF METHODS: BASE CASE
CONCEPTUAL/FEED LAYOUT METHODOLOGY

Different layout designers and companies may have their own specific procedures, but the summary outlined in Table 27.1, based on Mecklenbergh's original methodology, is the base case on which I discuss conceptual/FEED design methodology.

DETAILED LAYOUT METHODOLOGY

Detailed design broadly replicates the stages outlined in Table 27.1 in more detail.

The detailed design stage for plants of any significance should include several formal safety studies, as well as less formal design reviews. These should address layout issues as well as the process control issues which may form the core of such studies.

Table 27.1 Summary of Conceptual/FEED Layout Methodology

Step	Description
1	Generate initial design, sizing and giving desired elevations of major equipment
2	Carry out initial hazard assessment or apply MOND index; consider all relevant codes and standards
3	Produce plan view GA of plant based on this data and spacing recommendations (it may help to cut out paper shapes to scale and arrange them on a large sheet of graph paper before proceeding with CAD)
4	Question elevation assumptions, consider and cost alternative layouts
5	Produce simple plan and elevation GAs of alternatives without structures and floor levels
6	Produce more detailed plan GAs based on decision for last stage
7	Use this drawing to consider operation, maintenance, construction, drainage, safety, etc. Consider and price potentially viable alternative options
8	Consider requirement for buildings critically. Minimize where possible
9	Produce more detailed GAs in plan and elevation based on deliberations to date
10	Carry out informal design review with civil engineering input based on this drawing
11	Revise design based on this review
12	Hazard assess the product of the design review. Determine safe separation distances for fire and toxic hazards, zoning, control room locations, etc. Consider off-site effects of releases
13	Revise design based on hazard assessment
14	Confirm all pipe and cable routes. Informal design review with electrical engineer would be helpful
15	Multidisciplinary design review considering ease and safety of operation, maintenance, construction, commissioning, emergency scenarios, environmental impact, and future expansion
16	Reconcile outputs of design review and hazard assessment, taking cost into consideration
17	If they will not reconcile, iterate as far back in these steps as required to reach reconciliation

Table 27.2 Summary of Detailed Design Layout Methodology

Step	Description
1	Compile the materials and utilities flowsheets for piping and conveyors as well as vehicle and pedestrian capacities and movements on and offsite
2	Lay out whole site, including areas for the various plots, buildings, utilities, etc.
3	Use the flowsheets to place plots and processes relative to each other, bearing in mind recommended minimum separation distances, sizes, and areas
4	Add in services where most convenient and safe from disasters
5	Place central services to minimize travel distance (but considering safety)
6	Consider detailed design of roads, rail etc., keeping traffic types segregated, and maintaining emergency access from two directions to all parts of the site
7	Identify and record positional relationships between parts of the plant/site which need to be maintained during design development
8	Hazard assess site layout, with special attention to the possibility of knock-on effects
9	Single discipline design review (Chemical; Electrical; Civil; Mechanical): representatives from design and construction functions should critically review the design from the point of view of their discipline
10	Multidisciplinary design review: the various disciplines should critically examine the design with respect to hazard containment, safety of employees and public; emergencies; transport and piping systems; access for construction and maintenance; environmental impact including air and water pollution and future expansion
11	If there is still more than one possible site location at this point, the candidate sites can be considered in the light of the detailed design, and one selected as favorite

If it is identified that the site will comprise several plots, interactions between these plots and with any existing ones on the site and those on surrounding sites need to be considered at detailed design stage.

Mecklenbergh's original detailed design procedure bringing together plot designs into a site-wide design for multi-plot sites is summarized at Table 27.2.

Reviews at this stage may be undertaken in a 3D model in the "richer" industries, such as oil and gas, but the water industry still mostly uses 2D hardcopy drawings.

"FOR CONSTRUCTION" LAYOUT METHODOLOGY

After site purchase, a detailed design to optimize the site to its chosen location can be undertaken.

Mecklenbergh originally recommended another stage of design review and optimization for the "for construction" phase, which involved gathering detailed data on the site, the market for the products, etc., and testing design assumptions from previous stages for a good match to the real site. He also recommended repeating the hazard and design reviews, culminating in consideration of the plant with its wider surroundings.

However, it is usual for there to be considerable pressure on resources at the design for construction stage. This, taken with the common requirement to run design and procurement together, works against the idea of a further detailed design review as a practical solution. In any case, the great cost of changing layout at this stage means that any design errors which are less than catastrophic may best be left as they are. There is no practical utility in expending resource to discover minor errors which will not be corrected.

FURTHER READING

Health and Safety Executive. (2015). COMAH guidance, technical measures document: plant layout. Available at: http://www.hse.gov.uk/comah/sragtech/techmeasplantlay.htm [accessed 24.11.17].

Moran, S. (2016). *Process plant layout* (2nd ed.). Oxford: Butterworth-Heinemann.

Scott, D., & Crawley, F. (1992). *Process plant design and operation: guidance to safe practice*. Oxford: IChemE/Butterworth-Heinemann.

Classic mistakes in water and effluent treatment plant design and operation

28

CHAPTER OUTLINE

Academic Approaches..317
Separation of Design and Costing..317
Following the HiPPO...318
Running from the Tiger...319
Believing Salespeople...319
The Plant Works ... Most of the Time 320
Water Treatment Plant Design is Easy 320
Let the Newbie Design It...321
Why Call in a Specialist? 321
Who Needs a Process Engineer?...322
Ye cannae change the laws of physics, Jim 323
Further Reading 323

ACADEMIC APPROACHES

The main problem I have observed with academic approaches to water treatment plant design is the lack of grasp of the key constraint: price. Engineers are never asked how to solve a problem as if money were no object. Balancing cost, safety, and robustness are the essence of engineering, though I discuss a practitioners' version of this error in more detail in the next section.

The next most important academic mistake is to assume that a full-scale plant will outperform a laboratory scale plant. This has happened just once in my entire professional experience, whereas the opposite situation is vastly more common. It is very common indeed that a full-scale version of a laboratory experiment will not work at all. Never assume that scale-up is even possible. I have experience of many cases from my expert witness practice where it was not.

SEPARATION OF DESIGN AND COSTING

I once worked for a large UK water company on the conceptual design of a major extension to a large municipal sewage treatment plant. This company had an approach where process scientists chose the "best process," civil engineers designed the plant around the selected process elements, and civil engineering estimators carried out a costing exercise based upon the design produced.

This resulted in the selection of a too-novel process which was both expensive and unreliable. The process scientists took no responsibility for this, as they assumed engineers would deal with the nuts and bolts of making the plant work. Their brief was to select the "best process," based upon a simple and naïve cost/benefit analysis.

The scientists had no access to accurate costing data, nor knowledge of how to generate it, as they were not engineers. They had no brief to ensure process robustness, and they lacked the engineer's conservatism and dislike of novelty. They were not trained in the analysis of safety as all engineers are. The civil engineers who used their process recommendations as the basis of their design were not able to examine them critically.

The outcome of this exercise will be unsurprising to experienced engineers. The novel process made it to the final decision stage, even though the process economics were bogus, and the process too novel to be used for the first time at such a large scale. The decision to use the wrong process was however sealed when a senior manager expressed an interest in the novel approach. The decision was made to go with the "HiPPO" (Highest Paid Person's Opinion).

FOLLOWING THE HIPPO

Professional engineers use a combination of explicit and tacit knowledge, founded in any given design case by as thorough an analysis of the design envelope as is practically possible to make their design decisions.

Their line management, the managers of various disciplines within the company they work for, and their fellow engineers and technicians within and outside that company may all have opinions on these decisions. Within the engineer's field of expertise, they should rely upon their own analysis, reasoning, and experience, neither unreasonably rejecting any suggestions nor wasting time on ones they know to be inappropriate.

It is common for a higher ranking engineer or manager to want to make their own mark on the process. Managers are generally less risk averse than professional engineers. Engineers from other disciplines or from outside the designer's organization have different priorities, and may well not grasp the full picture. The process designer cannot see the whole picture themselves unless they are willing to listen to these opinions, but they should insist on making decisions rationally.

While relevant experience should be respected, it is important not to follow the HiPPO blindly. Sometimes, however, it may be wise to stand aside from HiPPOs to avoid being trampled.

RUNNING FROM THE TIGER

I recently undertook an expert witness case where an undergraduate had been allowed to design a full-scale plant worth millions of dollars. None of the several layers of senior engineers who should have been checking the student's work corrected it, nor even asked why a student was leading the job and offering such an unusual design. When I questioned this, I was told "She is like a tiger!"

This was clearly quite a formidable student. However, no amount of force of personality can be an effective substitute for a grasp of the facts, together with the training and experience to know what to do with them. It is our job as engineers to ask the difficult questions, and many tigers are just paper.

BELIEVING SALESPEOPLE

Mostly, technical sales staff in firms selling well-established products are the designer's best source of reliable information for detailed design of plant. They will also know how their equipment has been used by others—information not in the public domain. They may even, in a competitive bidding situation, know how your competitors are using their equipment.

Good engineering salespeople do not tend to lie, though what they say may be incorrect for several reasons.

They may pass on—in good faith—bad information which they have been supplied with by their company. Such information is more likely to be selective or statistically insignificant than false. Never take someone else's data analysis at face value.

Salespeople may also omit key facts. They may genuinely lack knowledge of superior alternative competing products or, more commonly, they may withhold any knowledge like this which they do have. The best salespeople will however give you this information if they have it, if only because they know a competent engineer will find it eventually and think less of them for withholding it.

Some salespeople may offer advice on matters they do not fully understand. Salespeople are commonly technicians, rather than professional engineers. Their knowledge of the manufacture, operation, and testing of their products and how they might be adapted to suit your design should be highly detailed. Their other knowledge may be secondhand, anecdotal, ill-understood, and unsupported by their training or experience. Always ask questions until you are sure that you understand where they are offering opinion as opposed to knowledge.

Then, there are the minority of salespeople whose ethics are questionable. These are most commonly found in the world of novel products and processes. From my experience of such companies, I can offer the following checklist:

- Is the salesperson a professional engineer?—this is a good sign
- Is the salesperson a time served tradesperson or technician? —this is also a good sign

- Does the salesperson have a PhD?—this, perhaps perversely, can be a warning sign
- Does the product or process appear to magically resolve a longstanding problem in a sort-of-obvious way?—This should raise alarm bells
- Do you have access to referees who have installed the product or process at full scale on a water like yours at least 5 years ago?—If references are not available, proceed with great caution
- Can you fully understand the mechanism of action and proven effectiveness of the product or process being offered, based on your own analysis of raw data?—A lack of availability of third-party verified data is a bad sign, as is a mechanism of action which flies in the face of established understanding

THE PLANT WORKS ... MOST OF THE TIME

I see this issue most commonly in troubleshooting and expert witness work. A client will maintain that their plant works, but the sampling data show that the plant does not always meet specification. It may even be that every sample taken has failed, but the client remains convinced that the plant is working the rest of the time, so it is working. Mostly.

This may be an operational failure, but it is most commonly an issue of underdesign. The design envelope must be set to have the plant working as much of the time as specified. This is most commonly 100% of the time, though there are exceptions. The costs of each incremental improvement in treatment efficiency may rise exponentially, so underdesign is tempting in as competitive an environment as water treatment.

A plant which has been underdesigned may "work" some of the time, but "working" means meeting specification to an engineer, and the specification usually includes acceptable ranges of availability and product variability. Working some of the time is more properly called "not working" from an engineer's point of view. Managers and other non-engineers may however have another opinion.

WATER TREATMENT PLANT DESIGN IS EASY

Water chemistry is perhaps less exciting than that of more energetic compounds, and much of the equipment looks archaic. Some would argue that there is nothing new or clever happening in water: we just follow the design manual blindly, price the equipment out of a catalogue, and build it as cheaply as possible.

There is some truth in this. It is indeed possible to build some sort of water treatment plant in this way and I have seen it done several times. Plants like this are inferior in two respects to those designed by experts: cost and robustness.

The water industry is savagely competitive and cyclical. Margins are always very tight, and tendering jobs at cost or even below cost is commonplace during downturns. Water process engineers design themselves ahead, rather than competing with discounters on a like for like basis.

The characterization of the chemical makeup of the feedstock to a water treatment plant is usually very poor, flowrates are variable, and treatability or pilot trials are very rare. Water process engineers have to make the best use of experience, heuristics, and statistical analysis to generate a number of scenarios which define the design envelope. This approach produces a robust design.

The fallacy that water treatment plant design is easy leads to another: the notion that it is an ideal starting point for a new designer.

LET THE NEWBIE DESIGN IT

I have carried out several expert witness engagements where fresh graduates or even undergraduates were allowed—by large international engineering firms without in-house water treatment process design expertise—to design water treatment plants.

This may have seemed like a cost-effective decision, but new graduate chemical engineers usually have no knowledge of water chemistry or the unit operations used for treatment. Universities tend to focus on oil and gas or bulk chemical design, to the extent that they teach design at all. Graduates have a very limited knowledge of biology or statistics, and are used only to the reaction chemistry of pure substances. They might be aspiring process engineers, but they are likely to know nothing about water treatment plant design, or even its underpinning sciences and math.

So, do not let the newbie design it, as they will quite possibly make every mistake in this section. By the time companies have paid for the correction of these mistakes, they may have spent many times what it would have cost to employ a specialist.

WHY CALL IN A SPECIALIST?

In my professional life I have seen many effluent treatment plants designed by experts in the main process (e.g., design of papermills, refineries, and various other kinds of manufacturing facility) rather than in effluent treatment. These plants are often needlessly expensive, lack robustness, and miss the basic tricks of the specialist.

For example, the designs of water treatment specialists will not carry water containing potentially pipe-blocking materials (especially any combination of grease and rags) in pipes. These blockages, of which the fatberg (Fig. 28.1) is a prime example, are the key reason water specialists make far more use of open channels than other process designers. The fatberg in Fig. 28.1 is far from large, however: the present UK record weighed 100 tons and was as long as a Boeing 747.

FIGURE 28.1

Fatberg[1].

Courtesy Arne Hendriks.

I have seen non-specialists learn the hard way why open channels are used on more than one occasion. Neither is it a good idea to filter water containing such materials, as I have seen nonspecialists try, as this simply deposits a fatberg on the filter.

Water specialists tend to add far more buffering capacity to our plants than nonspecialists, to deal with the variability in flow and composition of feed, and the increased possibility of unscheduled maintenance.

Most importantly, specialists know how to account economically for the variability of feed. Nonspecialists may try to surmount this issue by simply making everything much bigger than it needs to be, but many items of plant can be too big as well as too small. Turndown ratios matter a lot in water treatment plant design.

WHO NEEDS A PROCESS ENGINEER?

Those cases I have been involved with—in which a process plant was built without process engineering input—are useful to demonstrate what process engineers do and how we add value to a project despite being expensive.

---

[1]Licensed under CC BY 2.0: https://creativecommons.org/licenses/by/2.0/deed.en.

Based on what happens when we are not involved, I have concluded the following are the key benefits of involving a process engineer:

- Our analysis of design data and the setting of the boundaries of the design envelope is crucial to the production of a robust design
- We make all the items of equipment work together to meet product specification across the full range of input flows and water quality by carrying out a mass balance and producing a process flow diagram to inform the specification of unit operations
- We ensure that the control system design is integrated with the process flow diagram by producing a piping and instrumentation diagram and functional design specification
- Our involvement in plant layout helps ensure that it produces a plant which is economical and safe to operate
- Our involvement in selection and sizing of unit operations helps to avoid excessive novelty

By process engineer, I mean a specialist chemical engineer. I have seen scientists and civil engineers called process engineers in some organizations. A job title does not make a scientist into an engineer, or give a civil engineer the required skillset. Process engineers must have with proven skills in the areas outlined earlier to carry out a process design.

YE CANNAE CHANGE THE LAWS OF PHYSICS, JIM

It is important not to be seduced by the claims of novel processes and their associated marketing breakthroughs, such as zero-sludge-yield assumptions.

Any process at laboratory scale fed with a constant flow of water of constant composition will exceed the performance of full-scale commercial water engineering equipment by a large factor. As discussed at the start of this section, this is not meaningful. Our design heuristics include very significant comfort margins. We will need strong evidence at full-scale before even considering changing them.

FURTHER READING

Moran, S. (2015). *An applied guide to process and plant design*. Oxford: Butterworth-Heinemann.

Moran, S. (2016). *Process plant layout* (2nd ed.). Oxford: Butterworth-Heinemann.

Troubleshooting wastewater treatment plant 29

CHAPTER OUTLINE

Introduction 325
Sick Process Syndrome 326
Data Analysis 327
    Too Little Data 327
    Too Much Data 328
Appropriate Statistics 328
The Rare Ideal 329
Site Visits 329
    Using All Your Senses 330
    Interviewing Operators 330
    The Operating and Maintenance Manual 331
    Maintenance Logs 331
    Operator Training 331
Design Verification 332
Putting It All Together 333
Further Reading 334

INTRODUCTION

I have carried out process troubleshooting on dozens of wastewater treatment plants over the last 20 years or so. Based on this experience, when a wastewater treatment plant does not work, it is most likely to be the fault of the owner in some way.

In the vast majority of cases, either they asked it to do something it was not intended to do, or it was badly specified, badly designed, badly constructed, badly commissioned, badly operated, badly maintained or, most commonly, several of these things.

Because people realize it is to some extent their fault (or perhaps despite this), they tend to believe that the plant "almost works," or "often works," but there is in fact no such thing. As with all process plants, it either meets the specification or it does not. "almost working" is not working.

Many of the most troublesome plants I have attended are onsite industrial effluent treatment plants. In every case in my experience, the operators of such plants come under pressure to accept whatever effluent is sent to them by the main production process, such that the capacity of their treatment plant is never a restriction on the main production process. Successful production plants tend to increase production volumes and product diversity over time, but owners often fail to upgrade their effluent treatment plants when they do so.

Even in plants which have not increased main production, effluent treatment plants can be used as a place to send troublesome liquids which the plant was never intended to treat, in the interests of getting the main process back on line as soon as possible and/or rather than paying the expense of having them taken away by tanker for proper disposal.

For example, I once witnessed the entire contents of hypochlorite scrubbers at a coking works being dumped directly to a biological effluent treatment plant by night shift crews, killing all life. I have known tons of out of specification concentrated detergent being dumped directly into the drains leading to oil–water separators, totally and irreversibly destroying the effectiveness of downstream granular-activated carbon filters. Both of these measures cost tens of thousands of dollars and several weeks to fix, in order to save operations teams a few minutes and a few hundred dollars.

SICK PROCESS SYNDROME

In recent years I carried out several hundred waste minimization analyses of manufacturing facilities in various UK process industries for a UK Government scheme called Envirowise. My fellow consultants and I saw many production processes which we characterized as, at best, "barely in control," and which we referred to as sick processes. These sick processes have many of the following characteristics:

- high and variable rates of process failure
- constant searching by staff for technical fixes
- taking control away from operators
- lots of different theories about the cause of the problem
- denial of problems
- a history of failed attempts to improve control
- the process being considered an art rather than a science

These signs and symptoms are apparent in many sick wastewater treatment processes. The way to cure the illness is to move beyond denial and folk wisdom, and to use statistical analysis to both accurately characterize the problem, and be sure whether any attempted fix has worked.

DATA ANALYSIS

When presented with problems with a wastewater treatment plant, we are usually provided with very little hard information. On the other hand, on some occasions there is so much data that the operators have been swamped by it, leading to "analysis paralysis." To treat these two cases in turn:

TOO LITTLE DATA

Very often, there will be fewer than 10 analyses of treated effluent quality available, and clients will be reluctant to spend anything on obtaining further analyses. In my experience, it is preferable to have at least 20 analyses, to increase the chances of identifying problems, but the troubleshooter will often have to make do with what is available. If necessary, a site test kit (Fig. 29.1) can be used during site visits to obtain a quick snapshot estimate of what is happening with respect to BOD, TSS, pH, and temperature. The kit in the illustration is made by a UK company, Palintest, though comparable suppliers no doubt exist in the United States and elsewhere.

However little data there is, I perform, at a minimum, some summary statistical analysis to determine the minimum, maximum and, where appropriate, the average and the upper and lower limits of the 95% confidence interval. Ideally, there will be sufficient data to do this for each of the basic "conventional

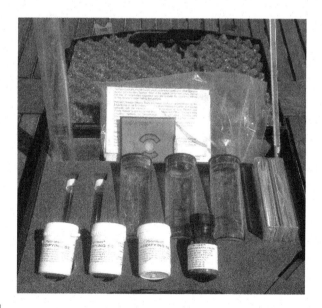

FIGURE 29.1

Portable site analytical test kit.

pollutants" (Total Suspended Solids (TSS); Biochemical Oxygen Demand (BOD); pH; Fats, Oils and Greases (FOG); and coliforms under the US Clean Water Act), as well as any site-specific parameters (Temperature (T); NH_4; Cl; FOGs; Heavy Metals (HM); Detergents, etc.), classed as "additional conventional pollutants," "toxic pollutants," or "nonconventional pollutants" by the Environmental Protection Agency or similarly by the UK Environment Agency.

Small numbers of analyses will often display a very wide range of values in the data (if they do not, be suspicious) and will also commonly show that operators do not always perform analyses or read instruments, and they do not always record some of the readings. It is not unheard of for operators not to record values which they think management would not like to see. Sometimes, missing data is shown by a blank in the table of readings, sometimes by a misleading zero, and sometimes by a suspiciously frequent figure, identical to three significant figures every time. Such a figure may also indicate an error in a laboratory analysis, a fault in instrumentation, or in the limit of detection of the measuring instrument or test. The troubleshooter faced with missing data needs to know which of these is the cause of missing data.

When undertaking statistical analysis, it can be helpful to check correlation coefficients between items, such as, for example, the FOG and TSS content and any pollutants which I would expect to partition to either oil, solids, or both, such as Polychlorinated biphenyls (PCBs). It is often the case that this process will yield "possible" rather than "probable" correlations (test statistic to standard error ratios of 2 and 3, respectively). In strict scientific research, these results would be of little use, but hints are of interest to the process troubleshooter. Weak negative evidence, in the form of items which have no sign of any correlation, can also be useful.

TOO MUCH DATA

I have encountered a few situations in which there is too much data or, more accurately, where there is a great deal of data (tens of thousands of data points), which have not been subjected to any simplifying analysis. This can result in "analysis paralysis," where operational staff are overwhelmed by the sheer amount of data.

Many clients restrict their own analysis of their data to plotting line charts of parameters against time in MS Excel. I find this approach unhelpful. Only when we use appropriate statistics can such large data sets allow useful conclusions to be drawn to a sufficiently high degree of certainty to guide remedial action.

APPROPRIATE STATISTICS

Many engineers are far keener on pure mathematics than they are statistics, which are essentially heuristics, only applicable under certain circumstances. Consequently, there is frequent misapplication of statistics by engineers who do not understand that you cannot, for example, produce a valid average of integers by adding them all together and dividing by the count of numbers.

Without getting too academic, the sort of statistics most people are familiar with, "parametric statistics," can only be applied if data points are continuous, independent, and distributed in accordance with a mathematic model, most commonly the normal distribution. They are more powerful (produce more certain results) than the nonparametric statistics which make fewer assumptions, but if statistics are used where their assumptions are not true, they can give a misleadingly certain answer.

Conversely, it is often claimed that we cannot do statistics with ranked lists of data, where all we know about parameter x is that it is associated with less of parameter y in one case, and with more in another. Nonparametric statistics allow us to work with such data. To give an example, if we had an approximate time to collapse of 10 nitrifying filters, we might see if the ranking of collapse times correlated with various rankings of water aggressivities using Spearman's rank correlation.

THE RARE IDEAL

Occasionally, people are willing to spend the necessary money to commission detailed investigations, taking and analyzing enough samples to draw statistically valid conclusions. In my experience, however, they only become willing to do this after an initial "quick and dirty" analysis has convinced them that there is probably a significant problem to be solved.

It is more frequently the case that the troubleshooter can only really characterize what is happening with a plant when testing whether the proposed fix works. It is vital to analyze a sufficient number of samples before and after making a change to show whether the change improved, worsened, or made no difference to the process. To do this, we must draw on experimental design theory to understand how many samples are needed to ensure that the statistics will yield conclusions to a useful degree of certainty.

In summary, only change one thing at a time, and take enough of the right kind of samples to be able to decide if each change has helped, hindered, or done nothing. A well-designed trial methodology is generated by working backwards from the requirements of the statistics which will be used to analyze its results.

Failure to carry out a rigorous trial before and after making changes is one of the commonest reasons for multiple failed attempts to fix problems.

SITE VISITS

I always insist on a site visit when troubleshooting. As Fig. 29.2 shows, much can be determined about the general condition of the plant by visual inspection. There

FIGURE 29.2

An unloved industrial effluent treatment plant.

are several things which cannot be determined remotely, and should always be personally and directly verified. I have listed the most important of these below. In my visit to site, I always ask to see the operating and maintenance (O + M) manual, and to interview as many operators, managers, and technical support staff involved in the process as possible.

USING ALL YOUR SENSES

When walking around the site, do not confine yourself to just looking at and listening to the plant. Odors can be highly revealing. A repulsive smell of fatty acids will lead me to suspect inefficient FOG handling and treatment. "Eggy" sulfides and thiols, cabbage-like mercaptans, fishy amines, and the stench of ammonia all tell their own tales, usually of failures in biological treatment plant design or operation.

INTERVIEWING OPERATORS

If possible, operators should be interviewed individually, and informally, using open-ended questioning to encourage them to talk fluently about day-to-day operations, as well as the occasional incidents, accidents, and emergencies.

Workers will often speak more freely about what is really happening when they are separated from management. You may however need to exercise discretion when reporting on how you found out about certain things.

THE OPERATING AND MAINTENANCE MANUAL

It is telling if nobody on site knows where the O + M manual is, if it is found dusty and neglected on top of some filing cabinet, or if it has been hand-amended by operators. I see all three of these situations frequently in troubleshooting work. They show that the plant is not being operated in accordance with its designers' intentions.

MAINTENANCE LOGS

While O + M manuals are frequently neglected, or even lost, all but the very worst sites have a maintenance log kept by the operators. This is usually kept in a logbook, most commonly by hand on templates in a loose-leaf folder. The hand-writing is frequently illegible, and a transcript will often be needed.

However, once deciphered, these logs are often very informative, and I like to have read them before interviewing the operators, as the logs tend to contain an unselfconscious account of what operators have actually been doing, rather than what the O + M manual states they should have been doing.

OPERATOR TRAINING

If an O + M manual can be found, I learn how many of the present operating and management team were trained in plant operation by the company which constructed the plant.

I learn how many worked alongside the commissioning crew. This is often a place where operators learn tricks or shortcuts used in commissioning which are inappropriate to everyday operation, or where they gain access codes for instrumentation, etc., which management did not intend them to have.

The use of such tricks and unauthorized access to control levels which operators are not supposed to alter have been the cause of several fatal accidents in process plants. In effluent treatment, it is more likely in my experience that such interventions are at the root of mysterious problems which only occur when management are absent or uninvolved.

That said, operators on both the main plant and the effluent treatment plant are often quite capable of inventing their own shortcuts to save themselves work. On one plant I saw that a high water content in thickened sludge was found to be due to the operators having changed the grade of polymer used, in order to produce a wetter cake which did not require periodical leveling in the dumpster. That this doubled the costs of waste disposal bills at the plant was of no concern to the operators.

Overcoming the Folk Wisdom Hurdle

It is commonly the case that staff will have strong opinions on what the problem is, and will tend to offer only the data which support their theory.

Detailed questions are required to get past these explanations to the basic signs and symptoms which allow an independent judgement on what might be happening.

The difference between a sign and a symptom is important. Signs are those things which can be observed directly, whereas symptoms are those things which operators (or in some cases management) feel *might* be happening. Troubleshooters should rely primarily on signs. Symptoms should be treated more cautiously.

Operators are not always wrong in their diagnoses, and their reports of symptoms can have evidential value. However, it is usually the case that if the problem really was understood well enough to fix it, an external troubleshooter would not have been instructed to look into it. External troubleshooters are rarely the first to try to fix a problem, so assume nothing when seeking the underlying cause.

Classic Mistakes

I will also look out for examples of design elements which I know from experience are hard to control well, or actively create operational or maintenance problems.

For example, mixing tanks, with hours of retention time, are used too often on industrial effluent treatment plants for pH control, frequently coupled with acid or alkali addition via actuated valves from elevated tanks by gravity.

While an overall process may only be required to produce effluent with pH in the range 5–9, far tighter pH control (often ± 0.1 pH units) may well be crucial to many downstream processes, including flocculation, coagulation, precipitation, disinfection, and both aerobic and anaerobic biological treatments.

Other common mistakes include a lack of flow and load balancing tanks, using high-shear centrifugal pumps on shear-sensitive material such as polymer flocs, comminutors instead of screens for gross solids, plastic piping on high-temperature effluent streams, poor layout details, and selecting various kinds of piping, fittings, equipment, and instrumentation incapable of handling entrained solids, gases, fats, oils, and greases in the effluent.

I will also be on the lookout for "kludges" such as hammers on a string for the "percussive maintenance" of diaphragm pumps and solids hoppers, or "hammer rash" (Fig. 29.3) on hoppers which have not yet acquired a hammer on a string.

DESIGN VERIFICATION

Always request (but do not expect) design data, design drawings and calculations, plant specifications, and procurement documentation which make clear what the plant was originally designed to do. The best chance of locating this data (or at least some of it) is usually within O + M manuals.

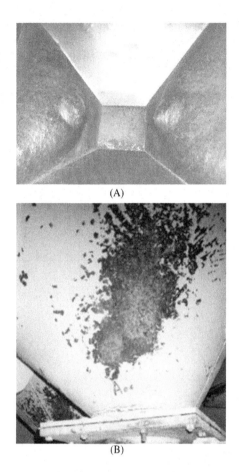

FIGURE 29.3

(A) and (B) Examples of percussive maintenance.

It can be helpful to measure key dimensions of all the equipment, and use the same design heuristics as are used in the early stages of plant design to estimate the capacity of each unit operation on the plant for yourself. This exercise can highlight potential design bottlenecks and often also design errors. In the rare cases where design information is available, it can be compared with what actually exists and has been measured on site.

PUTTING IT ALL TOGETHER

After gathering this information, it is usually possible to start to generate some candidate theories about what might be going wrong. These might involve problems with design, construction, operation or maintenance, or a combination of these.

Troubleshooters need to work out how to rationally decide what the problems are, or come up with ways to determine what they are if we cannot decide based on information in hand. Once we know what the problems are, we need to devise ways to fix them and ways to be sure that the fixes work without generating any new problems.

All of this happens under tight cost constraints, and often time constraints as well. No troubleshooter is commissioned to conduct a 5 year, multi-million-dollar research program into exactly what is going on in their plant. Clients simply want their plants to reliably meet specification in a cost-effective, safe, and robust manner.

FURTHER READING

Bonem, J. M. (2011). *Problem solving for process operators and specialists*. Hoboken, NJ: Wiley.

Water feature design

30

CHAPTER OUTLINE

Introduction 335
   Tender Stage Design 337
Water Quality 339
   Chemical Composition 339
   Clarity 339
   Biological Quality 340
Detailed Hydraulic Design 342
   Nozzles 342
   Weirs 342
   Water Quality 343

INTRODUCTION

Large water features are significant engineering projects and the treatment and handling of the large volumes of water used in such features (pipes for large features might be more than 1 m in diameter) requires input from a water engineering specialist. I have been involved in several complex designs, the largest of which was the Grand Cascade at Alnwick Castle, Northumberland, United Kingdom (where the Harry Potter movies were filmed) (Fig. 30.1).

There are five key aspects to such design work. Translating the artistic vision of architects, garden designers, and stakeholders into practical engineering questions; maintaining the aesthetics of the water; making sure that the engineering is hidden; maintaining water quality to bathing water standards (children love to paddle in water features); and making sure that spray from the feature does not harm the plants that commonly grow close to features.

To simplify, the engineer's key input to the design of a water feature may be considered in two parts: (1) maintenance of an acceptable quality of water and (2) provision of an acceptable head of water to operate water features.

Water quality considerations in these applications are firstly health and safety issues, such as avoidance of transmission of disease; secondly, aesthetic factors of clarity and absence of odor; and thirdly, the corrosive and other qualities that impact on the vessels, pipework, and so on used to produce the feature.

(A)

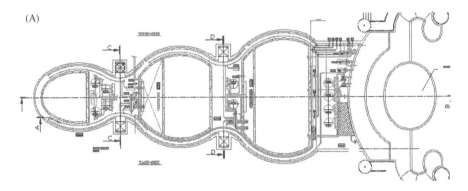

(B)

FIGURE 30.1

Grand Cascade, Alnwick Castle, Northumberland, United Kingdom (A) General Arrangement. (B) As constructed[1].

(A) Courtesy: Expertise Ltd. (B) Copyright image courtesy: Russel Wills.

Provision of an acceptable head of water and prediction of the behavior of water under differing conditions is dealt with by means of hydraulic calculations.

This chapter offers two levels of calculation: firstly, quick and simple techniques to allow a proposal to proceed with a reasonable degree of comfort in the integrity of the design and secondly, more detailed techniques that allow for more accurate design predictions at contract stage. It should be noted that hydraulics is never an exact science, and while the calculations given are acceptable to most

---

practitioners, if it is desired to offer guaranteed figures, the services of a specialist should be employed.

Anybody carrying out tender stage design should be as familiar with the contract stage calculation requirements as possible, to allow an appreciation of the requirements of the engineer carrying out the more detailed calculations.

TENDER STAGE DESIGN

Preliminary hydraulic design

Pipework

Rigorous calculations of straight run headlosses for circular pipework running full are usually unnecessary for tender stage designs. Nomograms for reckoning headloss per 100 m of pipe are available from many of the manufacturers of plastic pipework (see, e.g., Fig. 10.2). These should be adequate for tender stage calculations for plastic pipework.

As a general rule, flow velocities for gravity, suction, and delivery lines of 1.0, 1.5, and 3 m/s, respectively should not be exceeded at this stage. Velocities tend to be lower than these given for pipework runs over 50 m, and/or those containing many fittings or valves.

Calculation of fitting headloss (where number and type of fittings is known) should also be carried out. The k-value method should be used for tender stage calculations: see Chapter 10 for details.

FIGURE 30.2

Nozzles.

Tankage

The feed to the pump should be via a buffering tank of capacity equivalent to 3 min of the pump's peak flow, to permit the disentrainment of air bubbles which might otherwise contribute to reducing pump or feature performance. The shape of this tank should be designed to minimize the chances of short-circuiting of incoming flows.

Channels

Calculation of channel headlosses is more complex than pipework calculations, not least because of variation in depth of flow.

For tender purposes, it usually suffices to assume depth of flow is set by the depth over the outlet weir, if an outlet weir is present. If there is no outlet weir, calculation of depth of flow is more complicated, and it is suggested that 100 mm of water depth be allowed for tender purposes.

Setting flow velocity at less than or equal to 1.5 m/s at this depth should result in reasonable losses down the channel.

Allowing a 50-mm fall from the weir edge, or invert of the channel to the next water surface is usually sufficient to account for head loss.[2]

Weirs

Calculation of depth of flow over a weir is mostly dependent on the shape of the weir. The previously unpublished results of experiments carried out as part of the design of the Alnwick Castle water feature yielded the recommendation that 6 L/s/m be allowed for broad-crested weirs, and 4 L/s/m for sharp-crested weirs. These flows gave depths of the order of 10−15 mm over the weir.

Nozzles (Fig. 30.2)

Experiments with PEM and OESA brands of nozzles have shown that variability between literature and actual values and between batches of nozzles can be considerable. Manufacturers also change specifications of nozzles without notice, and may supply nozzles to differing specifications from those in their catalogues. It is therefore recommended that at least 25% be added to manufacturers' recommended head requirements for nozzles at the tender stage.

---

[2]Previous practice was to use a nomogram of this type with flow velocities of approximately 0.6, 1.4, and 2.25 m/s for Gravity, Suction, and Delivery pipework, respectively. It must be noted that use of these velocities did not guarantee hydraulic performance will be adequate, and conversely, exceeding these velocities is not necessarily problematic. It should also be noted that nomograms of this type only apply to the pipe materials stated, and that the plastic pipe nomogram often used will give large underestimates for rougher pipe materials.

WATER QUALITY
CHEMICAL COMPOSITION

The following substances are undesirable components of water destined for water features:

- Sulfides or other odorous compounds
- Iron, manganese, and other colored inorganic compounds
- Organic contaminants, especially humic and fulvic acids, which either lead to significant oxygen demand, or add color and/or odor

Water should not have a highly aggressive nature, as measured by Langelier index (see Appendix 6: Water aggressiveness indexes for details).

It may generally be assumed that supplies from potable sources are suitable in these respects. Any concerns over the level of any of these contaminants should be expressed at tender stage.

CLARITY

The most important parameter with respect to the visual impact of the feature is the clarity of the water. Unlike industrial applications, however, no performance standards are stated, and the clarity of the water is an aesthetic rather than a scientific measurement.

The water used should normally be free of colored compounds, whether organic, or inorganic. The suspended solids level is assumed to be the factor determining clarity of the water. The filters will remove only gross suspended solids. Any fine or colloidal solids will require additional treatment, as will any of the previous undesirable contaminants.

Sand filtration equipment to reduce suspended solids content is usually provided to address the need to maintain clarity of feature water. Manufacturers usually specify the type and size of unit to be used, but some rules of thumb may prove useful for giving an idea of likely sizes for these filters, to allow layout to proceed.

The volume of water to be treated in an hour may be determined by allowing all the water in the system to be passed through treatment in a period of between 2 and 6 h.

To determine this flowrate, firstly calculate the volume of all the water vessels, channels, and pipes in the system. If there are a lot of weirs in the system the head of water over the weirs may be a significant proportion of this volume. If there are many nozzles, water in the air may provide a significant part of the volume. These two items are, however, usually insignificant contributors to the total volume.

This total volume of water (v) is to be treated in x hours, and filter flowrate is therefore v/x.

x can be of the order of 2−6 h for tender purposes. How quickly we turn the system around is a factor of degree of contamination, amongst other things—outdoor systems receiving leaf debris, or street litter will require quicker turnovers.

High-rate sand filter sizes may be roughly estimated by allowing a flow per unit area of the filter of say 30 $m^3/m^2/h$, a figure far in excess of those used for more conventional rapid sand filters, but conservative in these applications.

Sand filters have to be cleaned periodically by means of backwashing. Filter backwashing is likely to result in wash water flowrates to drain of the order of the feed flowrate at the flow per unit area given. Where air scour is used to supplement the water, similar flowrates of air (in FAD m^3/h) are used.

Filters are backwashed much less frequently than in municipal water treatment, usually at weekly, rather than daily intervals. The actual washing frequency is, however, dependent on incoming water quality, flowrates per unit area, and dosing rates. Any guarantee figures should be referred to an expert.

There are a number of other filter types that are used for this application, rapid sand filters and precoat filters being the most common alternatives. These have the disadvantage of increased space requirements, and more troublesome operation, respectively. Note that (confusingly) rapid sand filters are not as rapid as high-rate filters, and are sometimes known in this sector as standard rate filters.

BIOLOGICAL QUALITY

In addition to the chemical and physical considerations outlined previously, prevention of the growth of organisms in the water is required. Of particular concern in some water features is the *Legionella* organism. This bacterium can cause serious pneumonia type illness in susceptible organisms, can survive in water kept below 60°C, and is transmitted well by any fine dispersion of water, such as those generated by sprays, jets, etc. Following the rules for maintenance and cleaning of features to prevent build-up of bacterial films, continuous disinfection of feature water, and periodic high-level disinfectant dosing are usually sufficient to control this hazard.

Lesser problems of water odor and appearance, as well as staining of water feature surfaces may also result from biological growth. The control measures suggested earlier would also inhibit the growth of organisms that may have these adverse aesthetic effects.

Filtration to maintain clarity also removes biological material from the system, and is the major factor in prevention of growth of algae within the feature, other than disinfectant dosing. Specific algaecides may be added to the feature, but they may well come with problems resulting from interaction with disinfectants, and the feature water. It is therefore highly unusual to add such algaecides on a regular basis.

The most commonly used disinfectant agents are bromine and chlorine. They have common advantages in that they are:

- highly lethal to the vast majority of organisms
- available in reasonably easy to handle forms

- relatively inexpensive
- capable of a persistent residual disinfectant action, preventing growth of organisms throughout the system

Disinfection by means of bromine is favored over chlorine, mainly because it is simpler to control, by virtue of its wider effective pH spectrum. A side effect of the use of bromine is oxidation of organic contaminants and the need for removal of ammonia from the system. Bromine dosing is usually based on systems dissolving bromine containing solid tablets. Manufacturers will be responsible for sizing the systems, but to allow for adequate feed pump capacity, one manufacturer recommends the provision of a flow to the brominator of 1 L/min per 10,000 L of feature capacity. Brominator capacities of approximately 1 kg of tablets per 7.5 m^3 of feature capacity are usual, to give reasonable filling intervals.

Ultraviolet light is also used as a disinfection method. High rates of disinfection are possible with UV, but it leaves no residual disinfectant in the water, unlike chlorine or bromine.

Ozone is a highly reactive form of oxygen that is gaining in popularity for swimming pool water feature treatment. Like UV, it leaves no residual disinfectant in the water, and therefore the possibility of growth of organisms within the body of the feature is a concern. This is often overcome in swimming pools by being used in tandem with chlorine dosing. The advantages of ozone in the swimming pool application do not apply to water feature use. Ozone is very toxic to humans, so stringent and costly provision for avoidance of ozone poisoning must therefore be incorporated within designs.

There are a couple of other minor techniques for disinfection that have come from developments to address the requirements of the United States and Soviet space programs for water recycling.

The US program devised a technique where disinfectant metal ions are introduced into the water by electrolysis of the water using precious metal electrodes. Use of this technique is practically limited to small domestic pools, as the effective agent is precious metal ions.

The Russians apparently developed electrolysis of salt through a semipermeable membrane. This system is sold as "Enigma" in the United Kingdom. The mode of action of this system is production of hypochlorous acid. There may be trace levels of other oxidized compounds, but these are insignificant with respect to the disinfecting effect of the system. There is no proof whatever of any effects over and above that explainable by the action of hypochlorous acid (the effective agent in standard chlorine disinfection). The system is more expensive than all other chlorine dosing systems.

DETAILED HYDRAULIC DESIGN
NOZZLES

If possible, testing of nozzles to determine their head requirements at the desired appearance is recommended. It should be noted that, for simple clear stream and aerated jet effects, plain piping can give effects equal to or better than far more expensive nozzles.

WEIRS

Different weir shapes may be used to give different effects. Experiments were carried out on the weir shapes shown in Fig. 30.3 as part of the Alnwick Castle design. (In all cases, flow went from right to left of the shapes as illustrated).

The experiments showed that a step overhang as in shapes (A) and (D) promotes nappe detachment, and radiusing of edges as in (A) and (C) tends to work against nappe detachment.

Type (A) gives an almost vertical curtain; the overhang prevents reattachment. Types (B) and (D) give a good curtain at low flowrates. Type (C) is very difficult to get a curtain of water over, and tends to have a clinging nappe at all flowrates.

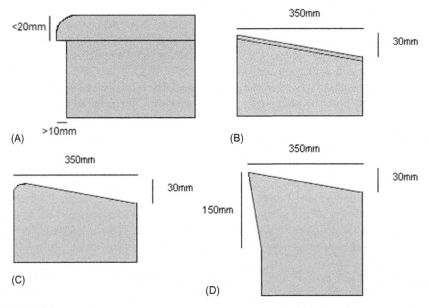

FIGURE 30.3

Broad-crested Weir Shapes (A) Step Weir, (B) Ramp Weir, (C) Radiused Ramp Weir, and (D) Gradual Overhang Ramp Weir.

WATER QUALITY

Chemical composition

It is assumed that, before the construction phase, adequate information has been made available on the compliance of supply water with the requirements laid out in the previous section on composition.

Plant should have been included to remedy any of the undesirable features of any proposed supply water.

Coagulation and filtration can remedy many organic and inorganic color problems. This may, however, involve construction of what is in effect a mini water treatment works, with metered dosing and controlled mixing of coagulants and pH correction chemicals, followed by additional filtration capacity.

It may be possible to remove odorous compounds, and oxygen demand in the water by means of oxidant chemicals, including some of the disinfectant chemicals.

Langelier indices in the aggressive range may be corrected by means of passive absorption of hardness from contactors containing magnesium and or calcium carbonates. It may be necessary to correct pH downstream of these contactors.

Clarity

Since client requirements for water quality will be aesthetic rather than scientific, it is recommended that filtration equipment be purchased from manufacturers with a guarantee of 95% removal of particles $>5\,\mu m$. While this will not guarantee overall water clarity, it is a common specification for filters producing drinking water.

Given the preceding specification for filters the clarity of the feature water is mostly dependent on the turnover, assuming all turbidity is caused by solid particles $>5\,\mu m$.

The turbidity of the water is the percentage of light that is absorbed as it passes through the sample water in a standard cell. This is therefore a measurement that corresponds inversely to the clarity of the water within the feature.

If we assume that a similar amount of turbidity is added each day, estimates are available on the percentage of this added turbidity that remains when the filters have done their work. Table 30.1 shows this relationship.

It is suggested that, for most applications, a turnover rate of 4 will suffice, removing 98% of turbidity at equilibrium. Higher rates would only be required when there is an unusually high load of incoming turbidity.

In the event that the turbidity is not filterable, advice should be sought from a water treatment expert on how to render the turbid components removable.

Table 30.1 Effect of Turnover Rate on Turbidity of Pool Water

Turnovers (per day)	1	2	3	4	5	6	7	8	9
Turbidity Remaining at Equilibrium (%)	58	16	5	2	1	0.5	0.08	0.02	0.01

Biological quality

The importance of the maintenance of water quality has already been emphasized. Common practice has been to leave this area to the manufacturers of chemical dosing systems. It may be, however, that on some bigger systems, concerns over areas such as pH stability have resulted in the use of pH dosing plant in addition to the disinfectant dosing kit.

The interaction between alkalinity, hardness, and pH is complex, and the interaction between bromine/chlorine and these factors adds more complexity. There is in addition to this a demand for disinfectant chemicals from oxidizable substances, especially ammonia, which makes it difficult to predict with accuracy the exact demand for dosing chemicals.

It is, however, necessary to be able to make a rational decision as to whether to install pH dosing equipment in addition to simple disinfectant dosing plant. It would be best to start from the analysis of water to be used within the feature in making this decision. The issue of size of feature is normally considered the major indicator as larger features have a greater chance of harboring stagnant areas where disinfectant chemicals may be destroyed by sunlight, or chemically used up, resulting in an area susceptible to biological growth. It may actually be best to address this feature by means of ensuring no such dead areas occur.

The pH and alkalinity of the water to be used both affect the need for pH correction equipment. Both chlorine and bromine disinfectants give best results at pH 7.2−7.8. Water that enters the system at a pH outside this range may be unsuitable for disinfection without pH correction. In addition to the loss of desirable disinfectant effect, there are a number of undesirable chemical reactions that are promoted by pH values far from the optimum.

Tablet brominators of the type commonly used for water features use chemicals that have very little effect on pH. Use of gaseous chlorine and sodium hypochlorite move pH into acid and alkaline ranges, respectively. If these chemicals are used, the advice of an expert should be sought in order to determine the pH that results from their addition.

If it is necessary to correct pH the alkalinity of the water must be taken into account in deciding which chemicals to use. Water with less than 50 ppm of alkalinity will have a very variable, and hard to control pH value.

APPENDICES

Appendix 1: General process plant design

INTRODUCTION

Coming into water and effluent treatment plant design after leaving university was something of a surprise. This was partly because of the transition between academic and practical approaches to design but there was a second factor.

Unlike chemical plants, which tend to be the focus of chemical engineering degree courses, water treatment plants have no control over the flow passing through them, or the composition of their feedstocks.

Plants that produce a clean water utility such as drinking or process water must match a variable demand for a product that must, in turn, always meet a certain specification. Plants that are designed to treat dirty water to a minimum standard allowing environmental release have to deal with a highly variable incoming flow and composition. This makes the specification of the design envelope especially important for water treatment plant process designers.

A further surprise to me was the wide number of people from different disciplines who considered themselves water process engineers, or sometimes process scientists. Until the 1980s, environmental engineers (a type of civil engineer) dominated dirty water treatment plant design engineering. There were also process chemists/scientists who tended to lead process design on the clean water side.

These two disciplines have distinctive approaches, which differ from that of the chemical process engineer such as myself. Although there are exceptions to the rule, environmental engineers tend to be weaker than chemical engineers in chemistry/biology, and process scientists tend to be weaker in engineering.

The ability of chemical engineers to work in both clean and dirty water process design gave us a competitive advantage, which means that nowadays, we are the go-to discipline for process design in the water sector, just as we are in others.

As the whole point of chemical engineering is that process design is process design, it should come as no surprise that chemical engineers make the best water process designers.

Chemical engineering is a kind of engineering, rather than a branch of chemistry. Professional engineering design practice has next to nothing to do with the activity called process design in many university chemical engineering departments. So let us start, as I did in my first book, by dispelling some confusion about what engineering is (and is not), and what design is all about.

WHAT IS ENGINEERING?

In academia, there is almost universal confusion between mathematics, applied mathematics, science, applied science, engineering science, and engineering.

Mathematics is a branch of philosophy. It is a human construction, with no empirical foundation. It is made of ideas, and has nothing to do with reality. It is "true" within its own conventions but there is no such thing in nature as a true circle, and even arithmetic, (despite its great utility) is not factually based.

Applied mathematics uses mathematical tools to address some real problem. This is the way engineers use mathematics, but many engineers use English too. Engineering is thus no more applied mathematics than it is applied English.

Science is the activity of trying to understand natural phenomena. The activity is rather less doctrinaire and rigid than philosophers of science would have us believe, and may well not follow what they call the scientific method, but it is about explaining and perhaps predicting natural phenomena.

Applied science is the application of scientific principles to natural phenomena to solve some real-world problem. Engineers might do this (though mostly they do not) but that does not make it engineering.

Engineering science is the application of scientific principles to the study of engineering artifacts. The classic example of this is thermodynamics, invented to explain the steam engine, which was developed without supporting science.

Science owes more to the steam engine than the steam engine owes to Science.
— **L.J. Henderson**

This is the kind of science that engineers tend to apply. It is the product of the application of science to the things engineers work with, artificial constructions rather than nature.

Engineering is a completely different kind of activity from all preceding categories. It is the profession of imagining and bringing into being a completely new artifact that achieves a specified aim safely, cost effectively, and robustly.

It may make use of mathematics and science, but so does medicine if we substitute the congruent "medical science" for "engineering science." If engineering were simply the application of these subjects, we could have a common first and second year on medical and engineering courses, never mind the various engineering disciplines.

WHAT IS DESIGN?

The ability to design is a natural human ability. Designers imagine an improvement on reality as it is, we think of several ways we might achieve the improvement, we select one of them, and we transmit our intention to those who are to realize our plan. The documents with which we transmit our intentions are, however, just a means to the ultimate end of design: the improvement on reality itself.

I discuss in this book a rather specialized version of this ability, but we should not lose sight of the fact that design is the same process, whether we are designing a water treatment plant, a vacuum cleaner, or a wedding cake.

Designers take a real-world problem on which someone is willing to expend resources to resolve. They imagine solutions to that problem, choose one of those solutions based on some set of criteria, and provide a description of the solution to the craftsmen who will realize it. If they miss this last stage and if the design is not realized, they will never know whether it would have worked as they had hoped.

All designers need to consider the resource implications of their choices, the likelihood that their solution will be fit for the purpose for which it is intended, and whether it will be safe even if it not used exactly as intended.

If engineers bring a little more rigor to their decision making than cake designers, it is because an engineer's design choices can have life and death implications, and almost always involve very large financial commitments.

So how does engineering design differ from other kinds of design?

ENGINEERING DESIGN

Like all designers, design engineers must dream up possible ways to solve problems and choose between them. Engineers differ from, say, fashion designers in that they have a wider variety of tools to help them choose between options.

Like all designers, the engineer's possible solutions will include approaches to similar or analogous problems that they have seen to work. One of the reasons why beginners are inferior to experts is their lack of qualitative knowledge of the many ways in which their kind of problems can be solved and, more important still, those ways that have been tried and found wanting.

Engineers need to make sure that they are answering the right question.

Petroski (see Further Reading) discusses the importance of avoiding failure in engineering design. Many of his examples of failure, however, were caused not by misspecification, but by designers who forgot that that the models used in design are only approximations, applicable in a narrow range of circumstances.

THE PROJECT LIFECYCLE

The niche or niches into which water and wastewater treatment plant design fits exist in the wider background of an engineering project lifecycle. The details of project lifecycles vary between industries, but there is a common core. For example, here is the product lifecycle for a water project:

1. Identify the problem—this stage is frequently overlooked because people think that they know what the problem is. Many solutions are for problems that do not exist (e.g., academic research often focuses on finding solutions without associated problems)
2. Define the problem in business, engineering, and science terms—often done poorly with the problem again being defined in terms of perceived solutions
3. Generate options that provide potential solutions to the problem
4. Review the options against agreed selection criteria and eliminate those options that clearly do not meet the selection criteria
5. Generate the outline process design for the selected options
6. Commence an engineering project to evaluate the possible locations, project time scale, and order of magnitude of cost
7. Based on the outcomes of step 6, reduce the number of options to those carried forward to the next level of detail
8. Develop the design of the remaining options to allow a sanction capital cost estimate to be generated and a refined project time scale
9. Based on the outcomes of step 8, select the lead option to be designed and installed
10. Carry out the detailed design of the lead option. A "design freeze" will almost certainly need to occur before the development work is complete
11. Construct the required infrastructure, buildings, etc., and install the required equipment
12. Commission the equipment
13. Commission the process and verify that the plant both performs as designed and produces product of the required quality
14. Commence routine operation
15. Improve process efficiency based on the data and experience gained during routine production
16. Increase the plant capacity, making use of process improvements and optimization based on the data and experience gained
17. Decommission the plant at the end of the product lifecycle

Where does process plant design fit into this? Consultants might consider stages 1−3 to be process plant design. Those with a background in design and build contracting usually think of process plant design as being predominantly what those in operating companies call "grassroots design"—broadly stages 3−10. Those who work for operating companies might call stages 15 and 16 process plant design.

Perhaps an argument can be made for all the stages, but it should be noted that, before step 14, very limited design information is available. The "design tools" popular in academia are used professionally only for stage 15/16 plant design, rather than stages 3−10 "grassroots" plant design for this reason.

I have used the term "water treatment plant design" in the sense of stages 3−10 design throughout this book, even though stages 1−3 and stages 15/16 are certainly related fields of professional activity, which I have been involved in. This is because this kind of design is closest to the meaning of the terms outside chemical engineering.

There are several reasons for this. Firstly, in the UK, the Construction Design and Management (CDM) regulations require a person or company to declare themselves the designer of a plant, responsible for the safety implications of the design. This designer is almost always the entity responsible for stages 3–10 design. Secondly, the process guarantee is almost always offered by the company responsible for stages 3–10 "grassroots" design. Thirdly, this definition of design is the one used by everyone involved in design activity.

PROCESS PLANT DESIGN

Water and wastewater treatment plant design is an art, whose practitioners use science and mathematics, models and simulations, drawings, and spreadsheets, but only to support their professional judgment. These things cannot supplant this judgment, since people are smarter than computers, and probably always will be. Our imagination, mental imagery, intuition, analogies and metaphors, ability to negotiate and communicate with others, knowledge of custom and practice and of past disasters, personalities, and experience are what designers bring to the table.

If more people understood the total nature of design, they would see the futility of attempts to replace skilled professional designers with technicians who punch numbers into computers. Any problem a computer can solve is not really a problem at all—the nontrivial problems of real-world design lie elsewhere.

Engineering problems will almost certainly always be far quicker to solve by asking an engineer, rather than by programming a computer, even if we had the data (which we can never have on a plant which has not yet been built), a computer smarter than a person, and a program that codes real engineering knowledge, instead of a simplified mathematical model with next to no input from professional designers.

Meadows (see Further Reading) has explained an intuitive system-level view that is identical in many ways to the professional engineer's view. We share this view with the kindergarteners who also excel at a design exercise called the marshmallow challenge that I use in my teaching. The roots of this system-level view are natural human insights, which we may be educating out of our students.

Meadows explains that she makes extensive use of diagrams in her book because the systems she discusses, like drawings, happen all at once, and are connected in many directions simultaneously, while words can only come one at a time in linear logical order.

Water and wastewater treatment plant design is system-level design, and drawings are its best expression—other than the plant itself—for the same reasons given by Meadows.

VARIATION AND SELECTION

Variation/creativity

When I examined myself and my methods of thought, I came to the conclusion that the gift of fantasy has meant more to me than my talent for absorbing positive knowledge.

— Einstein

The engineering design process consists of the generation of candidate solutions and of then selecting, from these, those most likely to be safe, cost-effective, and robust.

Coming up with the candidates is a creative process, involving the use of imagination, analogy from natural or artificial systems, knowledge of the state of the art, etc. Selection of candidates is, however, often more of a grind.

Engineers will almost always make use of mathematical tools to help them with selection, though the academic study of engineering can overemphasize these tools, which are supposed to inform professional judgment rather than supplant it.

Selection/analysis

Modeling and simulation programs only address this part of the design process, and may allow thoughtless processing of options rather than making a genuine considered choice between well-understood options.

Tools that help the understanding of a system are good. Tools that allow bypassing of understanding to get to a decision (even if our understanding is "mere" intuition rather than science) are not. Our best students will still understand, but the generality will not and there are risky implications associated with this.

Professional chemical engineers do not usually write simulation programs. Numerate graduates with no plant design or operating experience write them, and they run simplified mathematical models on machines with a great deal less processing power than people.

Professional engineers validate simulation programs against real plants before giving their outputs any credence. Therefore, the use of such programs by professional engineers is usually limited to modification of existing plant rather than the whole-plant grassroots design largely addressed by this book.

The key to a successful design is to understand the problems just well enough to be able to predict that the desired outcome will be reliably attainable. Science and mathematics are certainly tools in the engineer's portfolio, but they are very often not the most important ones, and they are not its basis.

Capturing nonscientific information about how past designs of the type being attempted have performed, and the factors associated with success, is often at least as important. The information is highly situation-specific, and is essentially a codification of experience.

The information may be presented as a design manual, or as standards, codes of practice, or rules of thumb. Such documents allow the very complex situations common in real-world design to be appropriately simplified, making theoretically impossible parts of the design practically possible. The main purpose of such documents is to control the design process such as to constrain innovation within the envelope of what is known to be likely to work based on experience.

Process plants versus castles in the air

An architect friend tells me that there are two kinds of architects, those who know how big a brick is, and those who do not. Castles in the air are of such stuff as dreams are made of, but real designs need to be built using things we can buy.

Process engineers do not often make things out of bricks, but we nevertheless make most of what we build from simple standardized subunits, such as lengths of pipe, and ex-stock valves, pumps, and so on. These items have types and sub-types that differ from each other in important ways. Choosing between them is no trivial matter.

If you do not know the dimensions and characteristics of the standard subunits from which you build something, your designs will be impractical to build. They will be likely to be less cost-effective, less robust, and less safe than the output of a more practically minded designer.

Commercially available components almost always represent the lower limit of the professional engineer's resolution. We do not care about the atomic structure of the metal in our pump, we care about those of its properties that affect the overall design.

WHAT WE DESIGN AND WHAT WE DO NOT

We do not tend to design things that we can buy, because having things specially made costs a lot more than buying stock items, and those who design things guarantee their efficacy. Engineers like to minimize both costs and risks.

Process engineers tend not to carry out design of mechanical, electrical, software, civil, or building works. They must, however, know about the constraints imposed on designers in these disciplines, and have a feel for the knock-on cost and other implications of their design choices.

STANDARDS AND SPECIFICATIONS

Standards and specifications exist to keep design parameters in the range where the final plant is most likely to be safe and to work. They also serve to keep design documentation comprehensible to fellow engineers. A brilliant design that no-one else understands is worthless in engineering.

There are several international standards organizations—ISO in Europe, DIN in Germany, ANSI, ASTM, and API in the United States, and so on. ("British Standards" in the United Kingdom are now officially a subset of ISO.)

I refer to British Standards in this book where available, most notably those governing engineering drawings. The use of British Standards (or any other used consistently and clearly) for drawings reduces the likelihood of miscommunication between engineers via their most important channel of information exchange.

The availability of interchangeable standard parts makes much of engineering design simple in one way, but introduces an extra stage that is frequently overlooked in academia. After an approximate theoretical design, practitioners redesign in detail using standardized subcomponents. We do not, for example, use 68.9 mm internal diameter pipe, we use 75 mm NB (nominal bore), because that is what is readily commercially available. NB and its US near-equivalent NPS (Nominal Pipe Size) are themselves specifications, rather than sizes.

DESIGN MANUALS

Companies frequently have in-house design manuals that are a formal way to share the company's experience of the processes it most often designs. These manuals are not in the public domain, because they contain a significant portion of the company's know-how. They are jealously guarded commercial secrets.

Some national and international standards are also essentially design manuals—for example, PD5500 used to be a British Standard and is now more of a design note or manual that embodies many years of experience of safe design of a safety-critical piece of equipment (namely, the unfired pressure vessel), despite being superseded in 2002 by a European standard (BS EN 13445).

RULES OF THUMB

Rules of thumb are a type of heuristic, and are usually very simple calculations that capture knowledge of what tends to work. Rules of thumb do not necessarily replace more rigorous (but still heuristic) analysis when it comes to detailed design, but their condensation of knowledge gained from experience provides a quick route to the "probably workable" region of the design space, especially at the conceptual design stage.

Rules of thumb are only ever good in a limited range of circumstances. These limitations must be understood and adhered to if they are to be valid. Rules of thumb encapsulate experience, and are therefore better than first-principles design.

First-principles design does not work, and published examples of "successful" first-principles design often turn out to be no such thing when investigated. My expert witness practice commonly centers on costly attempts by beginners to use first-principles design in the real world.

APPROXIMATIONS

Everything is approximation in engineering. Engineers who grew up in the era of the slide rule know that anything after the third significant figure is at best science, rather than engineering.

We need to know how precise and certain we must be in our answers to know how rigorous to be in our calculation. Very often, in troubleshooting exercises, knowing the usual interrelationships between a few measurements is all that is needed to spot the most likely source of problems.

Coarse approximations will steer us to the right general area for our answers, allowing design to proceed by greater and greater degrees of rigor as it homes in on the area of plausible design.

PROFESSIONAL JUDGMENT

Douglas (see Further Reading) gives a figure of $10^5 - 10^8$ possible variations for any new process plant design, and even this leaves out many of the important variables. This is far lower than the number of possible positions in chess, but it is akin to the number of patterns memorized by an experienced chess player.

An engineer's professional judgment allows them to semiintuitively discern approaches to problems, which might work in a similar way to the experienced chess player. They will summarily discard many blind alleys that a beginner would waste time exploring, and include options that beginners would be unlikely to think of. They will know that simple calculations will allow them to choose quickly between classes of solution.

Consequently, experts can quickly achieve outcomes that less experienced practitioners might never arrive at. This judgment takes many years of practice to develop.

STATE OF THE ART AND BEST ENGINEERING PRACTICE

When I serve as an expert witness, I need to differentiate between *facts* (perceptible directly to human senses or detectable with a suitably calibrated instrument), *current consensus opinion* (as held by most suitable qualified professionals), and *other opinion*. These are considered to represent progressively weaker evidence.

Best engineering practice is always changing, so no single engineer is an entirely reliable source of advice on what is best engineering practice. My experience as an expert witness has, however, taught me that professional engineers understand the gap between our personal practice and common practice. We know whether each of our heuristics is on the fringes or at the heart of consensus/best practice. Even those maverick professionals who hold fringe ideas that they think better than best practice know that these opinions are not commonly held.

This means that only current practitioners who regularly engage professionally with other practitioners are in touch with the moving goalposts of best practice. It also means that those closest to the heart of a discipline tend to have the greatest concordance between personal heuristics and consensus heuristics. This does not prevent people far from the heart of the discipline from holding fringe opinions strongly, it just keeps them from understanding that they are fringe opinions.

THE USE AND ABUSE OF COMPUTERS

Back in 1999 the IChemE's Computer Aided Process Engineering (CAPE) working group produced Good Practice Guidelines for the Use of Computers by Chemical Engineers. These have never been superseded, and are if anything increasingly relevant.

The guidelines emphasize the legal and moral responsibility of the professional engineer to ensure the quality and plausibility of the inputs to and outputs from the system, to understand fully the applicability, limitations, and embedded assumptions in any software used. It also emphasizes the importance of being properly trained to use the software, and only using fully documented software validated for your particular application.

They emphasize that the primary importance of understanding the problem one is trying to solve, working within proper engineering limits, of taking into consideration not just dynamic but transient conditions and, most of all, of applying sensitivity analysis to the results produced, especially for those areas identified as the most important to a successful outcome. The severity of the consequences of being wrong should also be taken into consideration.

They warn users to beware of the assumptions implicit in software, and to know where default values exist, and to be able to back-trace any data used to a validated source. This allows the documented accuracy of the data in the range used to be known, as well as whether it is valid in that range.

They recommend contacting more senior engineers and/or the suppliers of the software in the event of any uncertainty at all about the correctness, accuracy, or fitness for purpose of any software or outputs.

In several places, they specifically recommend suspicion about the outputs of computer programs, and an assumption of guilt until proven innocent.

They say repeatedly, in different ways, that software cannot be a substitute for engineering judgment, and that its use without understanding is a dangerous abandonment of professional responsibility.

However, I have seen many times, in both academic circles and in recent graduates, a failure to understand this truth. Computers may be a little faster than they were in 1999, but they are no closer to being people. All the potential problems of software are still there, and the awareness of their limitations has decreased.

FURTHER READING

Douglas, J. (1988). *Conceptual design of chemical processes*. New York, NY: McGraw-Hill.

IChemE Computer Aided Process Engineering (CAPE)Working Group *Good practice guidelines for the use of computers by chemical engineers* (1999) <http://www.icheme.org/communities/special-interest-groups/cape/resources/publications.aspx> Accessed 12.10.17.

Meadows, D. H. (2009). *Thinking in systems*. New York, NY: Routledge.

Petroski, H. (1992). *To engineer is human: The role of failure in successful design.* New York, NY: Vintage.

Wujec, T. *Build a tower, build a team (the Marshmallow problem)*, TED Talk. <https://www.ted.com/talks/tom_wujec_build_a_tower> Accessed 12.10.17.

Appendix 2: The water treatment plant design process

INTRODUCTION

Like all process plants, water treatment plants are made up of several main plant items or unit operations.

Main plant items are designed, specified, and purchased by the process contracting company that installs, commissions, and sets to work the water and wastewater treatment plant in such a way as to integrate the operation of the main plant items.

Process or chemical engineering is the discipline which ensures that all parts of a water and wastewater treatment plant can work together harmoniously, with consideration of cost, safety, operability, and performance.

There is far more to such a plant than the main plant items. The site must be prepared; a concrete slab laid down; the main plant items fixed down, interconnected with pipes and valves, connected to electrical cables, powered, and controlled by a control panel and instrumentation; and the whole usually enclosed in a suitable building.

A combination of different engineering disciplines usually designs water and effluent treatment plants. Process or chemical engineers are responsible for specification and sizing of the unit operations required to turn dirty water into clean water via several process stages. They usually work for specialist design consultants or process contractors.

There are many processes used in water treatment but, in summary, solid particles may be settled or filtered out. There may be modifications to water chemistry via chemical addition. There may be physical agitation to promote mixing, control particle size, and so on.

Process engineers know how to achieve the required effects using a range of specially designed or specified machines. Every plant is bespoke, because every site and its water are different.

Process engineers are usually also involved in determining the layout of the plant, selecting suitable materials of construction, as well as process control and instrumentation.

Process engineers normally produce certain key documents to help them do this. A process flow diagram (PFD) is normally developed early in the design process, summarizing an essential exercise known as a mass and energy balance.

Process engineers also produce a piping and instrumentation diagram (P&ID), which shows how the plant's components work together. They either produce or are involved in producing the general arrangement drawing (GA) that shows how the plant is laid out in space.

Process engineers also know how to specify the pumps, valves, pipes, etc., that control the flow of materials around the plant. Part of this process is hydraulic design, which allows the effect of pumps, pipes, valves, etc., on fluid flow to be calculated, in order to specify them. Hydraulic design requires, as a prerequisite, a detailed GA to allow pipe lengths to be known.

Mechanical engineers may also know how to lay out plant, carry out hydraulic design, and select materials, but they are mainly concerned in this field with other matters, such as design of pipework supports, decking, platforms, and so on, to suit the process engineer's design.

Electrical engineers size and lay out cabling and control panels to suit the process engineer's design. They are also involved in issues such as lighting and power arrangements on the plant. They may also have input on other issues such as choice of instrumentation.

Civil and structural engineers also must be involved, most notably to design and ultimately build the slab that the plant sits on. The weight and locations of the various parts of the plant need to be known to allow this design work to proceed.

The remainder of this Appendix is adapted from my first book *An Applied Guide to Process and Plant Design*, which is a more general process plant design text. See also Appendixes 7, Clean water treatment by numbers; Appendix 8, Sewage treatment by numbers; and Appendix 9, Industrial effluent treatment by numbers.

STAGES OF DESIGN

Plant design proceeds by stages. These are commonly known as Conceptual, Front End Engineering, Detailed, and "For Construction" design.

The stages of water treatment plant design (and indeed almost all designs) seem not so much to be conventional as having evolved to fit a niche. The commercial nature of the process means that minimal resources are expended to get a project to the next approval point. This results in design being broken into stages leading to three approval points: feasibility, purchase, and construction.

The basically invariant demands of the process are the reason everyone who designs something professionally does it in essentially the same way, even though chemical engineers are often nowadays explicitly taught a radically different approach in university (if they are taught any approach at all).

CONCEPTUAL DESIGN

Conceptual design of water and wastewater treatment plants is sometimes carried out in an ultimate client company, more frequently in a contracting organization, and most commonly in an engineering consultancy.

In this first stage of design, we need to understand and ideally quantify the constraints under which we will be operating, the sufficiency and quality of design data available, and produce several rough designs based on the most plausibly successful approaches.

Practicing engineers tend to be conservative, (none more so than water engineers) and will only consider a novel process if it offers great advantages over well-proven approaches, or if there are no proven approaches. Reviews of the scientific literature are very rarely part of the design process. Practicing engineers very rarely have free access to scientific papers, and are highly unlikely in any case to be able to convince their colleagues to accept a proposal based on a design that has not been tried at full scale several times, preferably in a very similar application to the one under consideration.

The conceptual stage will identify several design cases, describing the outer limits of the plant's foreseeable operating conditions. Even at this initial stage, designs will consider the full expected operating range, or design envelope.

The documents identified in Appendix 3, Design deliverables, are produced for the two or three options most likely to meet the client's requirements (usually economy and robustness). This will almost always be done using rules of thumb, since detailed design of a range of options, most which will be discarded, is uneconomic.

This outline design can be used to generate electrical and civil engineering designs and prices. These are important, since designs may be optimal in terms of pure "process design" issues like sludge yield or energy recovery, but too expensive when the demands of other disciplines are considered.

At the end of the process, it should be possible to decide rationally which of the design options is the best candidate to take forward to the next stage. Very rarely, it will be decided that pilot plant work is required, and economically justifiable, but this is very much the exception; design normally proceeds to the next stage without any trial work.

The key factor in conceptual studies is usually to get an understanding of the economic and technical feasibility of several options as quickly and cheaply as possible. Ninety eight per cent of conceptual designs are never built, so it is not sensible to spend a fortune investigating them.

Client companies have advantages over contractors in carrying out conceptual designs, as they may possess operating data unavailable to contractors; however, they usually lack real design experience. Contractors are in the opposite situation, while most staff employed by consultancies tend to have neither hands-on design experience nor operational knowledge. In an ideal world, therefore, client companies would collaborate with contractors to carry out conceptual design. In the real world, this cooperation/information sharing is less than optimal.

Modeling as "Conceptual Design"

Much work thought of as conceptual design, especially in the case of modifications to existing plant, takes place in large consultancies and operating companies where staff tend to lack whole-plant design experience. Their approach may therefore focus on a small number of unit operations rather than a whole plant, an approach commonly followed in academia. However, this work differs from the design techniques of professional water and wastewater treatment plant designers in several ways, most notably:

The "conceptually designed" plant will not be built by those carrying out the conceptual design exercise. The contractors who will build it may use this "design" as the basis of their design process, but since it is they who will be offering the process guarantee, they may redesign from the ground up.

Genuine information from pilot studies on the actual working plant would allow a computer model to be fed with a specific and realistic design envelope, and for its outputs to be validated on the real plant. This is very different from using the modeling program in an unvalidated state.

Thus, such conceptual design studies, while potentially very valuable as a debottlenecking or optimization exercise, are not a true plant design exercise. The modeling, simulation, and network analysis used in academia are of their greatest utility in this area, due to the availability of the large quantities of specific data that are needed to yield meaningful results.

However, those carrying out such exercises should nonetheless take into account the suggestions from contracting companies if they would like to arrive at a safe, robust, and cost-effective solution. There is no substitute for experience.

It is important to understand that professional judgment is still superior to the output of computer programs; however much effort went into pilot trials and modeling exercises, but human nature leads people to wish to benefit from costs that have been irreversibly incurred. If a great deal of resource has been spent on these studies, it may be hard to accept that simply engaging a contractor would have led to a suitable plant design. The sunk cost fallacy needs to be avoided.

FEED/BASIC DESIGN

If the design passes the conceptual stage, a more detailed design will be produced, most commonly by a contracting organization.

Competent designers will not use the static, steady state model used in educational establishments, but will devise several representative scenarios that encompass the range of combined process conditions that define the outer limits of the design envelope. All process conditions in real plants are dynamic; they do not operate at "steady state." They may approximate the steady state during normal operation if the control system is good enough, but they must be designed to cope with all reasonably foreseeable scenarios.

The process engineer normally commences by setting up (most commonly in MS Excel) a process mass and energy balance model linking together all unit operations, and an associated PFD for each of the identified scenarios. It is usually possible to set up the spreadsheets so that models of the various scenarios are produced by small modifications to the base case.

In certain industries a process modeling system might be used for this stage, but in the water sector this is the exception rather than the rule, because business licenses for such programs are very expensive.

A more accurate version of the deliverables from the previous stage is produced, based on this more detailed design/model and, wherever possible, bespoke design items are substituted with their closest commercially available alternatives, and the design modified to suit.

Process designers normally avoid designing unit operations from scratch, preferring to subcontract out such design to specialists who have the know-how to supply equipment embodying the specialist's repeated experience with that unit operation.

Drawings at this stage should show the actual items proposed, as supplied by chosen specialist suppliers and subcontractors. Even such seemingly trivial items as the pipework and flanges selected should be shown on the drawings as they will be supplied by a particular manufacturer, and pricing should be based on firm quotes from named suppliers.

The drawings should form the basis of discussions with, at a minimum, civil and electrical engineering designers, and a firm pricing for civil, electrical, and software costs should be obtained.

All drawings and calculations produced are checked and signed off at this stage by a more experienced—ideally chartered—practicing chemical engineer.

Once that is done, a design review or reviews can be carried out, considering layout, value engineering, safety, and robustness issues. Where necessary, modifications to the process design to safely give overall best value should be made.

DETAILED DESIGN

The detailed design stage is also known as "Design for Construction." This virtually always takes place in a contracting organization. The detailed design will be sent to the construction team, who may wish to review the design once more with a view to modifying it to reflect their experience in construction and commissioning.

Many additional detailed subdrawings are now generated to allow detailed control of the construction of the plant. The process engineer would normally not have much to do with production of such mechanical installation drawings, other than participation in any design reviews or hazard and operability studies (HAZOPs) that are carried out.

The junior process engineer will, however, have much to do with producing documents such as the datasheets, the valve, drive and other equipment schedules, the installation and commissioning schedules, and the project program.

This painstaking work is required to allow the procurement of the items that are described in them by nonengineers. It is not so much design work as contract documentation.

SITE REDESIGN

This does not usually feature much in textbooks on design, but it is common for designs to pass through all the previous stages of scrutiny and still be missing many items required for commissioning or subsequent operation.

It is far cheaper to move a line on a drawing than it is to reroute a process line on site. If communication from site to designers is managed very poorly in a company, expensive site modifications may be required on many projects before the problem comes to the attention of management.

Commissioning and site engineers are rarely involved in the design process (though they should be involved in any HAZOP) but often find these omissions when they review the design they have been given, or worse still, find the error on the plant as built. This is the most expensive stage at which to modify a design, but in the absence of perfect communication from site to design office, it will continue to occur.

The following items should be included in designs at the earliest stage, to facilitate the commissioning process:

- Tank drains that will empty a tank under gravity in less than 1 hour, and somewhere for the drained content to go to safely
- Suitable sample points in the line after each unit operation (there may be far more to this than a tee with a manual valve on it)
- Service systems adequately sized for commissioning, maintenance, and turnaround conditions
- Connection points for the temporary equipment required to bring the plant to steady state operation from cold, especially where process integration has been undertaken
- Water and air pollution control measures suitable for use under construction and commissioning conditions
- Access and lifting equipment suitable for use under construction, commissioning, maintenance, and turnaround conditions

See also Chapter 28, Classic mistakes in water and effluent treatment plant design and operation, for further examples of mistakes to avoid.

POSTHANDOVER REDESIGN

The commissioning engineer may have tweaked or even partially redesigned the plant to make it easier for it to pass the performance trials that are used to judge the success of the design, but the nature of the design process means that, while the unit operations have been tuned to work together, they have different maximum capacities.

When the spare capacity in the system is analyzed, it is usually the case that the output of the entire process is limited by the unit operation with the smallest capacity. This is a restriction of capacity or a "bottleneck." Uprating this rate-limiting step (or several of them) can lead to an economical increase in plant capacity.

Similarly, it might be that services that were slightly overdesigned to ensure that the plant would work under all foreseeable circumstances, and optimized for lowest capital cost rather than lowest running cost, can be integrated with each other in such a way as to minimize cost per unit of product.

It should be noted that the operational teams carrying out this kind of operation have at their disposal significantly more data than whole-process designers.

FAST TRACKING

Mixing the natural stages of design to accelerate a program is a reasonably common approach, which may be used in professional practice, though it has a downside. It is telling that contractors who are asked to make a program move faster (or "crash the critical path") are usually given acceleration costs to compensate them for the inherent inefficiency of the fact-tracked process.

As well as costing more, it is commonly held that fast tracking always increases speed at the cost of quality, whether that be design optimization quality, and/or quality of design documentation.

The standard approach practically minimizes the amount of abortive work undertaken, since each stage proceeds based on an established design envelope and approach. Each stage refines the output of the preceding stage, requires more effort, and comes to a more rigorous set of conclusions than the preceding stage.

When stages are mixed, the more rigorous steps are carried out earlier in the program, when the design has a larger number of variables. We may well therefore need to have a larger team of more experienced people working on the project. Even with the improved feel for engineering given by using more experienced engineers, it is much more likely in fast tracking that a design developed to quite a late stage will need to be rejected, and the process restarted from the beginning of the blind alley which the design went down. In highly profitable sectors, this will be less of a concern, but water and environmental engineering is one of the least profitable sectors, which explains much about its processes and procedures.

CONCEPTUAL/FEED FAST TRACKING

In a contracting company or design house, conceptual design can be very rapid indeed. Senior process engineers know from experience that is likely to be the best approach to an engineering problem. A few man-hours may be all that is required to produce a rough conceptual design.

However, this means that a client who wishes to skip the stage where they or a consultant develop the initial designs is ceding control of the design to a contractor. This is not without a downside—senior process engineers in contracting organizations will almost certainly come up with a design that will be safe, cost-effective, and robust, but a design based entirely in generic experience is unlikely to be the most innovative practical approach.

FEED/DETAILED DESIGN FAST TRACKING

If a project is to go ahead, and the contractor has been appointed, FEED and detailed design can be combined, and the operating company/ultimate client can take part in the design process. This is a traditional approach based in the United Kingdom on IChemE's "Green Book" contract conditions.

This approach can produce a very good quality plant, which is unusually fit for purpose, but the FEED study is normally used as the basis of competitive tendering. The removal of a discrete FEED study, along with that of the usually adversarial approach between client and contractor in water and wastewater treatment plant procurement, does seem to inflate costs somewhat.

DESIGN/PROCUREMENT FAST TRACKING

Time is generally lost from the originally planned design program at each stage of design such that the later stages—which produce the greatest number of documents, employ the most resources, and are most crucial to get right—often proceed, unhelpfully, under the greatest time pressure.

There are barriers to communication between the various stages of design that need to be well managed if conceptual design is to lead naturally to detailed design and from there to design for construction. If the communication process is managed poorly, unneeded redesign may be carried out by "detailed" or "for construction" design teams who do not understand the assumptions and philosophy underlying the previous design stage. Alternatively, designs may have to be extensively modified on site during construction and commissioning stages, usually at the expense of the contractor.

Design/procurement fast tracking is popular as a response to time lost in earlier stages for projects where the end date cannot move. As soon as an item's design has been fixed, the procurement process is started, especially with long lead time items such as large compressors.

The nature of design being what it is, this can result in variations to specifications for equipment design after procurement has started, and this usually carries a cost penalty, as does the high peak manpower loading, duplication of designs, and backing-out of design blind alleys inherent in this approach.

GENERATING THE DESIGN ENVELOPE

A water treatment plant operates for most of its life within bounds set by the effectiveness of its control system; for a significant part of its life in commissioning/maintenance conditions outside these bounds; and must be sufficiently safe when operating well out of bounds in an emergency.

Although a plant will spend most of its life operating within bounds (though not actually at steady state), the maintenance and emergency conditions we need to accommodate may well have a larger effect on the limits of plant design than the requirements of quasi-steady state normal operation.

So, we must design a plant that will handle normal variations for extended periods of time, maintenance conditions for shorter periods of time, and extreme conditions for short, but crucial periods.

Each parameter therefore has a range of values, rather than the single value it has in the steady state case. So, plant design may have dozens of parameters at each stage, each of which has a range of values. The range of values will be associated with a range of probabilities, like a confidence interval. In a good design, extreme values will be experienced with a low probability, and average values will be commonplace.

We frequently use confidence intervals to decide on the upper and lower bounds of incoming and outgoing concentrations of key chemicals. Performance trials are frequently statistically based, so we are in effect already working to a confidence interval in our product specification. Engineers normally, however, work beyond the specified confidence interval, such that a 95% confidence interval specification from the client might lead an engineer to produce a 99% confidence interval design.

We then need to permutate these ranges of parameters to generate design cases, representing the best, average, and worst cases we can imagine across important permutations.

For example, a sewage treatment work needs to work in the middle of the night when no one is flushing a toilet and most factories are closed, as well as at peak loadings. It needs to work during dry weather, and during a 100-year return period storm. It needs to work sufficiently well when crucial equipment has failed.

When it rains hard, we may initially see a sharp increase in incoming solids and biological material due to sewer flushing, and afterward we may see large volumes of weak sewage, mostly rainwater.

Thus, it is necessary to design to a low probability/high strength/high flow scenario, a medium probability/high flow/low strength scenario, a medium probability/low flow/high strength scenario, and a high probability/medium flow/ strength condition.

We need to construct a range of realistic design scenarios, and our designs need to work in all of them. It may be that our provision for less probable scenarios has a shorter service life or requires more operator attention than that for the normal running case, but the plant must be safe and operable under all foreseeable conditions.

Even bought-in chemicals have a specified range of properties, rather than a single value. There is, in short, no such thing as steady state.

In real design scenarios, we often have too little data to generate statistically significant design limits. We get a feel for the data by generating summary statistics: means, maxima, minima, confidence intervals, etc. There is a plug-in tool for MS Excel that will helpfully generate this set of statistics but even plugging in the formulae and functions manually is not very time consuming.

The lack of a formally statistically valid data set does not absolve us from the responsibility of designing the plant. Sensitivity analysis can be used to see how much it matters if the data are unrepresentative. The all-too-likely lack of rigorously valid data as a design basis is something the designer should be conscious of throughout the design.

For example, sample analyses are limited in number, and will not be fully representative for many reasons. Types of sampling methods, source condition, insufficient conditioning prior to sample collection, inappropriate sample collection methods, limitations of laboratory testing, and contamination of samples may contribute to error. Different laboratories have been shown to produce wide inconsistency in compositional analysis, even for the same sample.

All these uncertainties may affect the facility design. Therefore, it is important that the design parameters are not too tight and a sufficient design margin is provided.

Appendix 3: Design deliverables

The key deliverables from a process designer's point of view for a water and wastewater treatment plant design project (in the approximate order in which they are first produced) are as follows:

DESIGN BASIS AND PHILOSOPHIES

The output from the conceptual design stage may sometimes be restricted to guidance on the approach which should be followed in subsequent design stages: a design basis or design philosophy. These terms are sometimes taken to be the same thing, but there are some differences between them.

A design basis will usually be a succinct (no more than a couple of sides of A4) written document which might define the broad limits of the FEED study, including such things as operating and environmental conditions, feedstock and product qualities, and the acceptable range of technologies.

Design philosophies, by contrast, may run to 40 pages. Clients often specify a design philosophy in their documentation, and individual designers and companies may have their own in-house approaches. It is good practice for a formal design philosophy to be written as one of the first documents on a design project. Similarly, a safety and loss prevention philosophy should, ideally, be produced early in the design process.

The design philosophy should record the various standards and philosophies used, together with the underlying assumptions and justifications for the choice of these. This is both to allow a basis for checking in the detailed design stage, and for legal and cost control purposes. In the absence of a written justified design philosophy, a different engineer working at the detailed design stage might attempt to apply an alternative, and the plant may consequently become subject to expensive redesign.

SPECIFICATIONS

There are several types of specification which are produced or introduced at various stages of the design process and which inform the definition of the design envelope.

The expected quantities and qualities of incoming water should be included, as well as a description of end-product quality and quantity. These descriptions will ideally be in the form of ranges of concentrations, flows, temperatures, pressures, and so on. There may be statistical information to allow the designer to understand the distribution of likelihood of various conditions.

There may be reference to standards, legislation, and so on likely to be critical to this specific design.

PROCESS FLOW DIAGRAM

The process flow diagram (PFD) is sometimes unhelpfully called a "flowsheet," but this term can have several different meanings. In the United Kingdom and Europe, the general British Standard for engineering diagrams (BS 5070) applies to the PFD, as does BS EN ISO 10628. The symbols used on the PFD should ideally be taken from BS EN ISO 10628, BS1646, and BS1553. In the United States (and in industries influenced by US regulation) ANSI/ASME Y14.1 "Engineering Drawing Practice" and ANSI/ISA S5.1-1984 (R 1992) "Instrumentation Symbols and Identification" apply.

The PFD treats unit operations more simply than the P&ID. Unit operations are shown using standard P&ID symbols or sometimes as simple blocks. Pumps are shown, as are main instruments, as shown in Fig. A3.1.

The lines on the PFD are labeled in such a way as to summarize the mass and energy balance, with flows, temperatures, and compositions of streams. The visual representation of the plant's interconnections and mass and energy balance is the main purpose of the drawing.

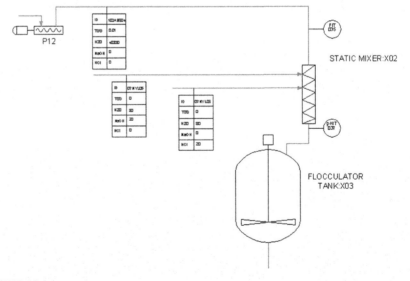

FIGURE A3.1

Process flow diagram for the pH correction section of a water treatment plant.

The sequential order of unit operations on the PFD (usually from top left to bottom right of the drawing) is a good starting point for their layout in space.

PIPING AND INSTRUMENTATION DIAGRAM

The Piping and Instrumentation Diagram (P&ID) is a drawing which shows all instrumentation, unit operations, valves, process piping (connections, size, and materials), flow direction, and line size changes both symbolically and topologically. An example of an extract from a P&ID is provided at Fig. A3.2.

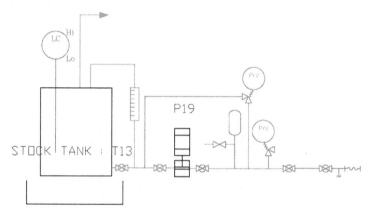

FIGURE A3.2

Extract from a piping and instrumentation diagram.

The P&ID is the process engineer's signature document, which develops during the design process. Its purpose is to show the physical and logical flows and interconnections of the proposed system. Recording these visually on the P&ID allows them to be discussed with software engineers, as well as other process engineers. It is useful that it shows pipe sizes, materials, and other detailed features which do not appear on the PFD.

There are a great many variants in additional features between industries, companies, and countries, but producing the drawing to a recognized standard makes it an unambiguous record of design intent, as well as a design development tool.

As with the PFD, the standards for the symbols which should ideally be used by British and EU engineers are BS EN ISO 10628, BS1646, and BS1553 and, like all engineering diagrams, it should be compliant with BS 5070. In the United States, ANSI/ASME Y14.1 "Engineering Drawing Practice" and ANSI/ISA S5.1-1984 (R 1992) "Instrumentation Symbols and Identification" apply. However, many companies and industries have their own internal standards for P&IDs. There are also several common P&ID conventions which do not appear in all standards.

- Flow comes in on the top left of the drawing, and goes out on the bottom right
- Process lines are straight and either horizontal or vertical
- Flow direction is marked on lines with an arrow
- Flow proceeds ideally from left to right, and pumps, etc., are also shown with flow running left to right
- Sizes of symbols bear some relation to their physical sizes: valves are smaller than pumps, which are smaller than vessels, and the drawn sizes of symbols reflect this
- Unit operations are tagged and labeled
- Symbols are shown correctly orientated: vertical vessels are shown as vertical, etc.
- Entries and exits to tanks connect to the correct part of the symbol—top entries at the top of the symbol, etc.

Less complex P&IDs produced during earlier design stages will normally come on a small number of ISO A1 or A0 drawings, but the P&IDs for a complex plant may be printed in the form of several bound volumes where every page carries a small P&ID section.

Every process line on the drawing should be tagged in such a way that its size, material of construction, and contents can be identified unambiguously, as in the following example:

Number Showing Nominal Bore (NB) in mm	Letter Code for Material of Construction	Unique Line Number	Letter Code for Contents
150	ABS	004	CA

In the example above, a line tagged 150ABS004CA would be a 150 mm NB line made of ABS (plastic), numbered 4, containing compressed air.

Main process line components should be numbered first, increasing from plant inlet to outlet. So, line 100ABS001CA would, e.g., normally be upstream of line 150ABS004CA in the example above. Design development can, however, mean that this becomes muddled on the "as-built" version of the drawing.

Every valve and unit operation on the P&ID will also be tagged with a unique code, a common key for which follows in Table A3.1. There are a variety of approaches to tag numbering, dependent on industry/company, so it is key to define this tagging format early in the design process.

Every instrument will be given a code. This is set out in BS1646 as follows:

First Letter: measured parameter

L = Level
P = Pressure
T = Temperature
F = Flow

Additional letters: what is done with the measurement (more than one of these is possible)

I = Indicator
T = Transmitter
C = Controller

The letter code will be followed by a unique number for that coded item; e.g., PIT1 would normally be the first pressure indicator/transmitter on the main process line. The British and other Standards cover these conventions in more detail.

Table A3.1 P&ID Tag Table

Valves	
MV	Manual valve
AV	Actuated valve
FCV	Flow control valve
CV	Control valve
ESV	Emergency Shutdown Valve
Unit Operations	
U	Unit
T	Tank
P	Pump
B	Blower
C	Compressor

The P&ID is a master document for HAZOP studies. It also frequently shows termination points between vendor and main EPC company, and between main EPC company and equipment supplier, in a way which makes contractual responsibility clear.

Even a conceptual design can be used to generate rough piping, electrical and civil engineering designs, and prices. These are important, since designs may be optimal in terms of pure "process design" issues such as yield or energy recovery, but too expensive when the demands of other disciplines are considered.

EQUIPMENT LIST/SCHEDULE

A schedule or table of all the equipment which makes up the plant is usually first produced at FEED stage (Fig. A3.3). Tag numbers from drawings are used as unique identifiers, and a description of each item accompanies them. There may be cross-referencing to P&IDs, datasheets, or other schedules.

Similar schedules are produced for all instrumentation, electrical drives, valves, and lines.

INSTRUMENT SCHEDULE

Project	Permanent Effluent/Groundwater Treatment Plant	Project Ref
Project Site		Document Ref
Client		
Client Ref		

	Rev 0	Rev 1	Rev 2	Rev 3	Rev 4
Prepared by	SMM				
Checked by	STM		CONFIRM		
Date	30/04/2004				
Approved by					
Date					

Inst No	Description	Supplier	Type	P&ID no	Line No	Size (mm)	Material of Constr	Design Fluid conditions Press (bar)	Design Fluid conditions Temp (C)	Operating range Min	Operating range Max	Alarm conditions H	Alarm conditions L	Notes	CSL /Equip supplier
LC001	MH 102 (PS1) Level Controller	Milltronics	Ultrasonic	90501 002		n/a	proprietary	atmos	ambient	0	5000 mm	Y	Y	Milltronics Multiranger with 2 sensors in sump. panel mounted indicator. Some alarms from PLC	CSL
PI002	Pressure Indicator	TBA	Bourdon	90501 002	0056	tbc	SS enclosure	5	ambient	0	2 bar	N	N	Standard 50/75/100 mm pressure gauge	CSL
PTx003	Pressure Transmitter	GEMS	Transducer	90501 002	0056	n/a	316 ss	6	ambient	0	2 bar	Y	Y	Pressure transmitter and panel mounted indicator	CSL
FTx004	Flow Transmitter	ABB	Electromagnetic	90501 002	0056	300	proprietary	16	ambient	0	120 m3/hr	Y	Y	AB Magmaster inline meter with panel mounted display	CSL
LC005	MH 92A (PS2) Level Controller	Milltronics	Ultrasonic	90501 002		n/a	proprietary	atmos	ambient	0	5000 mm	Y	Y	Milltronics Multiranger with 2 sensors in sump. panel mounted indicator. Some alarms from PLC	CSL
PI006	Pressure Indicator	TBA	Bourdon	90501 002	0062	tbc	SS enclosure	5	ambient	0	2 bar	N	N	Standard 50/75/100 mm pressure gauge	CSL
PTx007	Pressure Transmitter	GEMS	Transducer	90501 002	0062	n/a	316 ss	6	ambient	0	2 bar	Y	Y	Pressure transmitter and panel mounted indicator	CSL
FTx008	Flow Transmitter	ABB	Electromagnetic	90501 002	0062	300	proprietary	16	ambient	0	240 m3/hr	Y	Y	AB Magmaster inline meter with panel mounted display	CSL
LC009	MH 55A (PS3) Level Controller	Milltronics	Ultrasonic	90501 002		n/a	proprietary	atmos	ambient	0	5000 mm	Y	Y	Milltronics Multiranger with 2 sensors in sump. panel mounted indicator. Some alarms from PLC	CSL
PI010	Pressure Indicator	TBA	Bourdon	90501 002	0066	tbc	SS enclosure	5	ambient	0	2 bar	N	N	Standard 50/75/100 mm pressure gauge	CSL
PTx011	Pressure Transmitter	GEMS	Transducer	90501 002	0066	n/a	316 ss	6	ambient	0	2 bar	Y	Y	Pressure transmitter and panel mounted indicator	CSL
FTx012	Flow Transmitter	ABB	Electromagnetic	90501 002	0066	300	proprietary	16	ambient	0	540 m3/hr	Y	Y	AB Magmaster inline meter with panel mounted display	CSL

FIGURE A3.3

Extract from an equipment schedule.

Some modern software packages promise to remove the necessarily onerous task of producing these schedules from the engineer's task list, by generating them from the databases created during P&ID development, or by 3D plant modeling software. This approach is however nowhere near as common in the water sector as it is in others.

FUNCTIONAL DESIGN SPECIFICATION

A functional design specification, or FDS, is sometimes called a control philosophy, although both terms have other meanings in different contexts. The document referred to in this text as the FDS describes (ultimately, in practice, for the benefit of the software author) what the process engineer wants the control system to do.

It starts with an overview of the purpose of the plant and proceeds to document, one control loop at a time, how the system should respond to various instrument states, including failure states. This description will be set out in clear and straightforward language, designed to be entirely unambiguous, and comprehensible by nonspecialists.

The FDS is read in conjunction with the P&ID and refers to P&ID components by tag number. It is used alongside the P&ID in HAZOP studies.

GENERAL ARRANGEMENT DRAWINGS

Plant, equipment, and pipework layout drawings of various kinds are the primary tools of the layout designer. These are given different names across industries and locations, but this text considers all such variants to be types of General Arrangement drawings or GAs.

A single General Arrangement (GA) drawing can show a small section of a water and wastewater treatment plant, or even a whole site, including sufficient details of pipework to allow its design and installation to be carried out (or, in the case of an "as-built" GA, how it was installed).

However, for larger plants, there may be two kinds of GA drawing. A "plot plan" in this context is purely an equipment layout and will be accompanied by a separate piping layout drawing known as a piping layout or study.

Where this approach is followed, piping layouts are usually drawn at a 1:30 (sometimes 1:50) scale, and plot plans are usually drawn at approximately a 1:500 to 1:1000 scale. Plot plan scales are far too small to show piping with any meaningful clarity for large plants.

In professional practice, a specialist piping or mechanical engineer may produce the finished piping layout and plot plan, but chemical engineers almost always lay out equipment in space, and may produce a single general arrangement for small plants or multiple plot plan drawings for large ones as part of their design process.

GA drawings should in the United Kingdom conform to BS 8888 and, in the United States, to ANSI/ASME Y14.1 "Engineering Drawing Practice." They should show (as a minimum), to scale, a plan and elevation of all mechanical equipment, pipework, and valves which form part of the design, laid out in space as intended by the designer. Where possible, the tag numbers used in the P&ID should be marked on to their corresponding items on the GA as well, to allow cross-referencing, as illustrated in Fig. A3.4.

The inclusion of key electrical and civil engineering details is normal, and there are also usually detailed versions for these disciplines which refer to common master GAs.

Ideally, the drawing will be produced to a commonly used scale (1:100 being the commonest scale for single GA small plants), and be marked with the weights of main plant items. Fractional or odd-numbered scale factors should be avoided. Sectional views demonstrating important design features are a desirable optional extra in 2D drawings.

There are invariably two (and frequently more) issues of 2D layout plans at each stage of the design process. The first are "issued for comment" and can be

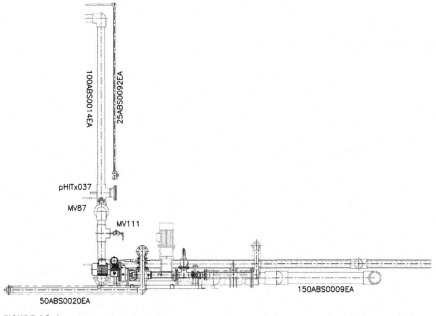

FIGURE A3.4

Section of a General Arrangement/plot plan drawing.

amended by members of the design teams. At the final "issue for design," the layout is frozen and financial sanction and planning permission will be sought.

When using 3D software on large plants, this is true of the plot plan, but piping layout comments are most often captured during model reviews.

COST ESTIMATE

If a company is going to accept a contract to construct a plant for a fixed sum of money, it needs to be certain that it can make a profit at the quoted price.

Equipment prices for specific items will be obtained from multiple sources. The specifications of these items will in turn come from a design which is sufficiently detailed to obtain an appropriate degree of certainty/control of risk. Civil, electrical, and mechanical equipment suppliers and installation EPC companies will also often be invited to tender for their part of the contract, although internal design and cost estimation resources may be used, especially early in the design process. Internal quotations are also usually obtained from discipline heads within the company for the internal costs of project management, commissioning, and detailed design. There may well be negotiations with these sources of information. Ideally, there will be multiple options for equipment supply and construction. A price based on a single quotation is far less robust than one which has a broader base.

Once there are prices for all parts and labor, residual risks are priced in. The insurances, process guarantees, defect liability periods, overhead contribution, profit mark-up, and so on are then added. This exercise can take a team of people weeks or months to complete, and the product is a $+/-1\%-5\%$ cost estimate.

Designers need to consider cost implications of each decision, although a key aspect of design expertise is not having to check back with cost estimators about which option is the cheaper.

DATASHEETS

Datasheets gather together all the pertinent information for an item of equipment in a single document, mainly so that nontechnical staff can purchase the correct items (Fig. A3.5). Process operating conditions, materials of construction, duty points, and so on are brought together into this document to explain to a vendor what is required.

Datasheets need to be cross-checked with several drawings, calculations, and schedules, and care must be taken to ensure that they are in accordance with the latest revisions of these documents. This is a more skilled task than the generation of schedules, and will therefore be likely to remain in the purview of young engineers for years to come.

Equipment Schedule and Pricing

Supplier

Client		Prepared by	STM		Date	Nov-03		
Site		Checked by			Date			
Project	Example	Contract ref		C2265	C2295	Rev	0	
Section	Skid mounted pumps	Section	X	of	Y			REV
Compressor						Item	P16,P17	

One air compressor, pressurised air storage reservoir and associated controls and ancillaries to provide air for oil water separator (if required) continuous sand filter (if required). DAF unit (if required), air actuated valves, and instruments. Current duty is estimated from preliminary calculations. Suppliers should satisfy themselves that the duty specified is correct and make any adjustments as necessary to the sizing. Supply should be for complete, free standing unit. Dryer to be provided if supplier considers this to be beneficial. System will be located inside control building

Any information not requested which the supplier believes will add to his offer to be provided on separate sheets

Performance parameter	Description	Pipework	Valves	Option1	Option 2	Option 3	Option 4
Min flow (Nm3/hr)				30	10	10	10
Max flow (Nm3/hr)				90	20	25	35
Pressure (bar)				8	8	8	8
Compressor							
Manufacturer							
Model							
Type							
Materials of construction							
Noise levels (dBA)							
Length (mm)							
Width (mm)							
Height (mm)							
Weight (kg)							
Inlet connection (mm dia)							
Outlet connection (mm dia)							
Motor type							
Enclosure class (IEC 34-5)							
Insulation Class (IEC 85)							
Motor Speed (rpm)							
Rated power (kW)							
Efficiency at design duty (%)							
Mains frequency (Hz)	50						
Rated voltage	415						
Rated current (amp)							
Starting (Star delta/DoL)							
Starting current (amp)							
Price for complete system, delivered Manchester							
Delivery period for complete, tested system (weeks)		Target <10 weeks					

Enter here any additional notes, details of additional equipment or facilities required

FIGURE A3.5

Example of an equipment datasheet.

HAZOP STUDY

A HAZOP study is a "what-if" safety study. It requires, as a minimum, a P&ID, FDS, process design calculations, and information on the specification of unit operations as well as perhaps eight professional engineers from several disciplines.

The report produced by the participants will usually, nowadays, include a full description of the line-by-line (or node-by-node) permutation of keywords and properties used in carrying out such a study. However, in the past, it was more usual to produce a summary document listing only those items which were identified by HAZOP as being problematic, what the problems were and how it was intended to avoid them.

In today's litigious environment, a full and permanent record of all that was discussed and considered in a HAZOP is best practice. This record may most conveniently be achieved by video recording the entire procedure.

ZONING/HAZARDOUS AREA CLASSIFICATION STUDY

Zoning the plant with respect to the potential for explosive atmospheres is not a strictly quantitative exercise. It is common for a small number of engineers to use design drawings to produce a zoning study or drawings showing the explosive atmosphere zoning which they consider appropriate for the various parts of the plant. Zoning can have a major impact on segregation and other issues in layout design.

ISOMETRIC PIPING DRAWINGS

At the detailed design stage, isometric piping drawings, or "isos", are produced for larger pipework, either by hand on "iso pads" or by CAD (Fig. A3.6).

The purpose of the iso is to facilitate shop fabrication and/or site construction. They are also used for costing exercises and stress analysis, as they conveniently carry all the necessary information on a single drawing.

Producing isos by hand or 2D CAD is time-consuming. Most 3D CAD plant layout software packages can automatically produce isometric drawings from model databases, but it is still usually thought prudent for these to be checked by experienced pipers, and the setup time of such systems is not worthwhile on smaller plants. 2D CAD is still, therefore, the norm on smaller projects.

Isometric piping drawings are not scale drawings, they are dimensioned. They are not realistic: pipes are shown as single lines, and symbols are used to represent pipe fittings, valves, pipe gradients, and welds. The lines, valves, etc., are tagged with the same codes used on the P&ID and GA. Process conditions such as temperature, pressure, and so on may also be marked on the iso.

It may well be that "clashes"—where more than one pipe or piece of equipment occupy the same space—are only identified at the stage of production of isos, so pipework design cannot be considered complete before isos are produced. Clash detection is a common functionality of 3D CAD and is carried out during design development.

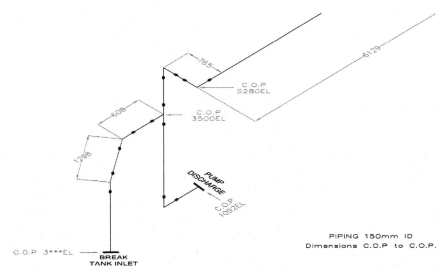

FIGURE A3.6

Isometric pipeline drawing.

FURTHER READING

Moran, S. (2015). *An applied guide to process and plant design*. Oxford: Butterworth-Heinemann.

Appendix 4: Selection and sizing of unit operations

INTRODUCTION

Detailed unit operation design is a job for those guaranteeing the performance of the unit operation. Water and wastewater treatment plant designers therefore very rarely carry out detailed design of unit operations, as they do not tend to offer the process guarantees.

The plant designer's job is to get their design sufficiently accurate that the design envelope for the equipment is robustly covered, because the process guarantee will limit its validity to that design envelope. Designers also frequently check that there are no misunderstandings in what vendors have offered or any design details incompatible with the broader design.

We will also usually make sure that the equipment weight, size, power, and other utilities requirements are in line with our expectations. If they are not, we may need to consider modifying the whole plant design to suit, or reject the item of equipment as it is less favored when these knock-on effects are considered.

MATCHING DESIGN RIGOR WITH STAGE OF DESIGN

We do not wish to spend any more time on design at each stage than is necessary to progress the overall design. Rule of thumb design is therefore the norm for water and wastewater treatment plant designers.

We do, however, use ever more onerous and accurate heuristics as design progresses, and it grows increasingly likely that the plant will be built.

RULE OF THUMB DESIGN

If you are working as a process designer, there will mostly (but not always) be a company design manual. This will set out the relevant rules of thumb for designing the plant you are working on, encapsulating the company's experience in the area.

Failing that, there will generally be a more experienced engineer who knows the rules of thumb. If there is such a person, they can be an invaluable source of support and information. The more experienced engineer is often the world's leading expert on the job you must do, as they don't just know a way to design the plant, but they know how to design the plant so that your company can build it.

If neither a design manual nor a more experienced engineer is available, this book and others like it contain general rules of thumb (see Further Reading).

There are some recent books offering "rules of thumb" generated by modeling and simulation programs. I would caution against these. Proper rules of thumb come from experience with multiple full-scale real-world plants, not first principles computer programs. First principles design doesn't work.

APPROACHES TO DESIGN OF UNIT OPERATIONS
FIRST PRINCIPLES DESIGN

First principles design has no place in everyday professional plant design practice. Even with sufficient data to design a unit operation from first principles, it would be at best a prototype, and your employer would be the one offering the process guarantee by making it part of the plant they were guaranteeing.

If you are designing unit operations for a living, you are not really a process plant designer, but you may have colleagues who have the necessary knowledge, or more likely still, a spreadsheet or program written by someone else which encapsulates the knowledge.

If you are asked to author such a spreadsheet or program, make sure it does reflect the experience held in the company, that the program is bug-free, and verify its accuracy across its range of operation by full-scale real-world experiments. It will have fewer bugs if you write it in MS Excel rather than compiled code, due to its increased transparency.

DESIGN FROM MANUFACTURERS' LITERATURE

Since detailed design involves putting together third party-designed unit operations, manufacturers' catalogues are a useful tool in the selection process. Updated catalogues also frequently include new items of equipment we might not have otherwise considered.

Nowadays these catalogues are more likely to be found on websites rather than in hardcopy, and this is very handy in academic settings, allowing us to bring realism to student designs without bothering manufacturers with enquires from students. Manufacturers' detailed information on products and their capabilities and limitations can allow plant designers to see new ways to design ourselves ahead.

SOURCES OF DESIGN DATA

For the professional designer, the strength of sources of design methodology information is generally as follows:

Strongest	Close to full-scale pilot plant trial
	Thoroughly developed and validated tailored modeling program directly based on many full-scale installations of exactly the type and size of plant proposed
	Supplier information
	Robust rules of thumb direct from a Chartered Engineer with lots of relevant experience
	Rules of thumb from a book by a Chartered Engineer
	Direct from a Chartered Chemical Engineer with little relevant experience
	First principles
	Guessing by beginner
Weakest	Less than thoroughly debugged and validated Simulation and Modeling program output

As far as sources of data required inputting to design methodologies is concerned, strengths are generally as follows for feedstock and product qualities and quantities:

Strongest	Statistically significant ranges of values for the exact type and scale of plant envisaged
	Statistically significant ranges of values for the type of plant envisaged
	Ranges of values for the exact type of plant envisaged, falling short of statistical significance
	Ranges of values for the broad type of plant envisaged, with or without statistical significance
Weakest	Guessing by beginner

As far as thermodynamic and other physical and chemical data is concerned, the test of validity should be whether the data is intended by those generating it to be valid over the range of physical conditions likely to be encountered under all reasonably foreseeable plant operating conditions.

SCALE-UP AND SCALE-OUT

In theory, there is no difference between theory and practice. But in practice, there is.

Yogi Berra

It is often assumed by academics that if a reaction works in a conical flask, getting it to work in a $100\ m^3$ reactor is a trivial thing. If this were true, there would be no chemical engineers. There is a lot more to chemical engineering than chemistry.

There are two basic approaches to making a bigger plant. We can have a lot of parallel streams of plants which we know to work at the given scale (scale-out), or we can have a smaller number of larger but at least slightly experimental plants (scale-up).

What we need to know as engineers to carry out a scale-up exercise is the critical variable or dimension. This variable is the thing we need to keep constant (or vary in a systematic way) to ensure the process works at the larger scale.

We might need to maintain the length, area, or volume of a process stage, or it might be more complex, such as several theoretical plates for a distillation column.

It should be noted that the key variable might change at different stages of scale-up as the balance of effects vary. Scale-up by a factor of more than about 10 from even a good pilot plant study should make us quite nervous as designers.

We might also need to revisit the original lab scale study, to make the reaction work with fewer hazardous reactants or solvents, etc., to offer a process which can be made to work in an economically viable plant.

NEGLECTED UNIT OPERATIONS 1: SEPARATION PROCESSES

Many students leave university with knowledge of only 6−10 separation processes, usually those most important in oil refining, and the knowledge they have of these processes is mostly at best engineering science.

Processes based in chemistry are overrepresented in university courses, as are liquid/liquid separations. The consequent lack of knowledge of options impoverishes the designer's imagination, and a lack of understanding of engineering design practice prevents practical use of technologies.

A list of the most important separation processes in water and wastewater treatment, arranged by phases separated is given in Table A4.1. This list is not however exhaustive, does not go beyond separations of two components, and does not mention the plethora of subtypes of the technologies listed.

Real-world separators differ from mathematical/theoretical ones in ways which usually make any first principles design unworkable.

Table A4.1 Separation Processes Used in Water Treatment

	Relative Price	Process Robustness	Safety Concerns	Separation Principle[a]	Product Recovery per Stage	Contaminant Removal per Stage
Gas/Gas						
Adsorption	M	M		Surface interaction		VH
Distillation	H	L/M	Flammable/toxic vapors, heat	Differential vaporization/condensation	L/M	M/H
Filtration	L	H		Physical exclusion of oversize particles	H	H
Gas/Liquid						
Distillation	H	L/M	Flammable/toxic vapors, heat	Differential vaporization/condensation	L/M	M/H
Adsorption	M	M		Surface interaction	H	VH
Stripping	L	H	Flammable/toxic vapors/liquids	Mass transfer from liquid to vapor phase	M	M
Filtration	L	H		Physical exclusion of oversize particles	H	H
Gas/Solid						
Adsorption	M	M		Surface interaction	H	VH
Filtration	L	H		Physical exclusion of oversize particles	H	H
Cyclones	M	H		Sedimentation	M/H	M/H
Drying	M	H	Heat	Evaporation, usually heat assisted	VH	VH

(Continued)

Table A4.1 Separation Processes Used in Water Treatment *Continued*

	Relative Price	Process Robustness	Safety Concerns	Separation Principle[a]	Product Recovery per Stage	Contaminant Removal per Stage
Liquid/Liquid						
Decantation	L	H		Sedimentation	M/H	H
Membranes	M	L/M		Physical exclusion of oversize droplets	H	H
Liquid/Solid						
Coagulation	L	M/H		Destabilization of a colloid	M	M
Precipitation	L	H		Exceeding solubility limit	M/H	M
Flotation	M	M/H		Sedimentation effected by reducing solid or liquid density with gas bubbles	H	H
Ion exchange	H	M		Differential affinity	VH	VH
Electrolysis	H	H	Electricity	Electrochemistry	H	H
Centrifugation	M	M	Rapidly rotating equipment	Sedimentation	M/H	H
Hydrocyclone	M	M		Sedimentation	M/H	M/H
Magnetic separation	H	M			VH	VH
Membranes	H	M		Physical exclusion of oversize particles for coarser membranes, diffusion for RO	VH	H
Solid/Solid						
Magnetic separation	H	M/H		Differential magnetic attraction	VH	H

[a]These principles are major influencers of the separation process, but they are at best tools for analysis and partial understanding.

Table A4.2 Pump Selection (General) (see also Table 10.1)

	Rotodynamic	Positive Displacement
Head	Low—up to a few bar	High—hundreds of bar
Solids tolerance	Low without efficiency losses	Very high for most types
Viscosity	Low viscosity fluids only	Low- and high-viscosity fluids
Sealing arrangements	Rotating shaft seal required	No rotating shaft seal
Volumetric capacity	High	Lower
Turndown	Limited	Excellent
Precision	Low discharge proportional to backpressure	Excellent discharge largely independent of backpressure
Pulsation	Smooth output	Pulsating output
Resistance to reverse flow	Very low	Very high
Reaction to closed valve downstream	No immediate damage to pump	Rapid pump damage likely

NEGLECTED UNIT OPERATIONS 2: PUMPS AND COMPRESSORS

Fluid moving equipment generally comes in two main varieties: rotodynamic or positive displacement. Each of these comes in many subtypes, but beginners have often not been taught the crucial differences between the two broad types.

Table A4.2 gives general data to help with the pump selection process, while Table A4.3 provides some ideas on how to choose between the commonest types of liquid and gas moving equipment.

We tend to use positive displacement pumps for metering duties, and centrifugal (rotodynamic) pumps for moving large volumes of flow at relatively low pressures. Table A4.3 can inform more detailed choices.

NEGLECTED UNIT OPERATIONS 3: CONTROL AND INSTRUMENTATION IN WATER AND EFFLUENT TREATMENT

AUTOMATIC CONTROL

We no longer build fully manual plants in the developed world nowadays. Computers are too cheap and reliable, and operators too expensive and human for routine operation activities to be best done by people.

Control is mostly achieved using a combination of programmable logic controllers (PLCs), PCs, and high-level control such as a Distributed Control System

Table A4.3 Pump Selection (Detailed) (see Also Table 10.2)

	Relative Price	Environmental/ Safety/ Operability Concerns	Robustness	Shear	Maximum Differential Pressure	Capacity Range	Solids Handling Capacity	Efficiency[a]	Seal In-Out	Fluids Handled[b]	Self-Priming?
Rotodynamic											
Radial flow	M	Cavitation	H	H	M/H	L–VH	M	H	M	Low viscosity/aggressiveness[b]	N
Mixed flow	M	Cavitation	H	H	M	M–VH	M	H	M	Low viscosity/aggressiveness[b]	N
Axial flow	M	Cavitation	H	H	M	M–VH	M	H	M	Low viscosity/aggressiveness[b]	N
Archimedean screw	H	Release of dissolved gases	VH	H	L	M–VH	H	L	N/A	Low viscosity/Aggressiveness[b]	N
Positive Displacement											
Diaphragm	M	Overpressure on blockage	VH	L	M/H	VL–M	H	L	H	Low/high viscosity/aggressiveness	Y
Piston diaphragm	H	Overpressure on blockage	M	L	VH	VL–M	L/M	M	H	Low/high viscosity/aggressiveness	Y
Ram	H	Overpressure on blockage	H	M	VH	M–H	H	M	H	Low/high viscosity/aggressiveness	Y
Progressing cavity	M	Overpressure on blockage	M	VL	H	M–H	H	H	H	Low/high viscosity/aggressiveness	Y
Peristaltic	H	Overpressure on blockage	L	VL	M	VL–M	H	M/H	M	Low/high viscosity/aggressiveness	N
Gear	L	Overpressure on blockage	L	M	VH	VL–L	L	L	H	Low/high viscosity, low aggressiveness	Y
Screw	L	Overpressure on blockage	L	M	VH	L–M	L	L	H	Low/high viscosity, low aggressiveness	Y

Other

Air lift	L	VH	VL	VL	L-M	H	L	L	N
Eductor	M (Blockage of eductor)	M	M	VL	L-H	M	H	M	Y
								Low viscosity, low/high aggressiveness	Low viscosity, low/high aggressiveness

[a] Centrifugal pump efficiency reduces as viscosity increases, but PD pump efficiency increases. Centrifugal pump efficiency is more to do with impeller type than anything else, impeller type is determined by process conditions such as any solids handling requirement.

[b] Aggressiveness is related to presence of abrasive particles, undissolved gases, or unfavorable LSI.

(DCS) or supervisory control and data acquisition (SCADA)), though there may be a few field-mounted controllers specified for several reasons.

Physical field-mounted PID controllers are declining in use, and the technologies most like them which we do still use have their own built in control algorithms.

Water and wastewater treatment plant designers do not write the software for these controllers. Tuning control loops during commissioning is mostly done nowadays by plugging in a laptop and pressing the "optimize" button on the manufacturer's dedicated software.

Commissioning, while very important, is not the subject of this book. Water and wastewater treatment plant designers need to know how to specify instrumentation and control hardware, populate their piping and instrumentation diagrams (P&IDs) with these items, and write functional design specifications (FDSs) so that software engineers can design and price their software.

Water and wastewater treatment plant designers need to have an idea of what neighboring disciplines do, and what they need to do their jobs but we don't, however, need to be able to do their jobs.

SPECIFICATION OF INSTRUMENTATION

Instrument engineers/technicians have their specialism, but we don't need to be one of them to specify instruments well enough to design a water and wastewater treatment plant. A selection table has been included in this chapter (see Table A4.5) to provide a sufficient overview.

We should be willing to be corrected by an instrumentation specialist at more detailed design stages on details of instrument choice, just as we should by other specialists on their respective areas of expertise.

PRECISION

Precision in mathematics is (confusingly to engineering students) related to what engineers call resolution. If I tell my students off for "spurious precision," this is the sense in which I am using it. Precision in engineering is different, and is to do with repeatability and reproducibility. It is not to do with how close the measured value is to the true value (accuracy) or the smallest change in the measured value with the instrument can detect (resolution). It is to do with whether the instrument will yield the same reading against the same true value the next time we test it.

We might further split engineering precision into reproducibility and repeatability, the first encompassing variability over time, and the second being precision under tightly controlled conditions over a short time.

To take a specific example, pH probes require regular recalibration against standards to maintain precision and accuracy. Over the period between calibrations, the accuracy (measured value for a given true value) varies. Eventually it is not possible to calibrate the instrument to give accurate readings against the standards, and a new probe must be substituted. There are gradual decreases in accuracy, precision, and response time during the periods between new probe installations.

ACCURACY

Accuracy is to do with the gap between the true value and the value indicated by the instrument. To take a specific example, in filter feed treatment, pH must be controlled within the range around the setpoint $+/-0.05$ pH units, therefore the accuracy of measurement needs, reliably, to be at least this good.

COST AND ROBUSTNESS

Increased instrument precision and accuracy both tend to cost money. Very precise and accurate (basically lab grade) instrumentation also tends to be less robust.

Lab instruments tend not to be suited to field mounting. We therefore tend not to specify any more accuracy or precision than is strictly needed, and we may treat manufacturers' lab test values for an instrument with skepticism.

All instrumentation needs to have a purpose to justify its cost. It may be true that "you can't manage what you don't measure," but it is also best not to measure things you don't need to manage.

SAFETY

Safety critical instrumentation requires a higher standard of evaluation than that which only affects operability or, less important still, process monitoring without associated control actions.

We might, e.g., specify for a process or safety critical reading the use of redundant cross-validated instruments (in which the reading most likely to be correct is determined by a voting system).

SPECIFICATION OF CONTROL SYSTEMS

The P&ID shows graphically, and the FDS describes in words what the process designer would like the software to do. Turning these deliverables into code is the job of the software engineer.

PLCs are the basis of many modern water and wastewater treatment plant control systems, with DCS or SCADA supporting the control interface or human—machine interface (HMI).

The true nature of the control system should be reflected on P&IDs and in FDSs. We should not expect to see a local control loop and field-mounted controller on a P&ID representing a loop which works via signals going out and back via PLC. There are appropriate symbols in the British Standard to show this correctly.

STANDARD CONTROL AND INSTRUMENTATION STRATEGIES

This section breaks down process control systems into some commonly used blocks, which should allow P&ID and control philosophies to be populated with the standard features which appear on almost every plant.

Many readers will know what feedback, feedforward, and cascade control are, but their university courses may have focused on mathematical software engineering such as transforms and algorithms. In 21st century process control, signal processing is built into the box, and the software engineer writes the algorithms, though they may well need input from the process engineer to do with required outcomes of control functions.

Water and wastewater treatment plant designers do need to understand what software engineers are going to need from them, so that they can design in controllability.

ALARMS, INHIBITS, STOPS, AND EMERGENCY STOPS

Water and wastewater treatment plant designers in Europe will need the assistance of electrical engineers to ensure compliance with the IEE regulations and the various European Union directives which apply to this area. Those elsewhere will have to conform to their own corresponding local regulations, codes, standards and laws.

However, I have included this section because beginners usually do not understand basic issues such as that all electrical equipment needs to be easy to switch off in an emergency, and very frequently comes with safety features which switch it off automatically in several potentially hazardous situations. These might include such things as motor winding overtemperature, motor overtorque, fluid ingress, and so on.

It is frequently the case (and it may be a legal requirement) that the more hazardous of these cases will be set up by the electrical/software engineers such that they require an operator to attend site to reset the "trip."

Less potentially serious conditions may stop motor operation only while the state is current, or if less serious still only prevent the motor from starting. Both conditions might be called inhibition.

These conditions will usually be set to generate local alarms in software. More serious ones may generate off-site alarms, or activate an alarm beacon on site.

A design which has an excessive number of alarms should be avoided. If there are too many alarms, operators will be subject to alarm flooding, and develop what is known in healthcare as "alarm fatigue" and either ignore them or find ways to disable them. Table A4.4 sets out the Engineering Equipment & Materials Users' Association (EEMUA)'s recommendations:

Table A4.4 EEMUA Criteria for Acceptability of Alarm Rate in Steady State Operation

Long-term average alarm rate in steady operation	Acceptability
More than one per minute	Very likely to be unacceptable
One per two minutes	Likely to be over-demanding
One per five minutes	Manageable
Less than one per ten minutes	Very likely to be acceptable

Adapted from "EEMUA Publication 191 Alarm systems - a guide to design, management and procurement" with permission from EEMUA.

It should be noted that commissioning engineers frequently disable alarms and interlocks during the early stages of commissioning, but this should be a planned aspect of a commissioning procedure, and suitable safety plans should be made.

Many of these signals, alarms, and interlocks must be handled by the control system, and leaving them out of the control system specification—if that is the case—is a classic beginner's mistake leading to cost overruns down the line.

Many are, however, wired directly into the motor starter, which eliminates a potential weak link in the chain. Hard-wiring is standard for safety critical interlocks.

European and many other standards also require the provision of emergency motor stop buttons immediately adjacent to motors. Resetting the emergency stop locally cannot incidentally allow the drive to restart; there must be a trip to reset on the MCC as well.

ELECTRICAL AND CONTROL EQUIPMENT

An understanding of the needs of electrical and control engineers is crucial to competent design. The most important items to consider are as follows:

Motor control centers

Electrical motors or "drives" require maybe six times their running power to start them up. Rather than uprating all the power cabling and so on to this starting current, motor starters are used to send a pulse of power to get the drive spinning. They also contain overload protection and so on.

Direct on Line (DOL) starters are very cheap, but they simply apply the full line current to the motor all at once in a way which usually limits their use to drives rated at less than 11 kW.

Star delta starters are more expensive. They apply current to the motor in two configurations in a way that reduces starting torque by a factor of 3. These are probably required above an 11 kW drive rating.

Soft starters are the most expensive type. They control voltage during drive startup in a way which avoids the torque and current peaks associated with DOL and star delta starters, and have some of the sophisticated control functions of the Variable Speed Drive (VSD).

Inverters or VSDs are perhaps a little more expensive than star delta starters, but they have far more flexibility. They allow very sophisticated patterns of power ramping to be applied to the drive on startup, as well as allowing variable frequency to be supplied in a way which allows drive rotational speed to be controlled. They always have a microprocessor on-board nowadays so that multiple, quite sophisticated control loops and interlocks can be run directly through them.

These drives are usually collected in a big box called an MCC, which will usually contain internal subdivisions housing starters or groups of starters. There are some other common features shown in Fig. A4.1.

Fig. A4.1 shows a "Form 2" panel, with some separation between controls for different aspects of a process. This is one of the four forms of panel in IEC 60439-1: 1999, Annex D and BS EN 60439-1: 1999. There are lettered subtypes but, broadly:

- Form 1 has no separation, and is often referred to as a wardrobe type. Failure of one component in a Form 1 panel can damage other components, and a single failure will take the whole process offline

FIGURE A4.1

MCC Unit, showing incomer, starters, marshaling cubicle and PLC/UPS sections.

Courtesy: Process Engineering Group, SLR Consulting Ltd.

- Form 2 separates the bus bars (big copper conductors which carry common mains power throughout the panel) from other components. Form 2 is not much better than Form 1
- Form 3 separates the bus bars from other components, and all components from each other. This is the minimum specification if sections of the plant need to run whilst one of them is offline
- Form 4 is as Form 3, but also separates terminals for external conductors from each other.

In the United States (described in standard UL 508A) the system takes a different approach, but addresses the same issues. In either case the process designer will be required to specify broadly which kind of panel they want, though clients may specify a minimum separation between control switchgear.

Panels will also have a specified degree of ingress protection for water and dust (IP55 is usually the minimum standard).

Consideration will also need to be given to the direction of cable entry, which can be from the top, bottom, or a combination of both. Bottom entry requires the panel to sit on a channel in the floor, so there are civil engineering implications.

There is great deal more to this, but the above is the minimum level of knowledge required of a process designer to integrate their MCC design.

Cabling

The absolute minimum number of things a process plant designer needs to know about cables is as follows:

Every hardwired instrument needs incoming power and outgoing signal cabling. Every electrical drive (motor) on the plant needs incoming power cabling. Every MCC needs incoming power cabling and outgoing power and signals cabling.

Power cable size is calculated by electrical engineers in a similar way to how process engineers calculate pipe size. The more current a cable carries, the thicker it must be to avoid overheating. A complicating factor is that cables can have a variable number of wires or "cores" inside them. Thus we can have a thick cable with a single set of large cores feeding a single large drive, or a similarly thick cable with multiple smaller cores to feed several smaller drives.

Instrument cabling needs to be so arranged as to be unaffected by electromagnetic fields from the power cabling. This is usually achieved by some combination of physical separation and shielding. 1 m of separation will usually suffice, but electrical engineers will advise.

The basic kind of power cable is the relatively inflexible steel wire armored PVC insulated type. There is also a more flexible, unarmored, and waterproof kind used to connect submersible pumps underwater. Instrument cabling also comes in several types, and is a lot smaller than power cabling as it is only carrying 5–240 V, and no real power.

Cables have a minimum bend radius. The thicker the cable, the bigger this is. There needs to be space in the design to accommodate this. It is approximately equal to the radius of a commercially available long radius bend in a pipe large enough to carry the cable.

All kinds of cable are carried underground through ducts, which are nowadays plastic pipes. These may be cast into concrete slabs, unlike pipes carrying fluids. Usually a few extra ducts are specified in a design to allow for future expansion. Cables are carried over ground on (usually elevated) cable tray. Power and signals cabling should ideally be in separate ducts and cable tray.

Instrumentation

There are many specialized process instruments, but the four commonly measured parameters are pressure, flow, temperature, and level. Process designers need to be able to choose between the most common types of instrumentation used for measuring these parameters. Table A4.5 provides an overview.

Table A4.5 Instrumentation Selection

Type	Relative Price	Robustness	Contact With Process Fluid?	Solids Handling Ability
Pressure Instruments				
Bourdon	L	L	Y	M
Capacitance	M	H	Y	H
Resistance	L	M	Y	H
Piezoelectric	M	H	N	H
Optical	H	M	N	M
Flow Instruments				
Variable area	L	L	Y	L/M
Mechanical	L	L	Y	L/M
Pressure	M	M	Y	L/M
Electronic	M/H	H	N	H
Radiation	H	H	N	H
Vortex	H	H	N	H
Doppler	M	H	N	VH
Temperature Instruments				
Thermocouple	L	H	Y	H
Bimetal	L	M	Y	M/H
Resistance	L	H	Y	H
Level				
Sonar/Radar	M	M	N	H
Float	L	M	Y	H
Pressure	M	M	Y	M/H

Control systems

There are several control systems used in water and wastewater treatment plant design. Selection between them for whole system or local design may be as much a matter of client preference, industry familiarity, and designer preference as inherent characteristics of the system type.

Local controllers

Historically, control loops were operated by mechanical, electromechanical, pneumatic, or electrically operated boxes which were mounted locally to the thing being controlled, but by the 1990s, these were most commonly solid state electronic PID (Proportional, Integral, Differential) controllers.

Dedicated field-mounted controllers are still used occasionally (for pH control, e.g.), but water and wastewater treatment plant designers never write the algorithms for these boxes. Field-mounted controllers are products, whose manufacturers have done this job for us. Their limited configurability also makes these controllers the most robust solution.

Programmable logic controllers

In the water industry, PLCs are commonly used for whole system control. PLCs are a kind of computer which is custom built out of components to suit a given duty.

Central processing units are available in a range of powers, and rack-mounted cards are added to this to provide a suitable number of input and output channels.

Direct interface with a PLC is via a HMI or PC program. These can vary in appearance from the program's green-screen and ladder logic to sophisticated simulation interfaces of SCADA systems.

It should be noted that, while PLCs themselves cannot be directly infected with computer malware, the intelligence community has produced a worm (Stuxnet) which can attack PLCs via their SCADA connected PCs (thought to have been written to target the Iranian nuclear program). Stuxnet was purely destructive, but a more recent virus called Duqu is a keystroke-logging spyware program.

Any systems which include a PC may be compromised, especially as PCs are almost always connected to the Internet nowadays, and site communications and signals are increasingly connected via Wi-Fi rather than hard wired.

Supervisory control and data acquisition

SCADA systems run on a PC. They can receive signals from one or more PLCs, or from remote telemetry outstations which convert 4−20 mA signals from field instruments into digital data.

Such signals may be carried by local or wide area networks using internet protocols, by telephone lines or satellite signals.

The SCADA system has a human machine interface (HMI), usually in the form of simulation screens which look rather like animated PFDs, alarm handling screens, "trends" screens which allow variation in parameters to be visualized as a graph against time, and input screens which allow process parameters to be changed (by authorized users).

DCS

DCS used to be more distinct from SCADA than it is nowadays. Historically SCADA used dumb field instrumentation, and had a centralized system "brain," whereas DCS had more control out in the field.

As it is becoming increasingly difficult to buy "dumb" instrumentation, or even dumb motor starters, the distinction is not as sharp as it once was, but DCS is more likely to involve field-mounted controllers. These are less likely to be simple PID controllers, but will be capable of more sophisticated control.

FURTHER READING

Couper, J. R., et al. (2012). *Chemical process equipment selection and design*. Oxford: Butterworth-Heinemann.

Appendix 5: Costing water and wastewater treatment plant

GENERAL

Costing methodology increases in rigor and accuracy as the design process proceeds.

It is possible to cost a plant at a whole-plant level very quickly and roughly, based on outturn costs for other similar plants. This will be most accurate when in-house figures are used, but a very rough estimate can be made from published data.

"Proposals engineers" in EPC companies (or others who produce proposals to build plant) will have access to many quotations for individual items of mechanical and electrical equipment from previous jobs, which will allow them to produce quite accurate estimates of cost rapidly. Their counterparts in civil engineering companies estimate by means of linear meters/cubic meters/tonnes, so they can rapidly produce estimates as soon as they have drawings of the plant.

Both M + E and Civils estimators need the design to have been progressed to a certain stage to produce a meaningful estimate. The first draft of a conceptual process design needs to have been completed for a M + E estimator to begin work, as they will need equipment sizes at the very least. The civil engineer will need to know approximate sizes of concrete tanks, slabs, and buildings as well as details of any enabling works.

MECHANICAL AND ELECTRICAL EQUIPMENT COSTING METHODOLOGY

CONCEPTUAL DESIGN STAGE

If you are working in the proposals department of an EPC company, you will be expected to be able to come up with "budget" pricing for all aspects of a plant very quickly (usually in far less than a working day). This will necessitate having several price estimation techniques. If there is time to do more than one of them, this can serve as a useful cross-check. If not, it is sensible to add a contingency to your price.

You should resist the impulse to return with a low price at this stage. It is always best to "underpromise and overdeliver." The contingency added to your price needs to be reflective of how uncertain your sources of pricing information are. Remember that beginners always underestimate costs, sometimes radically (i.e., by a factor of 2 or more) so caution is advised.

Approaches normally followed in practice are:

- Estimation based on in-house knowledge of whole-plant costs, either as individual prices or far less frequently, formulated as cost curves. Factors for inflation since plant completion date and plant size will normally need to be applied
- Estimation based on in-house knowledge of M + E plant costs. Factors for civil and other costs, as well as inflation since plant completion date and plant size will normally need to be applied
- Estimation based on published information on whole or partial plant costs. Factors for civil and other costs, as well as inflation since plant completion date and plant size, will normally need to be applied. Individual plant prices may be readily found on the internet, and equipment sizes for existing plant can often be estimated using Google Earth. Cost curves for water and wastewater treatment plant can be found in the "Cost Estimating Manual for Water & Wastewater Treatment Facilities" by William McGivney and Susumu Kawamura (see Further Reading)

Estimation by 3D CAD software can also be used, but the effort required to get to the point where such an estimate is meaningful is so much greater than the time available to a proposals engineer that it is not used at this stage in practice.

FEED STAGE

The degree of detail required at FEED stage varies. There is, however, usually enough design effort in this stage that you can start to out-design your competitors, and reflect the cost savings in your pricing, if the savings are large enough to exceed the margin of error in your pricing technique.

FEED stage pricing should ideally be based on current quotations for the exact equipment and services required by the design. Some items of equipment and services may, for practical reasons, need to be estimated using the techniques listed in the last section but, as before, these need to have margins added to reflect the risk of such an approach.

The risks of cost estimation are not just that the equipment may cost more than expected. A key risk is of not having discussed the design with the equipment supplier sufficiently. The equipment or service may be entirely unavailable, unavailable where needed, unavailable when needed, or unavailable at the specified size. Even if it is available it may have utility, layout, or other requirements, which were not considered in the conceptual design, or it may not be able to meet the required performance specification (all too commonly because the specification was based on optimistic or naive reading of promotional materials).

"Quotations" in this instance are, ideally, legally binding offers from a subcontractor or equipment vendor to supply equipment or services to meet a given specification to a stated timescale under certain contract conditions. Any vagueness about any of these points represents a risk that must be priced in. Suppliers will often produce offers under their own terms of sale, rather than those requested. They will commonly leave out the price of delivery and offloading to site. Their actual quotations should be checked carefully against the requests for quotation which they were sent. Any variance from what was asked for should ideally be removed, and if it cannot be eliminated, priced for. Variations to contract conditions can cost a great deal of money.

DETAILED DESIGN

The detailed design stage involves generating costs that an EPC company is willing to stand by, for a plant which they believe will work.

All M + E prices at this stage are normally supported by at least three valid quotations for an accurately sized piece of equipment. Ideally, electrical, mechanical, and civil engineering contractors will have offered firm quotations for the site works.

The level of design detail at this stage should be such that even the required number of bolts to secure the pipework flanges and their surface finish will have been determined. This leads to an ability to price all the individual items on documents supplied to the EPC company by the plant purchaser for this purpose.

M + E equipment is broken down item by item in documents called "Cost Schedules," and Civil engineering items on what are known as "Bills of Quantities." This approach is better suited to the pricing methodology of civil engineering firms, as I will explain later.

From my own professional experience the costing methodology proceeds broadly as follows:

1. Read documentation
2. Possibly appoint local "agent" or "sponsor" for overseas job
3. Visit site and/or client and/or "agent"
4. Carry out or expand upon process design
5. Produce datasheets for all mechanical equipment
6. Produce GA showing locations of all concrete tanks and structures, process equipment (marked with weights), buildings, ducts, etc. for civil engineering estimator
7. Produce GA showing locations of MCC, all drives and instruments, process equipment, ducts, etc. for electrical installation company
8. Produce drive schedule for electrical installation company

9. Produce GA showing locations of all tanks and structures, process equipment, piping and valves, buildings, ducts, etc. for mechanical installation company
10. Negotiate with equipment suppliers as well as civils, electrical, and mechanical companies about numbers, size, and location of equipment, etc.
11. Redesign plant, issue revised documentation
12. Request estimates from internal discipline managers for in-house engineering time
13. Carry out in-house design review or reviews
14. Redesign plant, issue revised documentation
15. Possibly carry out design review with purchaser
16. Redesign plant, issue revised documentation
17. Negotiate with all the above parties on cost, timescale, performance guarantees, contract conditions
18. If lead contractor, pull together all the specifications from suppliers into tender submission documents
19. Chase up the suppliers who have failed to return quotations in time
20. Produce (in MS Excel) a document that adds up all the costs for the "bought-in items," adds on in-house engineering time estimates, cost of required bonds and insurance, a percentage for the defects liability period, one for contingency, and another for profit. This spreadsheet needs then to work out what the overall percentage markup on bought-in items is, and apply this percentage markup to the bought-in cost of each individual item, which has been listed in the spreadsheet as per the clients cost schedule.
21. Attend a tender settling meeting to negotiate with management how much markup is going to be applied to the bought-in items and to ensure management are satisfied that the plant will work, that all the necessary equipment has been included etc.
22. Revise the MS Excel document to suit the tender settling meeting recommendations
23. Insert all the individual marked-up prices into the price schedules
24. Package up and deliver the completed tender document with very little time to spare
25. If subcontractor, send Price Schedules to the lead contractor with very little time to spare

There are, as I have described elsewhere, many other parallel activities related to design development in the case of the process/M + E contractor's duties, but I have omitted them here for the sake of clarity. Anyone who has ever done this job will understand the great degree of resource constraint, the importance of accuracy, and the importance of complementary skills of communication, persuasion, negotiation, and compromise.

COMMERCIAL KNOWHOW

This section is written from the perspective of those working for EPC companies, which is by far the most common situation. Even if you ask a consultant for a price, they will probably in turn ask an EPC company to generate it for them.

Margins in water and effluent treatment are very tight (unless you are a single bidder to a naïve client for a one-off industrial effluent treatment plant).

A competitive markup (including all in-house resources) on bought-in items and services might be around 22% for M + E bids, but civil contractors might add no markup, or even negative markup, gambling on their ability to make profit on "variation orders" once the job starts. Doing this requires a strict "letter of the contract" approach by the civil company that can annoy clients, and often ends up in court. Process/M + E contractors normally deplore such an approach, and are consequently frustrated when outbid by a company hoping to "screw it back on VOs" as the practice is known.

Tight margins do not allow a lot of scope for pricing in risk. Those pricing plant are forced to take a view on how risky the job is, based on sufficiency of design data, track record of technology, expertise of client, litigiousness of client, risks associated with plant location, and so on. They may recommend adding a lump sum or percentage contingency (undeclared or declared) to all or part of the cost of the job.

There may also be "Prime Cost" (PC) sums added to the tender document, sums of money that will be paid out if the client decided to proceed in a certain way, sometimes contingent on whether a risk crystallizes. These may be in the original tender documents, or the bidder may ask that they be included.

Bidders are normally required to submit a fully compliant bid to avoid disqualification, though they are often allowed to submit an alternative bid alongside the compliant one. Only if it is possible to argue that a compliant bid would not work it is usually possible to submit an alternative as the main bid. Doing so is a calculated risk, as it can offend clients to the extent that many would rather learn that you were right the hard way.

If the job is too risky, they may recommend to management that they decline to quote entirely, especially if there are many bidders.

Some clients invite 8 or 10 bidders, but this is only feasible for very large companies regularly issuing such tenders. Producing bids is time consuming and involves a large team of contributors. It is unreasonable to expect 10 companies to spend thousands of dollars each to produce bids, for which only one of them will be awarded any work. Unless the client is particularly large or prestigious the companies which will decline to quote will tend to be the good ones, who put signfiicant work into their bids.

FURTHER READING

Hall, S. M. (2012). *Rules of thumb for chemical engineers*. Oxford: Butterworth-Heinemann.

Kawamura, S., & McGivney, W. (2008). *Cost estimating manual for water treatment facilities*. Chichester: Wiley.

Moran, S. (2015). *An applied guide to process and plant design*. Oxford: Butterworth-Heinemann.

Peters, M. S., Timmerhaus, K. D., & West, R. E. (2003). *Plant design and economics for chemical engineers*. New York, NY: McGraw-Hill.

Woods, D. R. (2007). *Rules of thumb in engineering practice*. Chichester: Wiley.

Appendix 6: Water aggressiveness indexes

INTRODUCTION

Water can affect wetted materials of construction in various ways, and it is commonplace for the chemical/process engineer to be consulted by civil and mechanical engineers for advice on materials selection.

Chemicals dissolved in water can interact with materials of construction, but pure water containing no dissolved chemicals can itself interact with materials of construction.

Water can be acidic or alkaline depending on whether it has acids or alkalis dissolved in it. Strongly acidic or alkaline waters can attack some materials of construction. The acidity (or alkalinity) of water can be measured in pH units though this is not how water engineers usually do it.

In water engineering, alkalinity is usually measured by titration with a strong acid and expressed as an equivalent amount of dissolved carbonate (because almost all of the alkali present in natural waters is carbonate or bicarbonate). Alkalinity is not affected by the addition or removal of carbon dioxide as it is measured by adding acid till the pH drops to the point when all carbonate species are converted to carbon dioxide (about 4.4).

High concentrations of dissolved carbonates and bicarbonates "buffer" the water against pH changes. Low concentrations of carbonates and bicarbonates in water mean that the addition of small amounts of acid or alkali will give a large pH change. Conversely, high levels of carbonates and bicarbonates in the water mean that the addition of small amounts of acid or alkali will not give a large pH change.

Dissolved calcium or magnesium salts are what makes water "hard" or "soft." Waters with high hardness tend to form deposits of calcium carbonate scale on heated surfaces, as can be seen, e.g., on the elements of domestic kettles in hard water areas. Conversely, soft waters tend to corrode wetted construction materials, and are therefore considered "aggressive."

To simplify matters greatly, corrosion may form a surface deposit, such as the familiar rust; it may be purely subtractive—removing material from a wetted surface; or it may pervade the construction material, changing its internal chemical composition.

All of these processes change the physical characteristics of a material. Rust is not as strong as steel. Thinner materials are weaker, and the chemical changes brought about by aggressive waters getting into construction material are rarely beneficial.

Engineers use various heuristic indices, such as the Langelier Saturation Index (LSI) in water and wastewater treatment plant design, to support professional judgment when predicting whether water is likely to form scale or to be aggressive.

LANGELIER SATURATION INDEX

The LSI compares two parameters: the pH at which a certain form of calcium carbonate (lime scale) would be theoretically expected to come out of solution to form scale (known as pH_s); and the actual pH value of the water.

If $pH_s < pH$, then water is in a "scaling" condition, and might be expected to have a lower tendency to cause corrosion. If $pH_s > pH$, then the water has an "aggressive" nature. The condition of the water may be determined using Eq. (A6.1).

LSI Calculation

$$LSI = pH + TF + CF + AF - 12.1 \qquad (A6.1)$$

where

pH = Measured pH value
TF = Temperature factor
CF = Calcium hardness factor
AF = Alkalinity factor

It should be noted that pH and alkalinity have separate terms in the LSI. The alkalinity used to determine AF is that which is measured by titration.

Table A6.1 allows the factors in the previous equation to be determined (though other factors are sometimes used):

Broadly:

- $pH + TF + CF + AF - 12.1 < -2$ implies very aggressive water
- $pH + TF + CF + AF - 12.1 < -0.5$ implies aggressive water
- $pH + TF + CF + AF - 12.1 > 0.5$ implies scaling water

Results in the range between ± 0.5 may be considered balanced, i.e., neither scaling nor aggressive.

In natural waters, calcium and magnesium are commonly found along with carbonates and bicarbonates, as the hardness and alkalinity both come from minerals such as limestone or chalk, which have dissolved in the water in the environment. Alkalinity and hardness are therefore commonly associated.

As mentioned previously, waters with low concentrations of alkalinity have a low buffering capacity, so also tend to have a low resistance to pH changes.

Waters with low concentrations of dissolved solids, alkalinity, hardness, and pH therefore tend to dissolve chemicals from metals and concrete with which they interact. The LSI can be used to estimate the tendency for waters to do this, a property that might be called aggressiveness.

The status of LSI amongst engineers is somewhat controversial. While it may be the consensus professional view that the LSI can be used as an indicator of likelihood of both scaling and aggressiveness, for both metals and concrete, there are some professional engineers who dispute this.

Table A6.1 Langelier Saturation Index Factors

Temperature (°C)	Temperature Factor (TF)	Ca$_2$ + Hardness (ppm CaCO$_3$)	Calcium Hardness Factor	Total Alkalinity (ppm CaCO$_3$)	Alkalinity Factor
0	0	5	0.3	5	0.7
2.7	0.1	25	1.0	25	1.4
7.7	0.2	50	1.3	50	1.7
11.6	0.3	75	1.5	75	1.9
15.5	0.4	100	1.6	100	2.0
19	0.5	150	1.8	150	2.2
24.5	0.6	200	1.9	200	2.3
29	0.7	300	2.1	300	2.5
34.5	0.8	400	2.2	400	2.6
40.5	0.9	-	-	-	-

AGGRESSIVE INDEX

The strongly related, simpler, but less accurate *Aggressive Index (AI)* (Eq. A6.2) was, however, developed specifically for use in determining the aggressiveness of waters toward cementitious materials (specifically those in asbestos cement pipes). The AI only considers pH, alkalinity, and hardness, and different factors are used in the equation from those used in the LSI.

Aggressive Index

$$AI = pH + AF + CF \tag{A6.2}$$

where

pH = Measured pH value
CF = Calcium hardness factor
AF = Alkalinity factor

AF and CF are factors obtained from the data in Table A6.2 based on the measured alkalinity and hardness of the sample.

- pH + CF + AF < 12 implies aggressive water
- pH + CF + AF < 9 implies very aggressive water

Lime (calcium hydroxide) can be added to aggressive waters, to increase hardness (by adding calcium), alkalinity (by adding hydroxide, which is an alkali), and pH (by adding alkali, which increases pH) simultaneously; and hence produce a pronounced increase in both LSI and AI, and hopefully reduce aggressiveness.

Table A6.2 Alkalinity and Calcium Hardness Factors for Use With the Aggressive Index

Hardness or Total Alkalinity (ppm CaCO₃)	Alkalinity Factor or Calcium Hardness Factor
10	1
20	1.3
30	1.48
40	1.6
50	1.7
60	1.78
70	1.84
80	1.9
100	2
200	2.3
300	2.48
400	2.6
500	1.7
600	2.78
700	2.84
800	2.9
900	2.95
1000	3

OTHER INDEXES

The *Larson-Skold Index* is based on observed corrosion of mild steel pipework carrying a single natural water type. It is calculated from the ratio of equivalents per million (epm) of sulfate and chloride to the epm of bicarbonate plus carbonate (an equivalent is molarity of an ion in a solution, multiplied by that ion's valency).

Despite its limited validity, it is reasonably commonly used to predict aggressiveness of once through cooling waters.

- L-SI < 0.8 implies balanced water
- 0.8 < L-SI < 1.2 aggressive
- L-SI > 1.2 very aggressive

The *Ryznar Stability Index (RSI)* (Eq. A6.3) is another one based on Tillmann's saturation level of pH concept (pH_s), calculated as:

Ryznar Stability Index (RSI)

$$RSI = 2(pH_s) - p \tag{A6.3}$$

It was generated by correlating observed scale thickness in municipal drinking water systems with saturation and measured pH.

- RSI < 6 implies scaling water
- RSI > 7 implies balanced water
- RSI > 8 implies aggressive water

The *PSI Index* (Eq. A6.4) is very like the Ryznar stability index, except that it uses a calculated equilibrium pH rather than system pH.

PSI Index

$$PSI = 2(pH_s) - pH_{eq} \tag{A6.4}$$

where

$$pH_{eq} = 1.465 \times \log_{10}([HCO3^-] + 2\,[CO_3^{2-}] + [OH^-]) + 4.54$$

- PSI < 6 implies scaling water
- PSI > 7 implies balanced water
- PSI > 8 implies aggressive water

Another very experienced designer in an operating company is of the view that current best practice is the use of Precipitation Potential, i.e., the actual amount of calcium carbonate that would precipitate (or dissolve) while reaching equilibrium. There is a British Standard based on this (BS EN 13577:2007) for concrete. While I agree that such testing is better than educated guessing using indexes, the problem for designers in contracting companies is that they are not provided with this information, nor do they have time or budget to gather it. Educated guesswork is therefore the order of the day.

Appendix 7: Clean water treatment by numbers

The following is a suggested generalized explicit design methodology for the commonest kinds of municipal drinking water treatment plant:

1. Analyze available raw water quality data for the max, min, mean, upper limit (UL) and lower limit (LL) of 95% confidence interval (CI) for the following parameters:
   * Aluminum
   * Color
   * Iron
   * Manganese
   * Suspended solids
   * Methyl orange alkalinity
   * Calcium hardness
   * pH
   * Temperature
        If there is more than one potential source, all potential combinations of the sources given by the specification must be considered in this way. This data can then be used as follows:
   * Design T for any temperature-dependent process calculations such as DAF = mean
   * Design T for hydraulic calculations other than NPSH = min
   * Design T for NPSH calculations = max
   * Design color for chemical dosing and sludge yield calculations = UL95% CI
   * Design SS for sludge yield calculations = UL95% CI
   * Design Fe for sludge yield calculations = UL95% CI (+ max added Fe as per next section)
   * Design Al for sludge yield calculations = UL95% CI (+ max added Al as per next section)
   * Design Mn range = min to max or Lower Level (LL)95%CI** to UL95% CI, whichever is greater
   * Design Alk range = min to max or LL95%CI to UL95%CI, whichever is greater
   * Design pH range = min to max or LL95%CI to UL95%CI, whichever is greater
   * Design Hardness range = min to max or LL95%CI to UL95%CI, whichever is greater

Engineering judgment should be used to check that this reasonably conservative approach is appropriate.

2. Determine the range of flows through the works (the maximum is normally given in the specification, but judgment may be required to determine the minimum)
3. Produce a proposed process PFD (or PFDs)
4. Size the main unit operations in turn by PFD order, updating the PFD and resizing as required to take into consideration recycles
5. Calculate the range of coagulant dose based on color (see Chapter 8: Clean water unit operation design: chemical processes) and PFD flowrates
6. Calculate the required range of acid or alkali addition quantities to give the coagulation pH at PFD maximum flow, considering possible variations in coagulant dose, incoming pH, and alkalinity (i.e., from maximum calculated coagulant dose/min design raw water pH/min design alkalinity to minimum calculated coagulant dose/max design RW pH/max design alkalinity). Size equipment for suitable turndown ratio
7. Calculate any interstage chemical dosing requirements such as Cl dosing for Mn^{2+} removal (see Chapter 8: Clean water unit operation design: chemical processes), along with that for any pH corrections required by such processes (e.g., Mn oxidation works best at pH 8.5, so it may be necessary to increase from the initial coagulation pH, and then reduce back to the best chlorine disinfection pH of 7). Size equipment for suitable turndown ratio
8. Calculate sludge yield (see Chapter 21: Sludge characterization and treatment objectives) at maximum PFD flowrates and color, SS, Al, Fe, and Mn figures
9. Design backwash water and sludge handling and treatment processes, and update PFD and preceding calculations as required
10. Calculate disinfection dosing requirement and size equipment
11. Calculate any additional dosing requirements for supply water, such as phosphate for solubility control, fluoride, etc
12. Produce layout drawings for the plant
13. Produce P + ID for the plant
14. Carry out hydraulic calculations
15. Discuss design from a commercial and technical point of view with civil and electrical engineers and equipment suppliers and revise accordingly
16. Complete design review and revise the design accordingly
17. Produce the documentation package for the client

Appendix 8: Sewage treatment by numbers

The following is a suggested generalized explicit design methodology for the commonest kinds of sewage treatment plant:

1. Analyze available data on incoming sewage for the max, min, mean, upper limit (UL), and lower limit(LL) of the 95% confidence interval (CI) for the following parameters:
   - Biochemical Oxygen Demand (BOD)
   - Chemical Oxygen Demand
   - pH
   - Suspended Solids (SS)
   - Ammonia
   - Alkalinity
   - Temperature (T)
        If there is more than one potential source, all potential combinations of the sources given by the specification must be considered in this way. This data can then be used as follows:
   - Design T for secondary treatment plant biological = min
   - Design T for secondary treatment plant aeration = max
   - Design T for hydraulic calculations other than NPSH = min
   - Design T for NPSH calculations = max
   - Design BOD = mean or UL95% CI (see later)
   - Design SS = mean or UL95% CI (see later)
   - Design Ammonia = mean or UL95% CI (see later)
   - Design Alkalinity range = minimum to maximum or LL95%CI to UL95%CI, whichever is greater
   - Design pH range = minimum to maximum or LL95%CI to UL95%CI, whichever is greater
        The overall plant design needs to be verified under several design conditions, at least encompassing the following two:
   - High Flow: Full Flow to Treatment (FFT) × mean BOD, Ammonia, SS
   - Low Flow: Dry Weather Flow (DWF) × UL95% CI BOD, Ammonia, SS
2. Analyze the available data on incoming flows.
3. Produce PFDs for average and high-flow cases.
4. Design the inlet works, storm tanks, and storm return arrangements. Normally 3−4 × DWF is FFT and Formula A—FFT is diverted to storm tanks.
5. Design the primary settlement tanks to treat DWF to FFT.

6. Design the secondary treatment: biological/aeration tank to treat DWF to FFT.
7. Design the secondary treatment: aeration systems including blowers to treat DWF to FFT.
8. Design the secondary treatment: solids removal stage to treat DWF to FFT.
9. Design the tertiary treatment plant where required to treat DWF to FFT.
10. Design the sludge treatment plant based on maximum yield.
11. Incorporate sludge liquors recycle into the previous stages of design (DWF to FFT).
12. Produce layout drawings for the plant.
13. Produce P + ID for the plant.
14. Carry out hydraulic calculations.
15. Discuss the design from a commercial and technical point of view with civil and electrical engineers and equipment suppliers and revise accordingly.
16. Complete the design review and revise the design accordingly.
17. Produce a documentation package for the client.

Appendix 9: Industrial effluent treatment by numbers

Industrial effluents differ far more from each other than sewage. They are also more individually variable over time, and usually far less well characterized. The following approach is, however, generally applicable:

1. Analyze available raw water quality data for the maximum, minimum, mean, upper limit (UL), and lower limit (LL) of the 95% confidence interval (CI) for the following parameters:
   * Fats, Oils, and Greases (FOG)
   * Total Petroleum Hydrocarbons (TPH)
   * Detergents
   * Heavy Metals
   * Cyanide
   * Chlorine
   * Sulfate
   * Color
   * Iron
   * Suspended Solids
   * Methyl Orange Alkalinity
   * Calcium Hardness
   * pH
   * Temperature (T)
   * Any other substances covered by treated water specification.
     If there is more than one potential source, all potential combinations of the sources given by the specification must be considered in this way. This data can be used as follows:
   * Design T for any temperature-dependent process calculations such as DAF = mean
   * Design T for hydraulic calculations other than NPSH = min
   * Design T for NPSH calculations = max
   * Design Color for chemical dosing and sludge yield calculations = UL95% CI
   * Design SS for sludge yield calculations = UL95% CI
   * Design Fe for sludge yield calculations = UL95% CI (+ maximum added Fe as per next section)
   * Design Al for sludge yield calculations = UL95% CI (+maximum added Al as per next section)

415

- Design Mn range = minimum to maximum or LL95%CI to UL95%CI, whichever is greater
- Design Alkalinity range = minimum to maximum or LL95%CI to UL95% CI, whichever is greater
- Design pH range = minimum to maximum or LL95%CI to UL95%CI, whichever is greater
- Design Hardness range = minimum to maximum or LL95%CI to UL95% CI, whichever is greater

  Engineering judgment should be used to check that this reasonably conservative approach is appropriate to the project. There is a very high probability in industrial effluent treatment that there will be insufficient data to draw statistically significant conclusions for all likely conditions. There may even be no data available at all. Ideally you would gather your own representative data, but in the common situation where this is not possible, some idea of the likely effluent composition in the industry considered may be gleaned from scientific papers, the Degremont manual (see Further Reading), or generally online. Clearly such data can have variable reliability, and must be treated with great caution. Variability throughout the year, week, and day must also be accounted for

2. Analyze available data on incoming flows. Such data will usually be rather sketchy, and flows may often have to be estimated based on water supply or effluent bills or even pipe or pump sizes. Variability throughout the year, week, and day must also be accounted for

3. Buffering tankage to control flows and loads to treatment must be designed first, based on the flows and loads data generated in the last two stages. If flow is even, but loads vary, a mixed tank of capacity greater than the period of variation that runs full will be required. If flows vary significantly an unmixed tank of capacity greater than the period of variation with fluid level rising and falling to accommodate peaks will be required. If both flows and loads vary greatly, one balancing tank of each type will be required, though hybrid approaches are often attempted to save money

The methodology from this point depends on whether the process is based on chemical dosing plant—like a drinking water plant; biological—like a sewage treatment plant; or a hybrid of the two. This will determine whether one or other of the methodologies outlined in Appendix 7, Clean Water Treatment by Numbers, and Appendix 8, Sewage Treatment by Numbers, may be followed or they should be hybridized as appropriate.

FURTHER READING

Degrémont Suez. (2007). *Water treatment handbook* (7th ed.). Cachan: Lavoisier.

Appendix 10: Useful information

INFORMATION ON COMMERCIAL GRADES OF CHEMICALS

Table A10.1 Commercial Grades of Chemicals: Approximate Data for Preliminary Design

Chemical	Solution (%w/w)	Specific Gravity or Relative Gas Density	Freezing Point	Usual Delivery Size
Aluminum sulfate	48.2 as 17% Alum	1.3	3–25°F	Road tanker, carboy, IBC
Anionic polymer	Solid	0.8	NA	Road tanker, FIBC, Sack
Calcium carbonate	Solid	2.7	NA	Road tanker, sack, IBC
Calcium hypochlorite	Solid	2.0	NA	Road tanker, sack, IBC
Carbon dioxide	Gas	1.5	NA	Road tanker, pressure cylinders
Cationic polymer	50	1.0	0°C	Road tanker, carboy, IBC
Chlorine	Gas	2.5	NA	Road tanker, tonne drums, cylinders
Ferric chloride	38–39	1.4	−20–0°C	Road tanker, carboy, IBC
Ferrous sulfate (Copperas)	42	1.5	< −28°C	Road tanker, carboy, IBC
Fluorosilicic acid	20.2 as H_2SiF_6 Or 16% F	1.2	< −15°C	Road tanker, carboy, IBC
GAC/PAC	Solid	2.3	NA	Road tanker, FIBC, Sack
30–36 °Tw degree Hydrochloric acid	29.4–35.4	1.15–1.2	−30–52°C (stronger solutions higher)	Road tanker, carboy, IBC
Milk of Lime	43–47	1.3–1.4	0°C	Road tanker, carboy, IBC
Monosodium phosphate	Solid	0.6–0.9	NA	Road tanker, FIBC, Sack

(Continued)

417

Table A10.1 Commercial Grades of Chemicals: Approximate Data for Preliminary Design *Continued*

Chemical	Solution (%w/w)	Specific Gravity or Relative Gas Density	Freezing Point	Usual Delivery Size
Ozone	Gas	1.7	NA	Locally produced
Poly aluminum chloride/poly aluminum silicate sulfate	10–24% Al_2O_3	1.2–1.3	<5°C	Road tanker, carboy, IBC
Phosphoric acid	75%	1.7	21°C	Road tanker, carboy, IBC
Potassium permanganate	3%	1.0	<0°C	Road tanker, carboy, IBC
Quicklime	Solid	3.0	NA	Road tanker, FIBC, Sack
Salt	Solid	2.2	NA	Road tanker, FIBC, Sack
Slaked lime	Solid	2.4	NA	Road tanker, FIBC, Sack
Sodium carbonate (Soda ash)	Solid	2.5	NA	Road tanker, FIBC, Sack
Sodium fluorosilicate	Solid	2.7	NA	Road tanker, FIBC, Sack
Sodium hydroxide	47%	1.5	6–8°C	Road tanker, carboy, IBC
Sodium hypochlorite	16% available Cl	1.3	−17°C	Road tanker, carboy, IBC
Sodium sulfite	Solid	2.6	NA	Road tanker, FIBC, Sack
Sodium thiosulfate	Solid	1.7	NA	Road tanker, FIBC, Sack
Sulfur dioxide	Gas	2.3	NA	Road tanker, tonne drums, cylinders
Sulfuric acid 98%	98%	1.8	0°C	Road tanker, carboy, IBC
Sulfuric acid 96%	96%	1.8	−10°C	Road tanker, carboy, IBC
Sulfuric acid 50%	50%	1.4	−35°C	Road tanker, carboy, IBC

Specific gravity of liquids and solids is the density relative to water (around 1000 kg/m^3).

Relative gas density is the density relative to dry air (around 1.2 kg/m^3).

STORAGE VESSEL SIZE

Allow one tanker plus 14 days of use + 10%

IDIOSYNCRATIC WATER INDUSTRY UNITS
SOLUTION DENSITY

Degrees Twaddell (°Tw) are commonly used in the UK water industry. They are based on the use of a hydrometer manufactured in Glasgow by Twaddell and Co. at the start of the 19th century.

The relationship between Twaddell and the normal units on a hydrometer (Specific gravity) is given in Eq. A10.1.

Relationship between degrees Twaddell and specific gravity

$$x = 1 + (0.005a)$$
$$a = \frac{(x-1)}{0.005} \qquad (A10.1)$$

where

a = degrees Twaddell
x = specific gravity (water taken as 1.000)

COLOR

There is only one color unit used in water treatment, but confusion can arise due to the wide range of names used for the same unit. The name preferred in both ASTM D1209 and ISO 6271 is "Platinum-Cobalt Color" or Pt-Co, but "APHA" or Hazen Unit is the name in most common use in industry.

TURBIDITY

Differences between the various units of turbidity can be attributed to differences in test methodology, but as with color they are all about the same.

$$1JTU = 1NTU = 1FTU = 1FNU$$

Nephelometric Turbidity Units (NTU), Formazin Turbidity Units (FTU), and FNU (Formazin Nephelometric Units) are all essentially measured in the same way, by the scattering of light passing through the sample at 90 degrees to path by an instrument known as a nephelometer.

Jackson Turbidity Units (JTU), on the other hand, are a measure of how long a vertical column of the sample we can view a candle flame through. It is a coarser method, best suited to field use, and samples >25 JTU.

Appendix 11: Worked example for chemical dosing estimation

A. COAGULANT DOSING

Coagulant Used	Bulk Liquid Aluminum
Required coagulation pH	6.1
Estimate dose rate from based on max color	$= 31°H$
	$= 2.5$ ppm (mg/L Al^{3+})
Use 7.5% alum	SG $= 1.31$: 52 kg Al^{3+}/m^3
kg/day alum required	$= 10.42$ MLD $\times 2.5$ ppm $= 26.05$ kg/day
30-Day storage requirement	$= 781.5$ kg
	781.5 kg@52 kg/m^3 $= 15.03$ m^3 of storage
	$2 \times 50\%$ tanks @ 7.51 m^3 each[a]
Let tank $= 2$ m Ø	CSA $= 3.14$ m^2 $\Rightarrow h = 7.51/3.14$
	$= 2.39$ m working depth
Pumping rate @ 7.5% @ max flow	$26.05/52 = 0.5$ m^3/day $= 0.021$ m^3/h
	Dosing pumps deliver 0.021 m^3/h at 1 bar g (nominal)

[a]NB less than 20.5 m^3 required for 18te tanker delivery.

B. pH ADJUSTMENT

B1 Neutralization of Existing Alkalinity

	$Al_2(SO_4)_3$	$: 3CaCO_3$
RMM	342	100

So: alum dose	$= 2.5$ mg/LAl^{3+}
	$\equiv \frac{1}{0.158} \times 2.5$mg/L $Al_2(SO_4)_3$
	$\equiv 15.82$ mg/L alum, i.e., 15.82/342 m molar
	$= 0.0463$ millimolar

(Continued)

Continued

B1 Neutralization of Existing Alkalinity

3:1 stoichiometry = $> 3 \times 0.0463$	= 0.139 m molar alkalinity required to neutralize alum present
Minimum alkalinity in raw water	= 76 mg/L$\equiv$76/100 = 0.76 millimolar

i.e., no caustic dosing will be required at this stage—we actually require acid dosing

B2 Neutralization of Remaining Alkalinity

Raw water pH	8.58 (max)
Alkalinity	110 (max)

At this pH (methyl orange alkalinity)/(free CO_2) is $\rightarrow \infty$

i.e., free CO_2 is theoretically $\rightarrow 0$

Note, however, that the design brief gives 7.2 mg/L at this pH

The Tillmann formula predicts poorly at $>> $ pH = 8, but the following approach should account for all required dosing:

B2.1 Assess Potential Alkalinity at Maximum and Operating pH Values

At pH 8.58, pot alk		= [Alk] + [CO_2] $\times$ 50/44
		= 110 + (7.2 $\times$ 50/44) = 118.18 mg/L
At pH 6.1	1. Pot alk	= [Alk] + [CO_2] $\times$ 50/44
	2. From Tillmann formula/nomogram	= [Alk]/[CO_2] = 0.625
Assume no CO_2 loss $\Rightarrow$ potential [Alk] still 118.18 mg/L		
So, at pH 6.1		[Alk] = 0.625 $\times$ [CO_2]
Substituting		0.625 [CO_2] + 50/44 [CO_2] = 118.18
		$\Rightarrow$ 1.76 [CO_2] = 118.18
		[CO_2] = 118.18/1.76 = 67.15 mg/L
Resubstitute		[Alk] + 50/44 [67.15] = 118.18
		$\Rightarrow$ [Alk] + 76.30 = 118.18
		$\Rightarrow$ [Alk] = 41.88 mg/L

B2.2 Calculate Alkalinity to be Neutralized

From pH 8.58 to pH 6.1, we have 118.18−41.88 mg/L of alkalinity to be neutralized	= 76.30 mg/L
	$\equiv$ 0.763 millimolar

B2.3 Neutralize with H_2SO_4

	H_2SO_4 + $CaCO_3 \rightarrow CaSo_4$ + H_2O + CO_2	
RMM	98 100	

So, 0.763 millimolar alkalinity, with 0.139 millimolar neutralized by the alum dose leaves 0.624 millimolar to be neutralized by acid

So, acid requirement = 1 mol/mol alkalinity	

(Continued)

Continued

B2 Neutralization of Remaining Alkalinity	
Let x = moles of acid	0.624/100 mol = x/98 mol
	$\Rightarrow x$ = 0.612 millimolar acid
	$\Rightarrow$ 59.98 mg/L
	$\equiv$ g/m^3
If we dose as 97% H_2SO_4	sg = 1.84
Dose rate	$\frac{434.17 \times 59.98}{1785}$ = 14.59 L/h max. dose
So, 30-day storage	= 10.51 m^3
	Say 2.5 m Ø $\times$ 2.3 m high

Index

Note: Page numbers followed by "*f*" and "*t*" refer to figures and tables, respectively.

A

"Absolute" filtration, 43
Academic misconceptions, 10–11
Acetic acid, 33*t*
Acidogenesis, 201–202
Acids and bases, 15–24
  buffering, 17–19
  catalysis, 23
  dechlorination, 23
  osmosis, 23–24
  oxidation and reduction, 21–23
  scaling and aggressiveness, 19–21
Acrylonitrile butadiene styrene, 5
Activated biofilter process, 177
Activated carbon, 113, 163, 234
Activated sludge (AS) process design, 179–185
  oxygenation capacity, 182
  oxygen requirements, 182–184
  physical facility design and selection, 178–179, 179*t*
  returned sludge, 184–185
  sludge age and mean cell residence time, 184
  sludge loading, 180–181
  tank geometry, 182
  tank volume, 181
  yield (aka sludge production factor), 180
Activated sludge (AS) system, 175
Activated sludge control chambers, 215–216
Actual flow data, 141
"Additional conventional pollutants", 327–328
Adenosine triphosphate, 22
Adsorption, 48–49, 94–95, 162–163, 234–235, 297–298
Advanced Water Purification Facility (AWPF), 146
Aerated lagoons, 175, 190
Aeration blowers, 295
Aeration systems, efficiency of, 162*t*
Aeration tanks/collection channels, 208
Aerobic/anaerobic environments, 27
Aerobic attached growth processes, 172–175
Aerobic inhibitors, 244, 245*t*
Aerobic sludge digestion, 175, 277, 286–287
  composting, 287
  disinfection, 287
Aerobic suspended growth processes, 175
Aerobic treatment design parameters, 247
Aerobic water treatment, biology of, 35

Aesthetics, plant layout and, 309–310
"Aggressive CO$_2$", 19–21
Aggressive Index (AI), 19–21, 407
Air lift percentage submergence, 151*f*
Airlifts, 149–151, 149*f*
  line size, 150, 150*f*
  required percentage submergence, 150
Air scour distribution, 85–86
Air stripping, 167
Aka moth flies, 298–299
"Alarm fatigue", 392
Algae, 28–29
Alkaline odorants, 298
Alkalinity, 15, 19, 63
  and anaerobic digestion, 281
  measurement, 405
Alnwick Castle design, 342
Alnwick Castle water feature, 338
Alternative bed sizing calculation, 190
Aluminum, 40
Aluminum coagulants, 105
Aluminum salts, 104, 166
Amino acids, 32
Ammonia, 140, 330
Anaerobic attached growth processes, 176
Anaerobic contact reactor, 283–284, 284*f*
Anaerobic digesters, 47, 266–267
  types, 283*t*
Anaerobic filter process, 201
Anaerobic filter reactors, 282–283, 283*f*
Anaerobic inhibitors, 244, 246*t*
Anaerobic sludge digestion, 256, 277–278
  digester design criteria, 279–282
    alkalinity and anaerobic digestion, 281
    digester temperature, 282
    rules of thumb, 280–281
    sludge mixing, 280
    types of heating, 280
    volatile matter destruction, 281–282
    zoning data, 281
  factors affecting, 278–279
  types, 278–282, 283*t*
    anaerobic contact reactor, 283–284, 284*f*
    anaerobic filter reactors, 282–283, 283*f*
    fluidized/expanded bed reactor, 285–286, 286*f*
    upflow anaerobic sludge blanket (UASB), 284–285, 285*f*

Anaerobic suspended growth processes, 176
Anaerobic treatment, 176
  advantages and disadvantages of, 173*t*
  biology of, 36−37
  design parameters, 247−248
Analysis paralysis, 327−328
Ancillary processes, 295
  fly control, 298−299
  noise control, 295−296
  volatile organic compound (VOC) and odor,
      296−298
    covers and ventilation, 297
    measurement and setting of acceptable
        concentrations, 297
    removal of, 297−298
Anoxic environments, 27
ANSI/ASME Y14.1 "Engineering Drawing
      Practice", 371−372, 376
ANSI/ISA S5.1-1984 (R 1992) "Instrumentation
      Symbols and Identification", 371−372
Antoine equation, 125−126, 126*t*
A/O process, 195
Applied mathematics, 348
Applied science, 348
Appropriate statistics, 328−329
Approximations, 354−355
Archaebacteria, 27
ASTM D5127, 67
Attached and suspended growth processes,
      comparison of, 178*t*
Autotrophs, 35
Average, defined, 6

B

Bazin's formula, 133, 212, 212*f*
BS1646, 372
Bacillus thurigiensis, 299
Bacteria, 27−28
Bacterial growth, 30−31, 31*f*
Band screens, 75
Bare carbon steel, 5
Bar screens, 152*f*, 153, 213−214
  more accurate head losses, 213−214
  rough head losses, 213
  velocities, 213
Base exchange softening, 106
Batch centrifugation, 271
Belt filter presses, 271, 272*f*
Bernoulli's equations, 56, 134
Best efficiency point (BEP), 129
Best engineering practice, 355
BET isotherms, 48−49, 50*f*
Bicarbonates, 18, 405

Bidders, 403
Bills of Quantities, 401
Bio-augmentation, 195−196
Biochemical oxygen demand (BOD), 10, 143, 155,
      243
Biochemistry, 31−35
  carbohydrates, 33
  fats, 32
  metabolism, 34−35
  nucleic acids, 33−34
  proteins, 32
    enzymes, 32
Biocides, 108−109
Biogas, 201−202, 256
Biological activated carbon (BAC), 115
Biological aerated flooded filters, 196−197
Biological filtration, 115−116
Biological fouling, 93
Biological nutrient removal, 192−195
  combined nitrogen and phosphorus removal, 195
  nitrogen removal, 192−194
  phosphorus removal, 195
Biological secondary treatment process, 172
Biology, 25
  aerobic water treatment, 35
  anaerobic water treatment, 36−37
    conversion of acetic acid to methane and
        carbon dioxide, 36−37
    conversion of higher fatty acids to acetic acid,
        36
    hydrolysis, 36
    volatile fatty acid production, 36
  bacterial growth, 30−31
  biochemistry, 31−35
    carbohydrates, 33
    fats, 32
    metabolism, 34−35
    nucleic acids, 33−34
    proteins, 32
  biology/biochemistry gap, 25−27
  important microorganisms, 27−30
    algae, 28−29
    bacteria, 27−28
    fungi, 30
    protozoa and rotifers, 28
Biomass, 201−202
Blending, 266−267, 305*f*
Blue green algae, 28−29
Boiler feed water, 66
  pharmaceutical water, 67
  semiconductor processing water, 67
Boiler water treatment, 109−110
  carryover, 109
  sludge and scale formation, 110

BOOT variant, 9
Borehole drinking water treatment, 21
Brackish waters, 66
Breakpoint chlorination, 167−168
Bridge scraper, 71, 156
British Retail Consortium (BRC), 261
British Standard "aggressive CO_2" test, 21
British Standard for engineering drawings (BS 5070), 370
British Standards, 353−354
Broad-crested weirs, 338, 342f
Broad weir calculation, 134, 212−213
Bromine dosing, 341
Bronze Roman hydraulic pump, 2, 2f
BS EN ISO 10628, 370
"Budget" pricing, 399
Buffering, 17−19
Buffering effects, in natural waters, 10
Buffer solutions, 17
Build Own Operate contract, 9
Butyric acid, 33t

C

Cabbage-like mercaptans, 330
Cable-raked screens, 153
Calcium carbonate, 343
Calcium chloride, 166
Calcium hardness, 15
Calcium hypochlorite, 22, 112
Calcium phosphate, 166
Calcium sulfate, 166, 226, 240
Calgon dosing, 105
"Cambi", 202
Capric acid, 33t
Caproic acid, 33t
Caprylic acid, 33t
Carbohydrates, 33
Carbonate-buffered system, 18
Carbonate buffering, 18−19
Carbon dioxide, 106−107
Carrier water systems, 102
Catalysis, 23
Catenary screens, 153
Cationic polymers, 40
Caustic (NaOH) solution, 298
Cellulose membranes, 91
Central processing units, 397
Central "stilling baffle", 156
Centrifugal units, 266
Chemical disinfection, 167, 242
Chemical dosing, 237
Chemical engineering, 122, 347
Chemical engineers, 25, 44, 347

Chemical oxidation, 169
Chemical oxygen demand (COD), 242−243
Chemical precipitation, 166
Chemical/process engineering misconceptions, 10
Chemically Enhanced Primary Treatment (CEPT), 200f
Chemoautotrophs, 35
Chloramines, 167−168
Chlorinated rubber, 5
Chlorination, 40, 51, 111−113, 167, 242
Chlorinator, 112−113
Chlorine, 22, 166−167, 242, 287, 341
    removal. *See* Dechlorination
Chlorine contact tanks (CCTs), 113
Chlorine dioxide, 22
Choking, 215
Chromic acid, 225
Chromium plating, 224
Circular channel headloss, 206
Circular tanks, 156−157
Civil and Structural Engineers, 360
Civil engineers, 3
Clarified filtrate, 271
Clarifier design, 71−73
    horizontal/radial flow clarification, 73
    proprietary systems, 73
    sludge blanket clarification, 72
Clean water characterization, 61−64
    alkalinity, 63
    color, 62−63
    iron, 63
    manganese, 64
    microbiological quality, 62
    suspended solids, 63
    total dissolved solids (TDS), 64
Clean water hydraulics, 117
    hydraulic profiles, 134
    open channel hydraulics, 132−134
        chambers and weirs, 133−134
        channels, 132−133
        flow control in open channels, 134
    pump selection and specification, 118−132, 118t, 120t
        pump sizing, 119−132
        pump types, 118
Clean water treatment by numbers, 411−412
Clean water treatment objectives, 64−67
    boiler feed water, 66
    cooling water, 66
    potable water, 65−66
    ultrapure water, 66−67
Clean water unit operation design
    biological processes in, 111
        biological filtration, 115−116

Clean water unit operation design (*Continued*)
  discouraging life, 111–115
  encouraging life, 115
  chemical processes in, 101
    boiler water treatment, 109–110
    cooling water treatment, 108–109
    drinking water treatment, 102–105
  physical processes in, 69
    distillation, 96
    flocculation, 70–71
    flotation, 73–74
    gas transfer, 97–98
    mixing, 70
    novel processes, 99–100
    physical disinfection, 98–99
    settlement, 71–73
    stripping, 98
Coagulants, 40, 104, 165
Coagulation, 104–105, 165–166, 343
Coagulation efficiency, 40
Coagulation/flocculation, 40
Coalescing media, 233–234
  hydrophilic media, 233–234
  hydrophobic media, 233
  removal of toxic and refractory compounds, 234
Coarse approximations, 355
Coarse bar screens, 152
Coarse surface filtration, 74–75
  band screens, 75
  trash racks, 75
Colebrook–White approximation, 122, 124
Coliforms, 50, 62
Colloidal fouling, 93
Colony forming units (CFUs), 50–51
Color, measurement of, 62–63
Color unit, 419
Combined aerobic treatment processes, 176–177
"Combined"/"firmly bound" CO_2, 18
Commercial grades of chemicals, 417*t*
Commercial membrane bioreactor systems, 198*t*
Comminution, 153–154
Comminutors, 153–154
Complete Mixing Activated Sludge (CMAS), 200*f*
Composting, 287
Compressors, pumps and, 387
Computer Aided Process Engineering (CAPE)
    working group, 356
Computers, use and abuse of, 356
Concentration polarization, 44–45
Conceptual design, 360–362, 399–400
  modeling as, 362
Conceptual/feed fast tracking, 366
Conceptual/FEED layout methodology, 312–313,
    313*t*

Conceptual layout, 310–312
  construction, commissioning, and maintenance,
    311
  emergency provision, 312
  indoor or outdoor, 311
  materials storage and transport, 311–312
  prevailing wind, 311
  security, 312
    central services, 312
    earthworks, 312
Conjugate base, 17
Constructed wetlands, 189–190
Construction Design and Management (CDM)
    regulations, 351
Continuous stirred tank reactor (CSTR) system,
    104, 176
Contractual arrangements, in water and effluent
    treatment, 8–9
Control of Major Accident Hazards (COMAH)
    Regulations, 307–308
Conventional/nonconventional pollutants,
    327–328
Conventional penstocks, 208
Cooling water, 66
Cooling water treatment, 108–109
Copperas, 40
Corona discharge, 113–114
Cost, plant layout and, 308–309
Cost estimate, 377
Costing water and wastewater treatment plant,
    399–404
  commercial knowhow, 403
  mechanical and electrical equipment costing
    methodology, 399–402
    conceptual design stage, 399–400
    detailed design, 401–402
    feed stage, 400–401
Cost Schedules, 401
Covers and ventilation, 297
Cryptosporidium, 51
Curve-fitting equations, 122
Cyanides, 169
  removal from wastewaters, 240–241
Cyclone degritters, 266

D

DAF designs, 271
Danish EA equation, 296
Darcy–Weisbach equation, 56–57, 124
Darcy's equation, 45, 46*f*
Data analysis, 327–328
Datasheets, 377, 378*f*
Death Phase, 30

Decanter centrifuges, 270, 270f
Dechlorination, 23, 51, 114, 167, 242
Dedicated field-mounted controllers, 397
Deep shaft process, 196
Degrees Hazen, 63
Degrees Twaddell, 419
Degritting, 266
Deionized/demineralized/softened water, 67
Denitrification, 192
Deoxyribose nucleic acid (DNA), 33
Department of Environment Transport and
       Regions (DETR), 261
Depth below sewer invert calculation, 217
Depth filters, 43, 233
Depth filtration, 45, 74—75
Design, defined, 348—356
Design basis and philosophies, 369
Design data, sources of, 383
Design deliverables, 369—380
    cost estimate, 377
    datasheets, 377
    design basis and philosophies, 369
    equipment list/schedule, 373—375
    functional design specification (FDS), 375
    General Arrangement (GA) drawings, 375—377
    HAZOP study, 378—379
    isometric piping drawings, 379
    Piping and Instrumentation Diagram (P&ID),
       371—373
    process flow diagram (PFD), 370—371
    specifications, 369—370
    zoning/hazardous area classification study, 379
Design envelope, generating, 367—368
"Design for Construction", 363—364
Design manuals, 354
Design mass loadings, selection of, 143—144
Design methodology information, sources of, 383
Design of water feature, 335
    detailed hydraulic design, 341—344
        nozzles, 341—342
        water quality, 342—344
        weirs, 342
    tender stage design, 337—338
        channels, 338
        nozzles, 338
        pipework, 337
        tankage, 338
        weirs, 338
    water quality, 339—341
        biological quality, 340—341
        chemical composition, 339
        clarity, 339—340
Design process, of water treatment plant, 359—368
    design envelope, generating, 367—368

fast tracking, 365—367
    conceptual/feed fast tracking, 366
    design/procurement fast tracking, 366—367
    feed/detailed design fast tracking, 366
stages, 360—365
    conceptual design, 360—362
    detailed design, 363—364
    feed/basic design, 362—363
    posthandover redesign, 365
    site redesign, 364
Design/procurement fast tracking, 366—367
Design verification, 332—333
Detailed design, 363—364, 401—402
Detailed layout methodology, 313—314, 314t
Diaphragm pumps, 265, 332
Digestate, 256
Digesters, 279—280
Direct on Line (DOL) starters, 393
Dirty water characterization and treatment
       objectives, 139
    flowrate and mass loading, 140—144
    treatment objectives, 144—146
        discharge to environment, 145
        industrial reuse, 146
        reuse as drinking water, 146
    wastewater characteristics, 139—140
Dirty water hydraulics, 203
    advanced open channel hydraulics, 214—218
        choking, 215
        control structures, 215—216
        critical and supercritical flow, 214
        hydraulic jumps, 214—215
        inlet works, 215
        sump flow presentation and baffles, 217—218
        sump volumes, 216—217
    minimum velocities, 204
    open channels, 204—211
        circular channel headloss, 206
        peripheral channels, 208—211
        shock losses in channels, 206—208
        straight channel headloss, 204—206
    plant layout, 218
    screens, 213—214
        bar screens, 213—214
    weirs, 211—213
        broad weirs, 212—213
        thin weirs, 211—212
Dirty water unit operation design
    biological processes, 171
        aerobic attached growth processes, 172—175
        aerobic suspended growth processes, 175
        anaerobic attached growth processes, 176
        anaerobic filter process, 201
        anaerobic suspended growth processes, 176

Dirty water unit operation design (*Continued*)
  bio-augmentation, 195–196
  biological aerated flooded filters, 196–197
  combined aerobic treatment processes, 176–177
  deep shaft process, 196
  expanded bed reactor, 201
  membrane bioreactors, 197–200
  natural systems, 188–191
  pure or enhanced oxygen processes, 196
  rotating biological contactors (RBCs), 188
  secondary treatment plant design, 177–185
  sequencing batch reactors, 197
  tertiary treatment plant design, 191–195
  trickling filters, 185–187
  two and three phase digestion, 201–202
  upflow anaerobic sludge blanket process, 200–201
 chemical processes, 165, 169
  chemical precipitation, 166
  coagulation, 165–166
  disinfection, 166
  nutrient removal, 167–168
  removal of dissolved inorganics, 168–169
 physical processes, 147, 163
  adsorption, 162–163
  filtration, 160–161
  final settlement tank design, 158–160
  flotation, 160
  flow equalization, 155–156
  flow measurement, 148
  gas transfer, 161–162
  grit removal, 154–155
  physical disinfection, 163
  primary sedimentation, 156–158
  pumping, 149–151
  racks and screens, 151–154
Disinfection, 49–51, 111–115, 166, 242, 341
  chemical disinfection, 242
  chlorination, 112–113
  dechlorination, 114
  dechlorination, 242
  membrane filtration, 115
  ozonation, 113–114
  of sewage sludges, 287
  ultraviolet light, 114
Dispersion number method, 182
Dissolved air flotation (DAF), 43, 73–74, 160, 223, 231
Dissolved calcium/magnesium salts, 405
Dissolved inorganics, removal of, 168–169
Dissolved oxygen (DO), 115
Distillation, 96

pretreatment, 96
types, 96
Domestic flows, 142
Dow/Mond indices, 308
Drain flies, 298–299
Dried sewage sludge, 256*f*
Drinking water treatment, 49, 102–105
  coagulation, 104–105
  pH and aggressiveness correction, 102–104
  precipitation, 105
  softening, 105–107
   base exchange softening, 106
   ion exchange softening, 106–107
   iron and manganese removal by oxidation, 107–108
   lime/soda softening, 106
   membrane treatment, 107
Drum screen, 152
Dry solids mass, 258*t*
Dry weather flow (DWF), 141, 186
"Dumb" instrumentation, 398
Duqu, 397

E

Economics, 4–5
Ecotherm, 202
Effluents, plating, 224–225
Eggy sulfides, 330
Electrical and control equipment, 393–398
  cabling, 395–396
  control systems, 397
  DCS, 398
  instrumentation, 396
  local controllers, 397
  motor control centers, 393–395
  programmable logic controllers, 397
  supervisory control and data acquisition, 397–398
Electrical engineer, 3, 125, 360, 392
Electrochlorination, 112
Electrodialysis, 99, 236
Electromagnetic radiation, 50, 98, 163
Empirical data, 141
"Empty Bed Contact Time" (EBCT), 49, 94, 94, 95
Enanthic acid, 33*t*
Engineering, 154, 348
Engineering common sense, 303
Engineering design, 349, 352
Engineering science, 1, 348
Engineering science of water treatment unit operations, 39

adsorption, 48–49
coagulants, 40
dechlorination, 51
disinfection and sterilization, 49–51
filtration, 43–45
    depth filtration, 45
    membrane filtration/screening, 44–45
flocculants, 40
mixing, 46–48
    combining substances, 46
    gas transfer, 47
    heat and mass transfer, 47
    promotion of flocculation, 46–47
    suspension of solids, 47
sedimentation/flotation, 40–43
    flotation, 42–43
    sedimentation, 40–42
Enhanced coagulation, 105
Enhanced High Rate Clarification (EHRC), 200f
Enhanced treated sludge, 261–262
Enigma, 341
Environment Agency (EA), 261, 306, 327–328
Environmental pollution, 295
Environmental Protection Agency, 296, 327–328
Environmental regulators, 8
"Environmental"/"sanitary" engineers, 2–3
Envirowise, 326
Enzymes, 32
Epoxy, 5
Equilibrating CO_2, 19
Equipment datasheet, 377, 378f
Equipment list/schedule, 373–375, 374f
Escherichia coli, 27f, 62, 98–99, 139, 145t
Estimating effluent flows, rules of thumb for, 142t
ETPs, 249–250
Eubacteria, 27
Eukaryotic organisms, 27–28
Eutrophication, 139–140, 192–193
Expanded bed reactor, 201
Expanded granular sludge bed design, 201
Explicit equation for headloss, 57

F

Facultative anaerobes, 36
Fast tracking, 365–367
Fatberg, 321–322, 322f
Fats, 32
Fats, oils, and greases (FOGs), 203, 415
Feed/basic design, 362–363
Feed/detailed design fast tracking, 366
FEED stage pricing, 400–401
Ferric chloride, 17, 40, 417t
Ferric salts, 40, 166

Ferric sulfate, 40, 417t
Ferrous salts, 40
Ferrous sulfate (copperas), 166
Filamentous organisms, 166, 169, 193, 241
Filter backwashing, 257, 340
Filter fly, 298–299, 298f
Filters, 85–86, 340
Filtration, 43–45, 74–95, 160–161, 340, 385t
Filtration media, 107–108
Final settlement tank design, 158–160, 181
Fine depth filtration, 74–87
Fine-inclined screens, 152
Finer screens, 152–153
Fine surface filtration, 87–95
    adsorption, 94–95
    design for fouling, 91–93
    design methodology, 93–94
    ion exchange, 95
    membrane materials selection, 90–91
    nanofiltration/reverse osmosis, 88–90
    solute rejection by reverse osmosis membranes, 93
    ultrafiltration (UF)/microfiltration (MF), 88
"Firmly bound" CO_2, 18
Fishy amines, 330
Fittings headloss, 121
    calculation of, 337
5-day BOD (BOD5), 140
5-stage Bardenpho process, 193, 195
Flash mixers, 70, 104, 114
Flocculants, 40
Flocculation, 70–71
    promotion of, 46–47
Flotation, 42–43, 73–74, 160
    coarse surface filtration, 75
        band screens, 75
        trash racks, 75
    dissolved air, 43
    fine depth filtration, 76–87
        air scour distribution, 85–86
        pressure filters, 86–87
        rapid gravity filters, 78–85
        slow sand filter design, 77
        slow sand filters, 76–77
    fine surface filtration, 87–95
        adsorption, 94–95
        design for fouling, 91–93
        design methodology, 93–94
        ion exchange, 95
        membrane materials selection, 90–91
        nanofiltration/reverse osmosis, 88–90
        solute rejection by reverse osmosis
            membranes, 93
        ultrafiltration (UF)/microfiltration (MF), 88
    gravity, 42–43

Flow balancing, 250–251
Flow control in open channels, 134
Flow equalization, 155–156
Flow measurement, 148
Flow measurement devices, 148
Flowrate and mass loading, 140–144
  domestic component, 142
  industrial component, 142
  infiltration/exfiltration, 142–143
  peaking factors, 143
  selection of design flowrates, 141–143
  selection of design mass loadings, 143–144
  upstream flow equalization, 143
Fluidized/expanded bed reactor, 283t, 285–286, 286f
Fluid mechanics, 53, 117
  Bernoulli's equations, 56
  Darcy-Wiesbach equation, 56–57
  less accurate explicit equation, 57
  pressure and pressure drop, head and headloss, 54
  rheology, 54–55
  Stokes Law, 57
Fluid moving equipment, 387
Flume, 148, 148f, 207
Fly control, 298–299
F:M ratio, 180
FNU (Formazin Nephelometric Units), 419
Foaming, 109
Food and Drink Federation (FDF), 261
Food Standards Agency (FSA), 261
"For construction" layout methodology, 314–315
"Form 4" panel, 394–395
Formazin Turbidity Units (FTU), 260, 419
Formic acid, 33t
"Formula A", 143
Forward osmosis, 100
4-stage Bardenpho process, 193
Free chlorine residual, 22
"Free CO₂", 18–19
Freundlich, 48–49
Frictional losses, determining, 121–122
Friction losses, 206–207
Froude number, 210, 214
"Full treatment flow" (FTF), 143
Functional design specification (FDS), 3, 323, 375, 390
Fungi, 30, 34–35, 66, 76, 108–109, 173–174

G

GAC adsorption, 298
Gallionella, 35
Gamma radiation, 98, 163
Gaseous chlorine, 344
Gaseous sulfur dioxide, 114
Gas handling equipment, 281
Gas production, 280–281
Gasification, 262
Gas stripping, 98
Gas transfer, 47–48, 97–98, 161–162
  stripping, 162
Gauckler–Manning–Strickler formula, 204
General Arrangement (GA) drawings, 359–360, 375–377, 376f
Glass-on-steel, 5
Glucose, 33f
Good Practice Guidelines for the Use of Computers by Chemical Engineers, 356
Gradual overhang ramp weir, 342f
Grand Cascade, 335, 336f
Grant and Moodie equation, 190
Granular activated carbon (GAC), 94, 115, 162–163, 234–235
  GAC adsorption, 298
  GAC filters, 49, 49f, 94
Grassroots design, 350–351
Gravity belt thickeners, 271
Gravity flotation, 42–43
Gravity sludge thickener, 269f
Gravity thickeners, 268
"Green Book" contract conditions, 366
Greensand filters, 23
Greenwash, 169
Grinding, 266
Grit removal, 154–155
  grit chambers, 154
  screens and filters for residual suspended solids removal, 155
  suspended solids removal, 154–155
  vortex separators, 154

H

Haaland equation, 124
"Half-bound" CO₂, 18
Halogens, 167, 242
Hammer rash, 332
Hand-cleaned screens, 152
Hand-raked screens, 153
Hardness, 15, 105
Hardness and alkalinity parameters conversion table for, 16t
Hardy Cross method, 127

Hazard and operability studies (HAZOPs), 307, 363, 378−379

Headloss, 54, 236
  calculation, 81, 122
    MS Excel spreadsheet for, 81f
    shape factors for cross-sections in, 214t
  estimation
    through clean screens, 213
    through partially blocked screens, 214
  explicit equation for, 57
  for sludges, 291

Heat transfer, 47

Heat treatment, 267, 275−276

Heavy metal inhibition, prediction of
  of digestion process, 244

Heavy metals, inhibitory effect of, 246t

Henry's law, 43

Heterotrophs, 35

Heuristic unit operation design, 39

High-rate biofiltration, performance chart for, 174f

High rate filters, 173−174, 247, 340

High-rate sand filter sizes, 340

High-shear centrifugal pumps, 332

High sulfate effluents, 225−227

HiPPO (Highest Paid Person's Opinion), 318

History of water treatment plant design, 2−3

Homologous series of stench, 32, 33t

Horizontal flow tanks, 73, 73t

HTH, 112

Human−machine interface (HMI), 391, 397−398

Humus tank, 174−175, 186

Hydraulic calculations, 4, 289, 335−336

Hydraulic design, 250

Hydraulic design, detailed, 341−344
  nozzles, 341−342
  water quality, 342−344
    biological quality, 343−344
    chemical composition, 342−343
    clarity, 343
  weirs, 342

Hydraulic jumps, 53, 214−215

Hydraulic loading rate (HLR), 186−187

Hydraulic networks, 126−127

Hydraulic profiles, 134, 135f

Hydraulic retention time (HRT), 70, 104, 114, 156, 176, 225, 247, 286

Hydraulics and layout, 249−250

Hydrocarbon-based chemical engineering, 5

Hydrochlorous acid, 22

Hydrogen sulfide dosing, 240

Hydrolysis, 36

Hydrophilic media, 233−234

Hydrophobic media, 233

Hydrostatic pressure, 24, 54

Hypochlorination, 22

Hypochlorite, 22, 166

Hypochlorite solutions, 113

I

Idiosyncratic water industry units, 419
  color, 419
  solution density, 419
  turbidity, 419

Impurities commonly found in water, 61−62

Incineration, 262, 267, 287

Inclined screen, 152, 153f

Indicator organisms, 50, 62, 139

Induced Air Flotation, 160

Industrial effluent characterization and treatment, 221
  calculating costs of, 222
  case studies, 223−228
    high sulfate effluents, 225−227
    paper mill effluents, 223−224
    petrochemical facility wastewaters, 225
    plating effluents, 224−225
    vegetable processing effluents, 227−228
  industrial wastewater composition, 222−223
  problems of, 228−229
    batching, 228
    changes in main process, 229
    nutrient balance, 228−229
    sludge consistency, 229
    toxic shocks, 228

Industrial effluent treatment, 11, 21, 70, 104, 250, 416
  calculating costs of, 222
  by numbers, 415−416
  problems of, 228−229

Industrial effluent treatment hydraulics, 249
  flow balancing, 250−251
  hydraulic design, 250
  hydraulics and layout, 249−250
  sludge handling, 251

Industrial effluent treatment plants, 10−11, 172, 249−250, 332

Industrial effluent treatment unit operation design
  biological processes, 243
    aerobic treatment design parameters, 247
    anaerobic treatment design parameters, 247−248
    inhibitory chemicals, 244
    microbiological requirements, 244−247
    nutrient requirements, 244
  chemical processes, 237, 241−242

Industrial effluent treatment unit operation design
  (*Continued*)
  cyanide removal, 240–241
  disinfection, 242
  metals removal, 239–240
  neutralization, 238–239
  organics removal, 241
  oxidants, 242
  sulfate removal, 240
  physical processes, 231
    adsorption, 234–235
    coalescing media, 233–234
    gravity oil/water separators, 231–233
    membrane technologies
Industrial wastewater, 221–222, 237, 247
  composition, 222–223
Infiltration, 141, 143
Infiltration/exfiltration, 142–143
Injection, water for, 67
Inline flow equalization system, 155
Inline systems, 156
Instrumentation selection, 396*t*
Interaction with engineering disciplines, 3
International standards organizations (ISO), 353
Inverters, 394
Invitation to Tender, 9
Ion exchange, 95
Ion exchange softening, 106–107
Iron, 40, 63, 166
Iron and manganese removal by oxidation, 107–108
Iron compounds, 169, 241
Iron oxidation, stoichiometry of, 107
Isometric piping drawings, 379, 380*f*

J

Jackson Turbidity Units (JTU), 419

K

Kill curves, 49–50, 50*f*, 261
Kill tanks, 287
KISS principle, 104
K_La, 48, 183
Kludges, 332
Kozeny-Carman equations, 45
K-value method, 121, 337

L

Lag Phase, 30
Langelier index, 92, 339
Langelier Saturation Index (LSI), 19–21, 107, 405–406
Langmuir isotherms, 48–49, 50*f*

Larson-Skold Index, 408
Lauric acid, 33*t*
Layout designers, 312–313, 375
Le Chatelier's principle, 17
Legionella bacteria, 109
Legionella organism, 340
Legislative drivers, in water and effluent treatment, 8–9
LEL (lower explosive limit), 297
Light nonaqueous phase liquids (LNAPLs), 231
Lime, 166, 275, 407
Lime/soda softening, 106
Literature of water treatment plant design, 3
Load balancing tanks, 332
Log Phase, 30
Lower explosive limit (LEL), 297
Low-rate trickling filter, 172–173, 173*f*
LSI versus Aggressive CO_2, 21
Luxury phosphate uptake, 195

M

M1E prices, 401
Magnesium, 15, 106, 406
Magnesium ammonium phosphate (struvite), 168
Magnesium carbonate, 343
Maintenance logs, 331
Manganese, in water, 64
Manganese removal by oxidation, 107–108
Manmade organic compounds, 172, 191
Manning equation, 132
Manning formula, 204–206, 205*t*
Manufacturers' literature, design from, 382–383
Marketing breakthroughs, 12, 12*f*, 169, 177, 244
Marshmallow challenge, 351
Mass and energy balance, 359, 370
Mass flux theory, 158
Mass transfer, 47
Mass transfer coefficient, 48, 183–184
Materials selection, 5
Mathematics, 6, 348, 351, 390
Matlock STW, 186*f*
Maximum organic loading rate, 187
  for stone media, 187*t*
MCC Unit, 394*f*
Meadows, 351
Mean, defined, 6
Mean cell residence time (MCRT), 184
Mechanical and electrical equipment costing methodology, 399–402
  conceptual design stage, 399–400
  detailed design, 401–402
  feed stage, 400–401
Mechanical coagulation/flocculation, 163

Mechanical Engineers, 360, 405
Mecklenbergh's Process Plant Layout, 301
Membrane bioreactors (MBRs), 175, 178,
    197—200, 236
  versus alternate technologies, 200*f*
  anaerobic, 199
  applications of, 200
  system design and optimization, 199
Membrane clean water treatment plant, 65*f*
Membrane distillation, 99—100
Membrane filtration, 44—45, 112, 115
  theoretical analysis of, 44—45
Membrane fluxes, 199
Membrane softening, 107
Membrane technologies
  oily water, 235—236
  removal of dissolved inorganics, 236
Memdos positive displacement pump, 119*f*
Mesophilic digester design criteria, 279*t*
Metabolism, 34—35
Metallic ions, 40
Metal oxide fouling, 92
Metals removal, 239—240
Methanogenesis, 36—37, 201—202
Microbiological quality, 62
Microfiltration (MF), 88, 115
Microsoft (MS) Excel, 6
Microstrainer, 77, 77*f*
Minimum rate of sludge return, 158
Minimum velocities, 204
Minworth sewage treatment works, AS treatment
    at, 179*f*
Misconceptions, in water and effluent treatment
    plant design, 10—12
Mistakes in water and effluent treatment plant
    design and operation, 317
  academic approaches, 317
  believing salespeople, 319—320
  HiPPO, 318
  newbie design, 321
  plant working, 320
  process engineer, 322—323
  running from the tiger, 319
  separation of design and costing, 317—318
  specialist, 321—322
  water treatment plant design, 320—321
Mixed liquor, 158, 178
Mixing, 46—48, 70
  high shear, 70
  low shear, 70
MLSS method, 185
Modeling as "conceptual design", 362
Mogden Formula, 221—222, 307
Moody diagram, 122

MS Excel, 122, 328, 368, 382
Multimedia filters, 83, 155
Myristic acid, 33*t*

N

Nanofiltration (NF)/reverse osmosis, 88—90
Natural systems, 188—191
  aerated lagoons, 190
  constructed wetlands, 189—190
  pond treatment, 188—189
  stabilization ponds, 191
Nephelometric Turbidity Units (NTU), 419
Net positive suction head (NPSHr), 125—126
Net positive suction head available (NPSHa), 125
Neutralization, 238—239, 247
Newbie design, 321
Newton, 54
Newtonian fluids, 54—55
"95% confidence interval", 6—7
Nitrate, 27, 115, 139—140, 192—193, 195
Nitrification, 116, 182—183, 192—194, 244
Nitrifying bacteria (Nitrosomonas), 35
Nitrifying biological filters, design of, 193—194
Nitrifying tricking filters (NTFs), 193
  design parameters for, 194*t*
Nitrites, 192—193
Nitrogen, 34—35, 140, 192, 278
  chemical removal of, 167—168
Noise control, 295—296
Nominal bore (NB), 150, 266
Non-Newtonian fluids, 26, 54—55, 289
Nonparametric statistics, 329
Novel laboratory test (SSV), 158
Nozzles, 337*f*, 338, 341—342
NTF recirculation rates, 194
Nucleic acids, 33—34, 34*f*
Nuisance organisms, 29
Nutrient removal, 167—168
  biological, 192—195
  chemical removal of nitrogen, 167—168
  chemical removal of phosphorus, 168
Nutshell filters, 233—234

O

Obligate anaerobes, 36—37
Octanol/Water Partition coefficient, 48—49
Odor abatement zone, 296, 297*f*
Offline systems, 156
Oil—water separator, 231—232, 232*f*
  three stage, 44*f*
Oily water, 235—236
On-site hypochlorination, 112
Open channel hydraulics, 132—134, 214—218

Open channel hydraulics (*Continued*)
  chambers and weirs, 133–134
  channels, 132–133
  choking, 215
  control structures, 215–216
    activated sludge control chamber,
      215–216
    water/sewage control chambers, 215
  critical and supercritical flow, 214
  flow control in open channels, 134
  hydraulic jumps, 214–215
  inlet works, 215
  sump flow presentation and baffles, 217–218
  sump volumes, 216–217
Open channels, 204–211
  circular channel headloss, 206
  flow control in, 134
  peripheral channels, 208–211
    aeration tanks/collection channels, 208
    storm overflows, 209–211
  shock losses in channels, 206–208
    outfalls, entries, and exits, 207–208
  straight channel headloss, 204–206
    rectangular channels, 204–206
Operating and maintenance manual,
  331
Operators, interviewing, 330
Operator training, 331–332
  classic mistakes, 332
  overcoming the folk wisdom hurdle,
    331–332
Organic coagulants, 40
Organic polymers, 168, 271
Organics removal, 241
O'Shaughnessey equation,
  281, 282
Osmosis, 23–24
Overland systems, 189
Oxidants, 241–242
Oxidation, 297–298
Oxidation and reduction, 21–23
  chlorine, 22
  chlorine dioxide, 22
  hypochlorite, 22
  ozone, 23
Oxidizing agents, 167, 240–242
Oxidizing disinfectants, 22–23
Oxygenation efficiency, 184, 196
  for common devices, 183*t*
Oxygen requirement calculation,
  182–183
Oxygen requirements vs F:M ratio, 183*f*
Ozonation, 113–114
Ozone, 23, 111–112, 167, 169, 242, 341

P

Packed beds, 233
Palmitic acid, 33*t*
Paper mill effluents, 223–224, 224*f*
Paramecium, 28
Paramecium aurelia, 29*f*
Parametric statistics, 329
Pascals (Pa), 54
Pasteurization, 202
PD5500, 354
Pelargonic acid, 33*t*
Penstock head calculation,
  215
Penstocks, 134, 208
Pentadecylic acid, 33*t*
Per Capita Contributions, 144*t*
Percussive maintenance, 332, 333*f*
Peripheral channels, 208–211, 209*f*
Petrochemical facility wastewaters, 225, 226*f*
Pharmaceutical water, 67, 114
pH control, 64, 103, 238–239, 332
pH dosing plant, 343
Phosphate, 34, 139–140, 195
Phosphorus, 35
  biological removal, 195
  chemical removal of, 168
Phostrip process, 195
Phragmites, 189
Physical field-mounted PID controllers, 390
Pie-blocking materials, 321
Pipe nomogram, 122, 123*f*
Piping and instrumentation diagram (P&ID), 122,
  359–360, 371–373, 371*f*, 390
"Pista" type grit separator, 154
Piston diaphragm pumps, 102–103, 118
Piston pumps, 265–266
Plant layout, 301
  and aesthetics, 309–310
  conceptual/FEED layout methodology, 312–313
  conceptual layout, 310–312
    construction, commissioning, and
      maintenance, 311
    emergency provision, 312
    indoor or outdoor, 311
    materials storage and transport, 311–312
    prevailing wind, 311
    security, 312
  and cost, 308–309
  detailed layout methodology, 313–314
  factors affecting, 303–310
    cost, 304
    health/safety/environment, 304
    robustness, 304

"for construction" layout methodology, 314–315
and safety, 307–308
site selection, 305–307
manmade environment, 306
natural environment, 306
regulatory environment, 307
Plate and frame filter presses, 272
Platinum-Cobalt Color, 419
Platinum Cobalt scale, 63
Plugging, 92–93
Poiseullie equation, 45
Polyacrylamides, 276
Polyaluminium silicate sulfate (PASS), 40
Polyamides, 91
Polyvinyl chloride, 5
Pond treatment, 188–189
Pool water, turbidity of
effect of turnover rate on, 343*t*
Population equivalent (PE), 143
Portable site analytical test kit, 327*f*
Positive displacement (PD), 118
Positive displacement pumps, 265, 289
Posthandover redesign, 365
Potable water, 65–66, 99, 114
Potable water sludges, 259–260
thickening, 260
water treatment sludge yield, 260
Potable water treatment clarifiers, 72*f*
Potential alkalinity, 103
Powdered activated carbon (PAC), 163, 234–235
Power cable size, 395
Practicing engineers, 361
Preaeration, 163
Precipitation
as hydroxides, 239
by reduction, 240
as sulfides, 239–240
Precipitation Potential, 409
Precoat filters, 340
Pressure and pressure drop, head and headloss, 54
Pressure filters, 86–87, 87*f*, 161
Primary sedimentation, 156–158
Primary settlement tanks, 157, 157*t*
Primary sewage sludges, 257
Primary tanks, 156, 157*f*
"Prime Cost" (PC), 403
Priming, 109
Process design, 3, 49, 199, 347, 363, 373
Process engineers, 3, 322–323, 353, 359–360
Process flow diagram (PFD), 359, 363, 370–371, 370*f*
Process plants versus castles in the air, 353
Professional engineers, 65, 203–204, 318, 352

Professional judgment, 303, 351, 355
Professional water process plant design practice, 39
Programmable logic controllers (PLCs), 387–390, 397
Project lifecycle, 349–351
Prokaryotes, 26
Propionic acid, 33*t*
"Proposals engineers" in EPC companies, 399
Protozoa, 28
PSI Index, 409
Pump affinity relationships, 129–131
Pump curves, 127–131, 127*f*
Pumping, 149–151
Pumps and compressors, 387
Pump selection, 118*t*, 120*t*, 387
detailed, 388*t*
general, 387*t*
Pure/enhanced oxygen processes, 196
Purified water, 67
Pyrolysis, 262, 267

Q

Quality, of water, 339–341
biological quality, 340–341, 343–344
chemical composition, 339, 342–343
clarity, 339–340, 343
microbiological, 62
Quantitative and verbal reason sills, in entrants to US academic programs, 26*f*
"Quick and dirty" analysis, 329
Quotations, 401

R

Racks and screens, 151–154
Radial flow clarification, 73
Radiused ramp weir, 342*f*
Raked gravity thickener, 268
Ramp weir, 342*f*
Rapid gravity filters (RGFs), 77–85, 79*f*, 161
Rapid sand filters, 340
Ready-made organic building blocks, 35
Reciprocating-rake screens, 153
Recirculation, 194
Rectangular channels, 204–206
Rectangular designs, 156–157
Redox reactions, 21. *See also* Oxidation and reduction
Reed beds, 190
Refractory compounds, 241
removal of, 234
Relative gas density, 418
Return Activated Sludge, 46

Reverse osmosis (RO) membranes, 24, 45
solute rejection by, 93
Rheology, 54–55, 55*f*
Rheopectic fluids, 55
Rigorous analysis, 131, 290
Rose's equation, 79, 84
Rotating biological contactors (RBCs), 174, 188
Rotifers, 28, 29*f*
Rotodynamic displacement (PD), 118
Rotodynamic pumps, 118, 149
Rough heuristics, 21
Roughing filters, 174, 177
Royal Commission effluent quality, 186–187
Royal Commission standard effluent, 228
Rules of thumb, 354, 381–382
Ryznar Stability Index (RSI), 409

S

Safe Sludge Matrix, 261
Safety, plant layout and, 307–308
Sales misconceptions, 11–12
Salespeople, 319–320
Saline waters, 66
Salmonella, 262
Sand filters, 161, 340
Sand filtration equipment, 339
Scaling, 92
Schmutzdecke, 76, 76*f*
Science, defined, 348
Screens, 44, 75, 153
Screens and filters for residual suspended solids
removal, 155
granular medium filtration, 155
microscreening, 155
Scrubbing, 297–298
Scum collection systems, 156–157
Secondary treatment plant design, 177–185
activated sludge physical facility design and
selection, 178–179
activated sludge process design, 179–185
suspended growth processes, 178
Sedimentation, 40–42
"Semicombined"/"half-bound" CO_2, 18
Semiconductor processing water, 67
SEMI F63, 67
Senior process engineers, 366
Sensitivity analysis, 368
Separation Area Calculation, 232
Separation processes, in water treatment, 385*t*
Sequencing batch reactors (SBRs), 175, 197
Setting flow velocity, 338
Settleability method, 185
Settlement tanks, 71–73, 158–160
clarifier design, 71–73

Sewage farms, 177
Sewage quality at different stages of treatment,
145*t*
Sewage sludges, 257–259, 265, 275
Sewage treatment by numbers, 413–414
Sewage treatment operators, 261
Sewage treatment process, 178
Sewerage pipes leak, 142–143
Sharp-crested weirs, 338
Shear-sensitive material, 332
Shock losses in channels, 206–208
Short-chain fatty acids, 32
Sick process syndrome, 326
Simple straining, 43
Single stage nitrification, 192
Site redesign, 364
Site selection, 305–307
manmade environment, 306
natural environment, 306
regulatory environment, 307
Site visits, 329–332
maintenance logs, 331
operating and maintenance manual, 331
operators, interviewing, 330
operator training, 331–332
using all your senses, 330
Slow rate systems, 189
Slow sand filters, 76–77, 76*f*, 161
design, 77, 78*t*
with *schmutzdecke* layer, 76, 76*f*
Slow sand filtration, 111, 115
Sludge, 47
beneficial uses of, 262
disposal of, 261–262
Sludge age, 184
Sludge blanket clarification, 72
Sludge blending, 266
Sludge characterization and treatment objectives,
255–256
beneficial uses of sludge, 262
disposal of sludge, 261–262
resource recovery, 256
sludge characteristics, 257–260
potable water sludges, 259–260
wastewater sludges, 257–259
solids destruction, 262
volume reduction, 255–256
Sludge conditioning, 276
Sludge handling, 251
Sludge headloss multiplication factor, 291
Sludge production rates, 259*t*, 267, 270
Sludge rake capacity design, 270
Sludge return rate, 185
Sludge thickening tanks, 268*f*
Sludge treatment hydraulics, 289

quick and dirty approaches, 290–291
    headlosses for sludges, 291
    sludge volume estimation, 290
    rigorous analysis, 290
Sludge treatment unit operation design
    biological processes, 277
        aerobic sludge digestion, 286–287
        anaerobic sludge digestion, 277–282
    conditioning, 276
    physical processes, 265
        blending, 266–267
        degritting, 266
        grinding, 266
        sludge and scum pumping, 265–266
        thermal treatments, 267
        volume reduction processes, 267–273
    stabilization, 275
Sludge volume estimation, 290
Sludge volume index (SVI), 158, 185
Sludge yield, 30–31, 175
Sluice gates, 134
Small RBC, 188f
Sodium acetate, 17
Sodium hypochlorite (NaOCl), 22, 112, 344
Sodium metabisulfite, 23, 51
Soft Starters, 394
Software engineer, 3
Solids destruction, 262
Solid state electronic PID (Proportional, Integral,
    Differential) controllers, 397
Spearman's rank correlation, 329
Specialist, 321–322
Specific gravity, 418–419
Spiral RO system, 91f
Spurious precision, 390
Stabilization, 275
Stabilization ponds, 191
Standard deviation, 7
Standards and specifications, 353–354
Star Delta starters, 393–394
Static flocculator, 71f
Static mixers, 48f, 70
Stationary Phase, 30
Statistically representative data, 7
Statistics, importance of, 5–8
    and discharge consents, 7–8
    in sewage treatment plant design and
        performance, 6–7
Steady state operation, acceptability of alarm rate
    in, 393t
Step Feed Activated Sludge (SFAS), 200f
Step weir, 342f
Sterilization, 49–51, 111, 115
Stokes Law, 41–43, 57

Storm overflows, 209–211, 209f
Storm tanks, 155–156
Storm waters, 141
Straight channel headloss, 204–206
Straight-run headloss, 121
    calculating, 122–124
Stripping, 98, 162
Stuxnet, 397
Submerged and side stream MBR units,
    comparison of, 199t
"Substances", 10
Sulfate precipitation, 168
Sulfate removal, 240
Sulfite, 21
Sulfolobus, 35
Sulfur dioxide, 23, 51
    and its salts, 225
Sump flow presentation and baffles, 217–218
    bellmouths, 218
    benching, 218
    incoming sewers and channels, 217
Sump working volume calculation, 216–217
Superchlorination, 113
Supernatant depth, 269
Supervisory control and data acquisition
    (SCADA), 387–390, 397–398
Surface aerators, 161, 162f
"Surface" filtration, 43
Surge analysis, 131–132
Surplus activated sludge (SAS), 157–158
Suspended growth processes, 175, 178
Suspended solid:biochemical oxygen demand (SS:
    BOD)
    sludge yield from domestic sewage by, 259t
Suspended solids, 10, 63, 84, 139, 172, 339
Suspended Solids Levels, 143
Suspended solids removal, 154–155
Synthetic gas (syngas) fuel, 267

T
Tablet brominators, 344
Technical Report (TR11), 158
Technical sales staff, 319
Temperature phased anaerobic digestion (TPAD),
    202
"Temple of storms" pumping station, 310f
Tender stage design, 337–338
    preliminary hydraulic design, 337–338
        channels, 338
        nozzles, 338
        pipework, 337
        tankage, 338
        weirs, 338

Tertiary treatment plant design, 191–195
  biological nutrient removal, 192–195
  need for, 191
  treatment technologies, 191–192
Thames water (United Kingdom), 77
Thermal hydrolysis, 262, 263*f*
Thermal hydrolysis of sewage sludge, plant for, 263*f*
Thermal hydrolysis pretreatment processes, 202
Thermal treatments, 267
Thermoplastics, 5
Thickened sludges, 255
Thickener Diameter Calculation, 268–269
Thickener working depth calculations, 269
Thin weir equation, 133
Thin weir flow rate calculation, 211–212
Thiobacillus, 35
Thiols, 330
Thixotropic fluids, 55
3D CAD software, 379, 400
Tillmann's formula, 19, 103, 103*f*
Time-independent fluids, rheology of, 55*f*
Top view pressure filters, 45*f*
Total digester gas flowrate calculation, 280
Total dissolved solid (TDS), 64, 94, 106
Total oxygenation capacity, 182*t*
"Total Recycle", 169
Total Viable Count, 62
Toxic and refractory compounds, removal of, 234
Toxic pollutants, 327–328
TR60, 113
Transient pressure equation
  in compressible conditions, 131–132
  in incompressible conditions, 131
Trash racks, 75
Trickling filters, 185–187, 186*f*
  hydraulic loading rate (HLR), 186–187
  maximum organic loading rate, 187
Tridecylic acid, 33*t*
Trihalomethanes (THMs), 111–112
Troubleshooters, 332, 334
Troubleshooting on wastewater treatment plant, 325
  appropriate statistics, 328–329
  data analysis, 327–328
    too little data, 327–328
    too much data, 328
  design verification, 332–333
  putting it all together, 333–334
  rare ideal, 329
  sick process syndrome, 326
  site visits, 329–332
    interviewing operators, 330
    maintenance logs, 331

    operating and maintenance manual, 331
    operator training, 331–332
    using all your senses, 330
Tubular systems, 89, 91*f*
Tuning control loops, 390
Turbidity, 419
Two and three phase digestion, 201–202
2D CAD software, 379
Type 1/"discrete" settling, 41
  relationship between time and depth in, 41*f*
Type 2/"flocculent" settling, 41
  relationship between time and depth in, 41*f*
Type 3/"hindered" settling, 41
  relationship between time and depth in, 42*f*
Type 4/"compressive" settling, 42
  relationship between time and depth in, 42*f*

U
UK Environment Agency, 327–328
UK Sewage Strength Classification, 144*t*
Ultrafiltration (UF) membranes, 88, 115, 236
Ultrapure water, 66–67
Ultrasonic sludge conditioning, 272–273
Ultraviolet light, 98–99, 114, 341
Undecylic acid, 33*t*
United Kingdom's Water Research Council, 158
United States and Soviet space programs, 341
Unit operations, 381–398
  approaches to design of, 382–383
    first principles design, 382
    manufacturers' literature, design from, 382–383
  control and instrumentation in water and effluent treatment, 387–391
    accuracy, 391
    automatic control, 387–390
    cost and robustness, 391
    precision, 390–391
    safety, 391
    specification of instrumentation, 390
  control systems, specification of, 391
  design data, sources of, 383
  pumps and compressors, 387
  rule of thumb design, 381–382
  scale-up and scale-out, 384
  separation processes, 384–386
  stage of design, 381
  standard control and instrumentation strategies, 392–398
    alarms, inhibits, stops, and emergency stops, 392–393
    electrical and control equipment, 393–398

Upflow anaerobic sludge blanket (UASB),
    284–285, 285*f*
    UASB process, 200–201
    UASB reactor, 176
Upstream flow equalization, 143
US Clean Water Act, 327–328
UV-activated peroxide, 169, 241

V

Vacuum filtration, 271
Valeric acid, 33*t*
Van der Waals forces, 32
Van Kleek method, 281
Variable Speed Drive (VSD), 394
Variation/creativity, 352
Variation orders, 403
VectoBac, 299
Vegetable processing effluents, 227–228
Velocity-associated "centrifugal" forces, 154
Veolia's Memthane process, 199
Viscosity, 289
Volatile fatty acids (VFAs), 36
    production, 36
Volatile matter destruction, 281–282
Volatile organic compounds (VOC) and odor
        control, 296–298
    covers and ventilation, 297
    measurement and setting of acceptable
        concentrations, 297
    removal of, 297–298
Volatile Suspended Solids, 63, 140
Volume reduction processes, 267–273
    dewatering, 271–272
    novel processes, 272–273
        ultrasonic sludge conditioning, 272–273
    thickening, 267–271
Vortex separators, 154

W

Waste activated sludge, 271
Waste and Resources Action programme (WRAP),
        222
Wastewater characteristics, 139–140
Wastewater filters, 161
Wastewater sludges, 257–259
    sewage sludge yields, 257–259
Wastewater treatment, 27, 47, 144–145, 165–166

Wastewater treatment plant, 30–31, 139, 148
    troubleshooting on. *See* Troubleshooting on
        wastewater treatment plant
Water, impurities commonly found in, 61–62
Water aggressiveness indexes, 405–410
    Aggressive Index, 407
    Langelier Saturation Index (LSI), 406
    Larson-Skold Index, 408
    PSI Index, 409
    Ryznar Stability Index (RSI), 409
Water and wastewater treatment plant design,
        349–351, 381, 390, 392
Water and wastewater treatment plant designers,
        390, 392
Water biology, 4
Water chemistry, 4, 15, 320
    acids and bases, 15–24
        buffering, 17–19
        catalysis, 23
        dechlorination, 23
        osmosis, 23–24
        oxidation and reduction, 21–23
        scaling and aggressiveness, 19–21
    hardness and alkalinity, 15
Water engineering, 2
Water engineers, 5, 18
Water for injection (WFI), 49, 64
Water hammer, 131–132
Water recycling, 341
Water/sewage control chambers, 215
Water soluble organic polymer coagulants, 40
Water supply figures, 141
Water treatment sludge yield, 260
Weir penstocks, 208
Weirs, 133, 211–213, 342
    broad weirs, 133–134, 212–213
    thin weirs, 133, 211–212
WRc publications, 271
WrC Technical Report TR189, 259–260

Z

"Zero Discharge", 169
Zero-sludge-yield assumptions, 323
Zigrang and Sylvester equation, 124
Zinc, 5
Zoning/hazardous area classification
        study, 379

Printed in the United States
By Bookmasters